Multifunctional Oxide-Based Materials

Multifunctional Oxide-Based Materials: From Synthesis to Application

Special Issue Editors

Teofil Jesionowski
Filip Ciesielczyk

MDPI • Basel • Beijing • Wuhan • Barcelona • Belgrade

Special Issue Editors
Teofil Jesionowski
Poznan University of Technology,
Poland

Filip Ciesielczyk
Poznan University of Technology,
Poland

Editorial Office
MDPI
St. Alban-Anlage 66
4052 Basel, Switzerland

This is a reprint of articles from the Special Issue published online in the open access journal *Materials* (ISSN 1996-1944) from 2018 to 2019 (available at: https://www.mdpi.com/journal/materials/special_issues/oxide_based_materials).

For citation purposes, cite each article independently as indicated on the article page online and as indicated below:

LastName, A.A.; LastName, B.B.; LastName, C.C. Article Title. *Journal Name* **Year**, *Article Number*, Page Range.

ISBN 978-3-03921-397-9 (Pbk)
ISBN 978-3-03921-398-6 (PDF)

Cover image courtesy of Teofil Jesionowski and Filip Ciesielczyk

© 2019 by the authors. Articles in this book are Open Access and distributed under the Creative Commons Attribution (CC BY) license, which allows users to download, copy and build upon published articles, as long as the author and publisher are properly credited, which ensures maximum dissemination and a wider impact of our publications.
The book as a whole is distributed by MDPI under the terms and conditions of the Creative Commons license CC BY-NC-ND.

Contents

About the Special Issue Editors vii

Preface to "Multifunctional Oxide-Based Materials: From Synthesis to Application" ix

Pavel Bazant, Tomas Sedlacek, Ivo Kuritka, David Podlipny and Pavlina Holcapkova
Synthesis and Effect of Hierarchically Structured Ag-ZnO Hybrid on the Surface Antibacterial Activity of a Propylene-Based Elastomer Blends
Reprinted from: *Materials* **2018**, *11*, 363, doi:10.3390/ma11030363 1

Katarzyna Siwińska-Stefańska, Adam Kubiak, Adam Piasecki, Joanna Goscianska, Grzegorz Nowaczyk, Stefan Jurga and Teofil Jesionowski
TiO_2-ZnO Binary Oxide Systems: Comprehensive Characterization and Tests of Photocatalytic Activity
Reprinted from: *Materials* **2018**, *11*, 841, doi:10.3390/ma11050841 15

Aleksandra Grzabka-Zasadzinska, Lukasz Klapiszewski, Sławomir Borysiak and Teofil Jesionowski
Thermal and Mechanical Properties of Silica–Lignin/Polylactide Composites Subjected to Biodegradation
Reprinted from: *Materials* **2018**, *11*, 2257, doi:10.3390/ma11112257 34

Adam Kubiak, Katarzyna Siwińska-Ciesielczyk and Teofil Jesionowski
Titania-Based Hybrid Materials with ZnO, ZrO_2 and MoS_2: A Review
Reprinted from: *Materials* **2018**, *11*, 2295, doi:10.3390/ma11112295 48

Chamila Gunathilake, Rohan S. Dassanayake, Chandrakantha S. Kalpage and Mietek Jaroniec
Development of Alumina–Mesoporous Organosilica Hybrid Materials for Carbon Dioxide Adsorption at 25 °C
Reprinted from: *Materials* **2018**, *11*, 2301, doi:10.3390/ma11112301 100

Kuibao Zhang, Dan Yin, Kai Xu and Haibin Zhang
Self-Propagating Synthesis and Characterization Studies of Gd-Bearing Hf-Zirconolite Ceramic Waste Forms
Reprinted from: *Materials* **2019**, *12*, 178, doi:10.3390/ma12010178 118

Ewelina Weidner and Filip Ciesielczyk
Removal of Hazardous Oxyanions from the Environment Using Metal-Oxide-Based Materials
Reprinted from: *Materials* **2019**, *12*, 927, doi:10.3390/ma12060927 127

Katarzyna Jankowska, Filip Ciesielczyk, Karolina Bachosz, Jakub Zdarta, Ewa Kaczorek and Teofil Jesionowski
Laccase Immobilized onto Zirconia–Silica Hybrid Doped with Cu^{2+} as an Effective Biocatalytic System for Decolorization of Dyes
Reprinted from: *Materials* **2019**, *12*, 1252, doi:10.3390/ma12081252 159

Mariusz Kłonica and Józef Kuczmaszewski
Modification of Ti6Al4V Titanium Alloy Surface Layer in the Ozone Atmosphere
Reprinted from: *Materials* **2019**, *12*, 2113, doi:10.3390/ma12132113 175

About the Special Issue Editors

Teofil Jesionowski Professor, Ph.D., D.Sc., Eng., is Vice-Rector of the Poznan University of Technology (PUT) and a member of the Commission for International Cooperation at the Conference of Rectors of Academic Schools in Poland. He also serves as head of the Chemical Technology Department. From 2008–2016, he was Vice-Dean of the Faculty of Chemical Technology of PUT. He received the title of Professor of Chemical Sciences in 2013. His research interests include the synthesis, characterization, and applications of advanced functional materials, functional fillers, and polymer composites; activators of rubber compounds, (bio)additives, and eco-friendly fillers; biomineralization-inspired syntheses and extreme biomimetics; biocomposites and biomaterials; removal of wastewater pollutants via adsorption, photocatalysis, or precipitation methods; pigment composites; enzyme immobilization; colloid chemistry and surface modification; and hybrid systems, biopolymers, and biosensors. He has published over 410 publications indexed by Thomson Reuters JCR. He is a cocreator of over 45 patents and patent applications. His publications have received over 5500 citations to date (H = 34). He is the coauthor of chapters in numerous monographs published by renowned publishing houses (Springer, Wiley, InTech, etc.) and a scholarship holder of the Foundation for Polish Science. He is also a member of the editorial boards of several scientific journals: *Scientific Reports*—Nature Publishing Group, *Dyes and Pigments*—Elsevier, and *Physicochemical Problems of Mineral Processing*.

Filip Ciesielczyk, Ph.D., D.Sc., Eng., has been an Assistant Professor at the Institute of Chemical Technology and Engineering at Poznan University of Technology (PUT) since 2010. In 2017, he successfully finalized his habilitation application. His research interests include materials chemistry, engineering and technology related with the synthesis and application of different type of metal oxides systems and inorganic/organic hybrids, surface chemistry, development of novel types of active adsorbents and catalysts (photocatalysts), catalytic processes, environmental protection concerning the removal of organic/inorganic impurities from water systems as well as from the atmosphere, and kinetic and thermodynamic aspects of adsorption processes. He has published almost 60 publications indexed by Thomson Reuters JCR. He is a cocreator of seven patents and patent applications. His publications have received almost 500 citations to date (H = 14). He is the coauthor of five chapters in monographs published by renowned publishing houses such as Springer. Additionally, he is a scholarship holder of The Kosciuszko Foundation, within which he had the opportunity to spend time at the Department of Chemistry and Biochemistry at Kent State University (Ohio, USA) with Professor Mietek Jaroniec's research group. He is also a member of the editorial boards of several scientific journals, such as *Materials* (MDPI), including the Special Issue entitled "Multifunctional Oxide-Based Materials: From Synthesis to Application", and *Physicochemical Problems of Mineral Processing*.

Preface to "Multifunctional Oxide-Based Materials: From Synthesis to Application"

The development of new technologies and improvement of those already existing requires not only up-to-date equipment, but novel products with specific physicochemical properties. Such a situation is observed especially in the case of oxide materials, which are utilized as active adsorbents, polymer fillers, plant protection agents, or substitutes of commonly used natural substances. The physicochemical properties of such systems mainly depend on the route of their synthesis and final treatment method (thermal treatment), and the fact that they are similar to physicochemical properties of naturally occurring systems confirms the possibility of their analogous application. During recent years, there has been increasing interest in the synthesis of different types of inorganic–inorganic or inorganic–organic combinations, owing to the possibility of obtaining hybrid materials which combine the properties of both components. Depending on the needs, the hybrid materials may be synthesized using different methods, and the selection of an appropriate method determines their physicochemical properties, such as morphology and dispersive character, electrokinetic and thermal stability, parameters of the porous structure, and hydrophilic–hydrophobic nature. A broad spectrum of methods of synthesis enables the design of hybrid materials characterized by diverse physicochemical and structural parameters, which is crucial in terms of the continuously increasing demand for such combinations. Each of the methods is characterized by advantages and disadvantages, and the selection of the appropriate one should be preceded by their detailed analysis. In particular, the considerations should include the economical aspect of the process, the possibility of its realization at the laboratory scale as well as a semitechnical scale, and the generation of waste substances. Another advantage is associated with the fact that the properties of hybrid oxide systems may be freely designed, by selection of the components which are included in the resulting materials as well as by means of further treatment using different surface modifications with multifunctional organic or bioorganic substances. Such technological procedures result in an enhanced activity of the synthesized hybrid materials, which is of importance when considering their potential fields of application, including in medical or environmental aspects, polymer or alloy processing, and catalysis (photocatalysis), among others.

The preparation, characterization, and application of a hybrid Ag–ZnO combination was reported by Bazant and coworkers (doi.org/10.3390/ma11030363), who proposed the so-called "drop technique" to obtain filler with hierarchical microparticles. The proposed methodology of synthesis enabled good distribution of the filler in the polymer matrix, the presence of which did not affect the properties of the resulting nanocomposite. The advantage of this oxide-based material is that it offers additional antibacterial activity, so it is important to analyze the potential application of such polymer composites in medicinal aspects, for example. The authors confirmed the excellent antibacterial activity against E. coli and very high activity against S. aureus of prepared polymer nanocomposites, which was supported mostly by the presence of Ag–ZnO filler.

During the last few years, there has been a dynamic increase of interest in studies focused on the selection of an appropriate synthesis method or modification of titanium dioxide, as well as the selection of proper components which, together with TiO_2, form advanced, multifunctional hybrid materials with strictly defined physicochemical properties, including photocatalytic or electrochemical activity. The development of methods for obtaining modified forms of TiO_2 is very complex and requires several complicated experimental procedures for proper implementation.

Titanium dioxide is a very attractive material that has the ability to absorb UV irradiation. Due to this fact, it offers both antibacterial as well as photocatalytic activity. A technological assumption concerning the fabrication of titania-based hybrid combinations with defined photocatalytic activity was presented by Siwinska-Stefanska and coworkers (doi.org/10.3390/ma11050841). The authors proposed to combine titania with zinc oxide via a sol–gel technique. The process conditions, including the molar ratio of components, among others, were optimized in order to obtain materials with enhanced photoactivity. Of key importance was the verification test concerning the photocatalytic degradation of selected, model organic impurities. The authors proved that the TiO_2–ZnO oxide hybrid obtained at a molar ratio of TiO_2:ZnO = 9:1 and additionally calcined showed the highest photocatalytic activity, which was attributed to the fact that this sample was composed of titania (with anatase being the dominant phase), zinc oxide, and $ZnTiO_3$ phases. So, it seems reasonable to design and fabricate hybrid oxide-based materials containing antibacterial or photocatalytically active elements in their structure, which furthermore ensures their specific application.

A comprehensive review concerning titania-based materials containing different oxide species in their structures was produced by Kubiak et al. (doi.org/10.3390/ma11112295). The current state of knowledge on the synthesis and practical utility of TiO_2–ZnO and TiO_2–ZrO_2 oxide systems and TiO_2–MoS_2 hybrid materials was presented. A key element of this review was to indicate the properties of titania-based hybrid systems synthesized with different methods, including hydrothermal, sol–gel, and electrospinning techniques, that affect their potential application. It has been shown that a proper selection of components included in the oxide or hybrid system, as well as their molar ratio, or properly designed thermal treatment enables obtaining a material with a strictly defined crystalline or textural structure. Additionally, it was confirmed that modified forms of TiO_2 synthesized by the sol–gel or hydrothermal methods can be successfully used as active photocatalysts, antibacterial materials, or even components of electrode materials.

Inorganic–organic hybrid materials are, as of recently, the most promising direction of studies focused on obtaining systems characterized by high quality and functionality. Lignin is the second most abundant natural renewable raw material from the biopolymer group and is therefore characterized by a notable application potential. Additionally, it is a valuable material for several chemical syntheses and modifications aimed at obtaining valuable polymer materials and biocomposites with special properties. The unique properties of lignin, including a high amount of functional groups (such as hydroxyl, ether, or carboxyl groups) in particular, enable its easy binding to mineral carriers. This group of carriers includes metal oxides, which are characterized by a well-developed surface area and high homogeneity as well as thermal and electrokinetic stability. The presence of hydroxyl groups on the surface of such systems determines their reactivity and facilitates their binding with a broader range of organic compounds, including lignin. The combination of lignin characterized by an "advanced" chemical structure and inorganic oxides results in the formation of multifunctional sorbents, carriers, or fillers. The justification for these types of connections is mainly associated with the "mutual benefit" for both components of the hybrid. This applies to the weakly developed porous structure of the biopolymer, which in combination with the oxide material gains notable benefits in this respect, and is also beneficial for the inorganic carrier, which gains notable surface activity due to the diversity of functional groups present in the structure of lignin. Detailed investigations related to a silica/lignin hybrid filler of a polylactide matrix were presented by Grzabka-Zasadzinska and coworkers (doi.org/10.3390/ma11112257). The authors found that the application of silica–lignin hybrids as fillers for a PLA matrix may be interesting not only in terms of increasing thermal stability, but also controlled biodegradation, which is of such importance nowadays.

The dynamic technological development generates increasing volume of waste substances which enter the natural environment. Aqueous solutions containing inorganic and/or organic contaminants (wastewater) are one of the major environmental issues. Wastewater containing hazardous metal ions is particularly cumbersome for the environment, and therefore several scientific institutes worldwide conduct intense studies focused on methods which would allow for their elimination from the environment. Adsorption seems to be the most versatile process due to its mild conditions, simplicity, limited costs, and most importantly, a broad range of sorption materials used for its realization. Studies regarding the adsorption process are mainly focused on the selection of appropriate materials used as adsorbents of different types of contaminants. An effective adsorbent should be characterized by defined porous structure, well-developed surface, and notable activity due to the presence of reactive functional groups. Therefore, the available literature reports describe several examples of materials which may be used as adsorbents, e.g., carbon materials (activated carbon), natural and waste materials (low-cost adsorbents), minerals, synthetic inorganic oxides, or biopolymers. Typical materials of natural origin (minerals), their counterparts (oxide systems), or hybrids synthesized in the laboratory using simple techniques, such as the coprecipitation (or, alternatively, sol–gel) method and further functionalization with a broad range of modifying agents, seem to be more attractive in this regard. The review prepared by Weidner and Ciesielczyk (doi.org/10.3390/ma12060927) concerns the application of different types of oxide-based adsorbents in the removal of hazardous metal oxyanions from wastewaters. Characteristics of oxyanions originating from As, V, B, W, and Mo, their probable adsorption mechanisms, and comparison of their sorption affinity for metal-oxide-based materials such as iron oxides, aluminum oxides, titanium dioxide, manganese dioxide, and various oxide minerals and their combinations were presented and discussed in detail.

Among the mentioned potential applications of metal oxides, their use as supports in the enzyme immobilization process is also of great importance, especially that resulting biocatalytic systems are also used in environmental protection aspects related to the removal of organic compounds, for example. The idea of the research presented by Jankowska et al. (doi.org/10.3390/ma12081252) concerned the immobilization of laccase onto ZrO_2–SiO_2 and Cu^{2+}-doped ZrO_2–SiO_2 systems in order to improve their stability in the reaction medium. The obtained materials were verified in the decolorization process of selected organic dyes. Through numerous analytical techniques, the authors proved the effective synthesis of a biocatalytic system which exhibited excellent activity in the decolorization processes. In addition, it was established that the presence of copper ions on the support material has a positive impact on laccase stability and activity. This fact also justified the possibility of the modification of metal oxides by various techniques to enhance their activity.

Another important contribution to the Special Issue was made by Gunathilake et al. (doi.org/10.3390/ma11112301), who belong to the scientific team of Professor Mietek Jaroniec—a globally recognized expert in material science, surface chemistry, and adsorption phenomena (his Hirsch index is 115). The presented research concerned the synthesis of alumina (Al_2O_3)—mesoporous organosilica (Al–MO) hybrid materials via a co-condensation method in the presence of Pluronic 123 triblock copolymer. The aim was to synthesize materials with well-developed porosity and high surface area that should enable effective CO_2 uptake. The authors proved that the proposed methodology of synthesis led to the formation of porous materials whose adsorption properties towards CO_2 depend mainly on the surface area of the sample, alumina precursor, and structure and functionality of the organosilica bridging group.

Metal oxides can be also found in the production of ceramic waste forms. Synthesis of a Gd-bearing Hf-zirconolite ceramic was proposed by Zhang and coworkers (doi.org/10.3390/ma12010178). The authors synthesized the Hf-zirconolite from an SHS/QP technique using CuO as the oxidant. The idea was to replace the Zr with Hf to obtain the chemical composition of $CaHfTi_2O_7$. Based on this, the Gd_2O_3 was introduced as the surrogate of trivalent actinides, which was designed to concurrently occupy the Ca and Hf sites of Hf-zirconolite. It was proved that the SHS/QP route is suitable for the preparation of zirconolite- and pyrochlore-based waste forms for HLW immobilization.

The final article included in this Special Issue is related to a study on the Ti_6Al_4V titanium alloy, including identification of the atoms forming the surface layer, after additional treatment with ozone. Kłonica and Kuczmaszewski (doi.org/10.3390/ma12132113) confirmed that the ozone treatment of the Ti_6Al_4V titanium alloy removes carbon and increases the concentration of Ti and V ions at higher oxidation states at the expense of metal atoms and lower-valence ions. Moreover, the study confirmed that ozone treatment, if performed under appropriate conditions, can be used in bonding technologies to shape surface microtopography and free energy, thus offering an alternative option to electrochemical methods. This fact is very important in analyzing the potential applications of such types of alloys.

The presented Special Issue is focused on the diversity of synthesis methods of inorganic oxide systems, as well as hybrid materials and their potential applications. A detailed analysis of this topic confirmed that the selection of an appropriate synthesis method of metal oxides and hybrid systems determines their physicochemical and functional properties. It was proven that the advantage of those materials is associated with the possibility to define their surface activity by additional functionalization with selected organofunctional compounds or biopolymers. This technological procedure clearly increases the possibilities of utilizing such materials and opens new perspectives for their practical use. The obtained and described experimental dependencies should be of high importance for the development technologies focused on synthetic materials with specific physicochemical and functional properties. They may contribute to the formation of novel technological conditions for obtaining next-generation oxide materials used in different branches of industry. Additionally, the literature reports published to date, which deal with the synthesis, characterization, and application of metal-oxide-based materials, will be supplemented with experimental data and technological conditions for realizing their fabrication with different type of components that offers unique properties for the final products

Teofil Jesionowski, Filip Ciesielczyk
Special Issue Editors

Article

Synthesis and Effect of Hierarchically Structured Ag-ZnO Hybrid on the Surface Antibacterial Activity of a Propylene-Based Elastomer Blends

Pavel Bazant *, Tomas Sedlacek, Ivo Kuritka, David Podlipny and Pavlina Holcapkova

Centre of Polymer Systems, Tomas Bata University in Zlin, Trida Tomase Bati 5678, 760 01 Zlin, Czech Republic; sedlacek@utb.cz (T.S.); kuritka@utb.cz (I.K.); podlipny@utb.cz (D.P.); holcapkova@utb.cz (P.H.)
* Correspondence: bazant@utb.cz; Tel.: +420-777-805-870

Received: 24 January 2018; Accepted: 26 February 2018; Published: 1 March 2018

Abstract: In this study, a hybrid Ag-ZnO nanostructured micro-filler was synthesized by the drop technique for used in plastic and medical industry. Furthermore, new antibacterial polymer nanocomposites comprising particles of Ag-ZnO up to 5 wt % and a blend of a thermoplastic polyolefin elastomer (TPO) with polypropylene were prepared using twin screw micro-compounder. The morphology and crystalline-phase structure of the hybrid Ag-ZnO nanostructured microparticles obtained was characterized by scanning electron microscopy and powder X-ray diffractometry. The specific surface area of this filler was investigated by means of nitrogen sorption via the Brunauer-Emmet-Teller method. A scanning electron microscope was used to conduct a morphological study of the polymer nanocomposites. Mechanical and electrical testing showed no adverse effects on the function of the polymer nanocomposites either due to the filler utilized or the given processing conditions, in comparison with the neat polymer matrix. The surface antibacterial activity of the compounded polymer nanocomposites was assessed against *Escherichia coli* ATCC 8739 and *Staphylococcus aureus* ATCC 6538P, according to ISO 22196:2007 (E). All the materials at virtually every filler-loading level were seen to be efficient against both species of bacteria.

Keywords: Ag-ZnO; thermoplastic elastomers; polypropylene; nanocomposites; hierarchical; antibacterial

1. Introduction

Recent years have witnessed a growing interest in developing polymer nanocomposites containing hybrid nanoparticles [1]. Advances made in polymeric nanocomposites have ushered in a new generation of macromolecular materials with low densities and multifunctional properties [2]. The primary advantage of nanocomposites is the tiny amount of filler needed to fulfil the given requirements, which could be one or even two orders of magnitude less than required by conventional micro-fillers [3].

Such advanced polymer systems incorporate a hybrid metal semiconductor materials that has attracted particular attention. This is not only due to the fact that co-joined metal and semiconductor nanoparticles possess a large specific surface area and high fraction of surface atoms, but also because they feature a unique electronic band structure which results in certain chemical activity [4–6]. For instance, hybrid Ag-ZnO nanoparticles constitute a filler of such materials. Indeed, Ag-ZnO hybrid nanoparticles are of particular interest since they exhibit biological, and photocatalytic activity, low toxicity, and exert a synergetic antibacterial effect.

In addition, adding said nanoparticles into a polymer gives rise to new, hybrid systems with combined properties and even some synergies. This was successfully proved for polymer melt compounded materials having further utilisation in medical, agricultural and catalytic applications [7–9]. Some research studies aim to eliminate bacterial growth and pathogen formation on various surfaces of subjects, examples being electrical equipment, walls, tables, floors, cars, food packaging, public and interior spaces and hospital facilities [10–16].

Nevertheless, nanoparticles may potentially cause problems due to safety issues. The occupational health risks associated with manufacturing and utilizing nanomaterials are not yet clearly understood [17–20]. Moreover, nanoparticles are difficult to process into thermoplastic polymers by conventional methods, e.g., extrusion, melt-blow, compression or injection moulding, as the filler forms micron-sized cluster-agglomerates and possesses a non-homogenous distribution. Therefore, dispersion and distribution of the nanoparticles in the polymer matrix can be spatially varied in the volume of prepared plastic article. Moreover, the surface of the product, if rendered as a thin film, can show serious surface and optical defects. In addition, during the thermoplastic processing of nanoparticles, hot spots can occur in the material adhering to heating elements, along with burns, and a further disadvantage is a greater amount of non-processable waste [21–25].

One way to avoid the problems described above is to use nanostructured hierarchical microparticles. These three-dimensional materials can offer advantages arising from their large surface area, as building blocks are often nanoscale in extent. Nonetheless, primary nanocrystals have been known to assemble themselves into complex structures, such as hierarchical flowers or spheres, these ranging in size from mesoscale to microscale; this is a phenomenon driven by reduction in overall system energy during synthesis. Therefore, multiscale hierarchical nanostructured materials are promising due to (i) properties gained when their building blocks are nanoscale in size and (ii) composite processability arising when at meso to micro scale [26,27].

A variety of synthesis methods for preparing Ag-ZnO have been reported. However, most are limited to research purposes as a consequence of the need for high temperature and pressure, expensive equipment, toxic reagents and a long reaction time [28–31]. In addition, many synthesis techniques utilize extremely dilute solutions as the resultant nanoparticles have to possess a specific structure [32–35]. These methods are not easily scalable to produce sufficient amounts of filler for typical commercial means of processing plastic, e.g., extrusion, melt-blow, compression and injection moulding.

In this paper, the authors offer up an original methodology for generating polymer nanocomposites by melt-mixing a pre-prepared filler, which involves adding Ag-ZnO hierarchical microparticles to the polymer matrix, in this case a blend of polypropylene and thermoplastic polyolefin elastomers (TPOs). In less than a decade, TPOs, defined as materials combining small enough semi-crystalline domains connected via amorphous elastomeric regions and renowned for their adjustable rubber-like characteristics, have emerged as a core material processed by standard thermoplastic processing equipment in automotive interiors and exteriors, buildings, wires and cables, film applications, medical devices, adhesives, footwear, foams, and other extruded and moulded goods [36–38].

Proposed methodology of compounding secured not only disagglomeration of processed micropartices into their building nano-blocks by shear forces during mixing but also ensured good distribution and dispersion of the resultant nanoparticles. Preparing the Ag-ZnO hierarchical particles merely calls for application of a facile hydrothermal method, even facilitating large-scale production of the same. It will be viable if the chemicals in the filtrate are recovered and reused, namely silver. Said polymer nanocomposites demonstrate excellent antibacterial properties against Staphylococcus aureus and Escherichia coli, as well as boasting unwavering mechanical and electrical properties.

2. Materials and Methods

2.1. Materials

Silver nitrate AgNO$_3$ (>99.5% purity) and Zinc nitrate hexahydrate (ZnNO$_3$)$_2$·6H$_2$O (>99% purity) were delivered by Penta (Praha, Czech Republic), while ammonium carbonate (NH4)$_2$·CO$_3$ (>99% purity) was supplied by Sigma-Aldrich (Praha, Czech Republic). Demineralized water was used throughout the experiments. Thermoplastic polyolefin elastomer of type Versify 3401 (Dow Europe GmbH; Rheinmünster,

Germany) and polypropylene random copolymer TOTAL PPR 6298 S01 (Total Petrochemicals & Refining S.A./N.V., Bruxelles, Belgium) were utilized as the polymer matrix or blends, respectively.

2.2. Synthesis of Ag-ZnO

Firstly, a solution of 0.1 mol $(ZnNO_3)_2 \cdot 6H_2O$ was mixed together with a solution of 0.01 mol $AgNO_3$; the total volume of water used to dissolve both salts was 100 mL; the amount of 0.125 mol $(NH_4)_2 \cdot CO_3$ was dissolved in 125 mL of water separately. The solution of ammonium carbonate was slowly added, during a period of 10 min, into the solution comprising $(ZnNO_3)_2 \cdot 6H_2O$ and $AgNO_3$, which was stirred vigorously throughout. The end product was washed by filtration and the powder obtained then dried at 100 °C for 8 h, then the intermediate was annealed at 450 °C for 2 h to obtain product.

2.3. Preparation of Nanocomposites

Pellets of the polypropylene (PP) and thermoplastic polyolefin elastomer (TPO) were premixed with the prepared filler and then the material was compounded by a co-rotating conical twin screw extruder Micro-Compounder Xplore MC15 (DSM Xplore Instruments BV, Sittard, The Netherlands). Melt treatment of a mixture at the speed of 50 rpm for 4 min at temperature set to 200 °C for the barrel, the die and the circular loop guaranteed achieving of a constant torque, thus demonstrating that stabilization had occurred, thereby indicating that the filler had been homogeneously mixed into the matrix. Afterwards, extruded strands were cooled in air and cut to the pellets. The filler concentrations 1, 3, and 5 wt % were chosen by the authors; hence, hereinafter the given filler containing compound is referred to as Ag-ZnO1, Ag-ZnO3 or Ag-ZnO5.

Consequently, sheets of the polymer nanocomposites, measuring 1 mm and 4 mm in thickness, were produced by the hot press method (Fontijne LabEcon 300, Barendrecht, The Netherlands); this involved preheating at 200 °C for 2 min, followed by compression for 4 min and subsequent cooling under pressure of 200 kPa. A pure PP and a PP/TPO blends without any filler were prepared in exactly the same way to obtain reference samples. The sheets obtained were applied as testing samples for evaluating mechanical properties, electrical resistivity and antibacterial activity.

The aforementioned pure PP and PP/TPO blends were processed with various PP/TPO weight contents, namely 100/0, 75/25 and 50/50 w/w, respectively. Hereinafter, these blends are referred to as PP100, PP75, PP50.

2.4. Structural Characterization

The crystalline structure of the powder obtained was characterized by X-ray diffraction (XRD), on a Rigaku Miniflex 600 X-ray diffractometer (Rigaku Europe SE, Neu-Isenburg, Germany) operated at 40 kV and 15 mA. Nickel-filtered Co Kα1 radiation (λ = 1.78892 Å) was applied over a 25–90° angular region, with the step size and rate set to 0.02° and 10°/min, respectively. The crystallite size (d_{diffr}) of the samples was estimated using the Scherrer Equation (1):

$$d_{diffr} = (0.9\lambda)/(\beta \cos\theta) \tag{1}$$

This involved gauging the line-broadening (β, full width at half-maximum, corrected by the response of the instrument) of the main peak intensity for zinc oxide (101) and (111) for silver, where λ is the wavelength for Co Kα1 radiation and 2θ is the Bragg angle. Phase composition was evaluated via Rigaku Miniflex 600 software, utilizing the normalized RIR method. The RIR is the ratio between the integrated intensities of the peak of interest and that of a known standard.

The micrographs of the prepared powder and nanocomposites were taken on scanning electron microscopes, a Phenom Pro (Phenom-World, Eindhoven, The Netherlands) unit for preliminary and overall observations, and a Nova NanoSEM 650 (FEI, Brno, Czech Republic) unit equipped with a backscattered electron (BSE) detector for detailed imaging. Foregoing coating of examined samples comprising a thin layer of gold/palladium had been applied by a SC 7640 sputter coater (Quorum Technologies, Lewes, UK).

The specific surface area A_{BET} was determined via multipoint Brunauer-Emmet-Teller (BET) analysis of the nitrogen adsorption/desorption isotherms at 77 K, recorded on Belsorp-mini II (BEL Japan Inc., Osaka, Japan) apparatus. Grain size is expressed as the mean diameter d_{BET}, according to Equation (2) [39]:

$$d_{BET} = 6/(\rho_s \cdot A_{BET}) \qquad (2)$$

where ρ_s is the density of adsorbent material. The equation is based on spherical approximation of the compact shape of grains.

The method of Barrett, Joyner, and Halenda (BJH) was employed for calculating pore size distributions from experimental isotherms using the Kelvin model of pore filling. Pore size distribution is calculated from desorption isotherm.

2.5. Investigation of Mechanical Properties

Tensile and three point flexural tests were carried out on a Testometric universal-testing machine of type M 350-5CT (Testometric Co. Ltd., Rochdale, UK), equipped with a load cell of 300 kN. The tensile properties of the polymer nanocomposites were investigated using a crosshead speed of 50 mm/min and the length of the gauge equalled 50 mm. The dumb-bell-shaped specimens of Type 2 (specified by ISO 527) of the thickness 1 mm were cut out from prepared sheets by the means of punching press. In total, six specimens of each material were tested.

Five individual samples of dimensions 80 mm × 10 mm × 4 mm, according to the ISO 178 standard [40], for all the compositions were milled out from the 4 mm pressed sheets using CNC milling machine Charly4U (Mecanumeric Co., Motta di Livenza, Italy) in order to define both the flexural strength and modulus of the same. A span of 64 mm was applied, pertaining to a span to depth ratio of 16:1. The samples were positioned in the middle of the supports, and the device was operated at a speed of 1.0 mm/min. under ambient temperature conditions.

2.6. Investigation of Electrical Properties

The surface and volume resistivity of the prepared nanocomposite films (according to the ASTM D257 standard [41]) was evaluated by means of a Keithley 8009 Resistivity Test Fixture, on a Keithley 6517A Electrometer/High Resistance Meter (Keithley Instruments Inc., Cleveland, OH, USA). Surface and volume resistivity was obtained under a DC voltage of 40 V after a bias time of 60 s. The testing specimens in the shape of discs of a diameter 65 mm and thickness 1 mm were cut out from the pressed sheets.

The electrical breakdown strengths of the polymer nanocomposites were gauged using a GLP1-g High-Voltage Tester (Schleich Co., Hemer, Germany), with measurements being taken eight times for each specimen for reasons of accuracy. A DC voltage was slowly applied to the sample(s) at a rate of approximately 0.1 kV per second, and each voltage was maintained for 10 s in order to evaluate the breakdown strength of said sample. The upper limit of current drawn during the experiment equalled 6 mA. All the tests were carried out at room temperature.

2.7. Evaluation of Antibacterial Activity

The surface antibacterial activity of the prepared compounds was assessed in vitro against *E. coli* ATCC 8739 and *S. aureus* ATCC 6538P, these comprising the representative strains of gram-negative and gram-positive bacteria, respectively; this was conducted according to ISO 22196:2007 (E) [42] "Plastics—measurement of antibacterial activity on plastics surfaces". The dimensions of the sample, square test pieces of sheet were 25 mm × 25 mm × 1 mm. In light of previous experience with the tests, modification to the original protocol was made that was adherent with ISO 22196:2007 (E) [42], the purpose being to reduce the risk of false results. The number of colonies grown from recovered cells was estimated after 48 h of cultivation (instead of 24 h as required by the original standard procedure) to ensure that all the colonies had developed to form countable sizes. Hence, the overall duration of the test after inoculation was 48 h at 35 °C [5,19]. An incubator, a HERAcell 150i model (Thermo Scientific, Waltham, MA, USA), was applied in this part of the work. Antibacterial activity, delineated as R, was calculated by Equation (3):

$$R = (U_t - U_0) - (A_t - U_0) = U_t - A_t \tag{3}$$

where R is antibacterial activity; U_0 is the average logarithm for the number of viable bacteria, in cells/cm^2, recovered from untreated test specimens immediately after inoculation; U_t is the average logarithm for the number of viable bacteria, in cells/cm^2, recovered from untreated test specimens after 48 h; and A_t is the average logarithm for the number of viable bacteria, in cells/cm^2, recovered from treated test specimens after 48 h [5,19]. Indeed, in several cases, some additional colonies were found by this procedure, which avoided incorrect overestimation of antibacterial activity as caused merely by the slower growth of the colonies.

Converting from logarithmic (base 10) reduction R to the percentage of reduction (P) is possible through applying the following formula: $P = 100 \times (1 - 10^{-R})$. This means that the R-values ($\log_{10}$ scale) of 1, 2, 3, 4, 5 and 6 correspond to reductions in microbial load of 90%, 99%, 99.9%, 99.99%, 99.999% and 99.9999%, respectively.

3. Results and Discussion

3.1. Crystal Structure and BET Characterization of the Filler

The powder XRD pattern for the prepared Ag-ZnO filler is shown in Figure 1. All diffraction peaks observed at 2θ = 37.06°, 40.19°, 42.35°, 55.80°, 66.78°, 74.49°, 78.97°, 80.94°, 82.38° and 86.86° are characteristic for the wurtzite ZnO structure (hexagonal phase, space group P63mc) and are in good agreement with the JCDD PDF-2 entry 01-079-0207. Diffraction peaks at 2θ = 44.60°, 51.99° and 76.53° correspond well with the fcc crystal structure of silver detailed in the JCDD PDF-2 entry 01-087-0720.

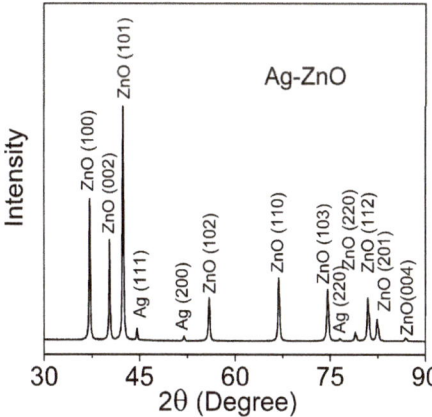

Figure 1. XRD pattern for the prepared Ag-ZnO hierarchical microparticles obtained by the synthetic method.

Quantitative analysis of the XRD pattern, performed via the reference intensity ratio (RIR) method (applying the Corundum standard [43]), showed that the content of silver in the Ag-ZnO sample was merely 1.3 wt %, with the remaining 98.7% constituting the wt % of ZnO. This confirms the presence of the target amount of metallic Ag on said ZnO. The fact that there is no shift in the peak position for the

ZnO phase in the sample indicates that Ag^+ ions either enter into the lattice of the ZnO or substitute a Zn site [44]. It is reasonable to expect, that metallic Ag particles are positioned on the surfaces or interfaces of the ZnO nanoparticles. Other Ag peaks were not observed in the XRD patterns, which is probably due to the small quantity of Ag nano-particles and their excellent dispersion.

The diffraction lines from the (101) and (111) planes in ZnO and Ag, respectively, as visible in the diffractogram images at 2θ angle 42.42° and 44.60°, were applied so as to deduce the diffracting area size d_{diffr}. Findings pertaining to the structural and morphological parameters obtained by XRD are displayed in Table 1. Assuming that the grains are spherical in shape and of uniform size, average particle size can be obtained via Equation 2 with respect to grain size d_{BET}. Thus, the analysis revealed the figure to be 64 nm. The average material density of Ag-ZnO was obtained from the material composition estimated by XRD assuming the tabular material density of silver (10.49 g/cm^3) and ZnO (5.61 g/cm^3). The specific surface area for a crystallite (A_{diffr}) is easily calculated from the size of the crystallites (d_{diffr}) and the material density of Ag-ZnO (5.64 g/cm^3). According to analysis of sorption/desorption isotherm, the pore sizes of our sample are expected to be approximately 17 nm. Although pores may not be truly cylindrical, we refer to their sizes as diameters consistent with Kelvin model, thus making the measurement error to be acceptable; therefore, we used the simple BJH method.

Table 1. Summary of XRD and BET analyses of Ag-ZnO. SSA stands for Specific surface area.

BET Method				XRD Method		
SSA, A_{BET} (m^2·g^{-1})	Grain Size, d_{BETm} (nm)	Pore Size, (BJH) (nm)	Phase	Crystallite Size, d_{diffr} (nm)	SSA for Pure Phase, A_{diffr} (m^2·g^{-1})	Contribution of Phases to SSA for Ag-ZnO (m^2·g^{-1})
16.6	64	17	Ag	51	11.2	0.15
			ZnO	49	21.8	21.52
			Total SSA of Ag-ZnO, A_{diffr} Ag-ZnO			21.7

Comparison can be made of results obtained from X-ray diffraction line-broadening analysis and those obtained from gas sorption BET analysis. The first and usually the most valuable information determined by BET concerns specific surface area A_{BET}. As can be expected, A_{BET} is lesser than the specific crystallite surface area. The X-ray diffraction characteristics are correctly obtained for coherently diffracting areas, i.e., pertaining to the size of the nanocrystalline domains, while gas sorption analysis examines the actual surface of the porous body accessible to N_2 molecules; the latter analysis characterises the surface of grains. Consequently, the average diameter of grain is bigger than that of nanocrystals. Hence, A_{BET} relates to the surfaces of grains that consist of one or more crystallites and to the amorphous phase, which should be present. Interface between congruent particles is not accessible to nitrogen adsorption becoming thus virtually invisible for BET measurement. It can be estimated from third powers of grain and nanocrystallite diameter that one grain comprises of approximately two nanocrystallites only. Such small packing level testifies for chain-like, weakly branched or twinning morphology of grains.

Higher activity of Ag-ZnO hierarchical particles with polymer matric against the microparticles due to relatively the large surface area, can be assumed, therefore some features of the composite have been improved.

3.2. Morphology of the Powder Filler

Microphotographs of the prepared filler obtained by SEM are shown in Figure 2. Here, Figure 2a,d shows the intermediate. An overall image in Figure 2a shows large agglomerates possessing diameter up to 70 µm. They exhibit complicated hierarchical morphology. A large agglomerate comprises aggregates of the layered zinc hydroxide nitrate complex containing most likely nanoparticles of silver oxide, as can be expected to be a product of mild basic precipitation conditions in the first stage of synthesis. The apparent size of these primary agglomerates is about several micrometers and they are arranged into grape-like assemblies that seems to be bound and covered by a network of fibres. Their morphology resembles a shrub covered by web produced by Spindle Ermine (*Yponomeuta cagnagella*) moth caterpillars. The diameters of fibers measure up to 200 nm, while their lengths are equal to tens of microns (Figure 2b,d).

It can be expected that these fine fiber structures are created in later stages of synthesis. During annealing of the zinc complex, conversion is made to the nanostructured hybrid Ag-ZnO microparticles resembling shape of original intermediate agglomerates and keeping their hierarchical morphology. Thermal decomposition of zinc hydroxide nitrates yields spherical, connected microparticles of ZnO, which go on to form long chains and two-dimensional networks of polyhedral to rounded nanoparticles, as seen in Figure 2e,h. The ZnO phase obtain this morphology during annealing as a legacy of original layered morphology of zinc hydroxide nitrates by topotactic transition [25,45]. The diameters of the ZnO nanoparticles range up to 100 nm in size. Figure 2h details the nanoparticle building blocks. The size and morphology of particles revealed by the SEM is in good agreement with results of BET analysis.

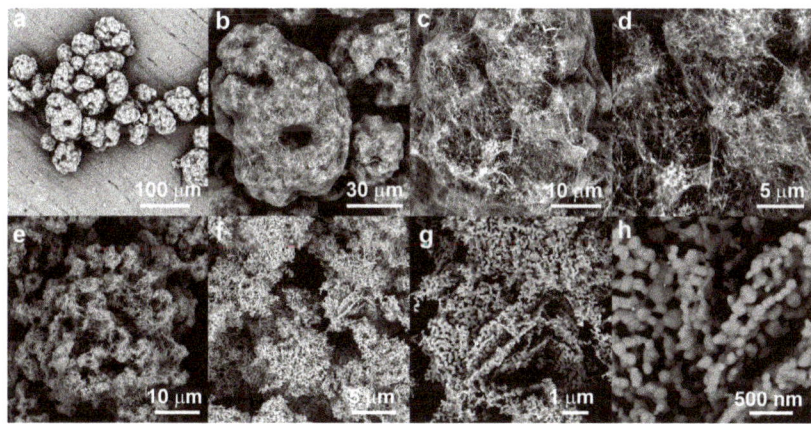

Figure 2. SEM microphotographs of the Ag-ZnO filler (**a–d**) before annealing and (**e–h**) after the annealing process.

3.3. Characterization of TPO Compounds

The morphology of the prepared polymer nanocomposites with 5 wt % of filler was analysed by SEM microscopy on the surfaces obtained by freeze fracturing in liquid nitrogen. Good dispersion and distribution of the filler in the polymer matrix is visible in Figure 3 recorded as representative images for typical sample fracture surface. As was demonstrated by SEM analysis, the filler is comprised from sparsely networked chains of nanoparticles that can be considered loosely bound. The interface between nanoparticles in long chains may be the weakest points and thus places where the cohesive forces of the synthesized ZnO microparticles can be overcome during compounding, a phenomenon which arises through high shear and elongation stresses, pertaining to reduction in the size of a component with sub-critical cohesive properties during compounding by mixing the molten phase. Therefore, large agglomerates of the filler were dispersed in the polymer matrix on individual particulates of the Ag-ZnO nanoparticles, said particulates measuring up to 100 nm in diameter and these single particulates dispersed in the polymer matrix are evident in Figure 3 for all compared nanocomposites. Separation of the silver nanoparticles from ZnO nanoparticles was neither confirmed nor excluded due to their very low concentration.

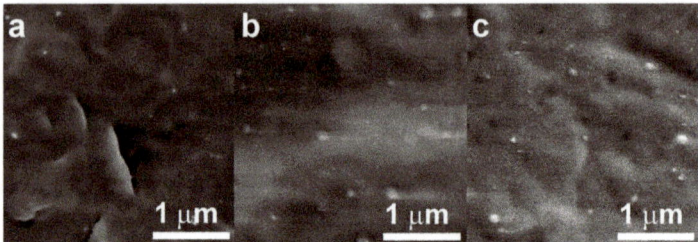

Figure 3. SEM microphotographs of fracture surfaces of polymer nanocomposites with Ag-ZnO particles: (**a**) PP100/Ag-ZnO5; (**b**) PP75/Ag-ZnO5; (**c**) PP50/Ag-ZnO5.

3.4. Mechanical Properties of the Polymer Nanocomposites

The authors chose to characterize the mechanical performance of the prepared nanocomposites according to their basic characteristics, i.e., yield stress, strain at break and flexural modulus, with adherence to ISO 527 [46] and ISO 178 [40]; the subsequent results are summarized in Table 2. Comparing the neat polymer matrix TPO and PP with all the prepared nanocomposites revealed that no adverse effects were exerted on the mechanical properties of the prepared material, either through using the fillers or the given weight percent of the same or through the processing conditions applied. Indeed, the mechanical properties e.g., flexural modulus, strain at break of the polymer nanocomposites and PP/TPO blends were only influenced by the amount of TPO material present dramatically. The yield stress and flexural modulus of the polymer nanocomposites slightly decreased alongside parallel increase in strain at break.

It cannot be expected, the mechanical properties will be dramatically influence by used spherical hierarchical microparticles of low-loading filler. On the other hand, the wettability and compatibility of hierarchical microfiller by thermoplastic polymer matrix plays important role for transport mechanical stress from polymer to filler.

Table 2. Mechanical properties of the compounds.

Samples	Yield Stress (MPa) Mean ± SD [a]	Strain at Break (%) Mean ± SD [a]	Flexural Modulus (MPa) Mean ± SD [a]
Neat PP100	30.9 ± 2.2	14.5 ± 1.3	850 ± 50
PP100/Ag-ZnO1	34.2 ± 2.5	12.3 ± 1.4	840 ± 40
PP100/Ag-ZnO3	32 ± 3	13.6 ± 1.3	840 ± 40
PP100/Ag-ZnO5	30 ± 5	17.4 ± 2.6	830 ± 40
Neat PP75	18.8 ± 1.3	1120 ± 120	183 ± 15
PP75/Ag-ZnO1	18.6 ± 1.4	1100 ± 110	180 ± 19
PP75/Ag-ZnO3	18.5 ± 1.2	1035 ± 150	178 ± 17
PP75/Ag-ZnO5	18.5 ± 1.6	1020 ± 130	181 ± 20
Neat PP50	15.5 ± 0.9	1300 ± 130	34 ± 7
PP50/Ag-ZnO1	15.2 ± 1.3	1280 ± 150	32 ± 10
PP50/Ag-ZnO3	14.8 ± 0.7	1260 ± 130	37 ± 7
PP50/Ag-ZnO5	14.5 ± 1.1	1250 ± 170	38 ± 8

[a] Standard deviation.

3.5. Electrical Properties of the Nanocomposites

Zinc oxide is a well-known semiconductor with a broad range of resistivity, which not only depends on the morphology of the materials, but especially on the type and concentration of a dopant. The resistivity can be varied within the range 10^{-4} to 10^9 $\Omega \cdot cm$ by doping. This means that ZnO can successfully be prepared even in a conductive state, although a heavily doped material and special conditions are required [47]. In contrast, silver is an excellent metallic conductor with a resistivity of 1.59×10^{-6} $\Omega \cdot$cm [48]. Therefore, a combination of these materials could produce a hybrid with reasonably high electrical conductivity, while Ag-ZnO as a powder filler would cause conductivity, even under circumstances of low content of the filler in the polymer matrix, if the critical threshold value is exceeded [6].

Nevertheless, due to their chemical nature, most TPO compounds and PP are fine electrical insulating materials; they boast good dielectric strength and do not absorb moisture. To check the performance of prepared nanocomposites, electrical properties were tested with special attention paid to the electrical strength as a critical phenomenon related with the material endurance and failure.

Table 3 summarizes the electrical surface, volume resistivity and dielectric strength for all the neat and prepared materials. The surface resistivity of the neat PP and PP/TPO blends lies in the region 3.2×10^{12} to 1.1×10^{13} Ω/sq, while volume resistivity ranges 7.2×10^{15} to 9.8×10^{15} $\Omega \cdot$cm. The surface resistivity and volume of the neat PP and PP/TPO blend correspond to literature values [49], whereas the hybrid, nanostructured Ag-ZnO microparticles (compressed into disc-shaped pellets) exhibited the surface resistivity value 6.6×10^{6} $\Omega \cdot$cm, which resembles the figure for semi-conductors. The values observed for resistivity are almost identical for the polymer nanocomposites filled with Ag-ZnO and the blend matrixes of neat PP or PP/TPO. Hence, it is evident that the resistivity of the composite material is fully governed by the properties of the given matrix.

Although the filler is not conductive enough to establish conductivity due long-range connectivity of the particles at high concentration, it may affect the critical behaviour of the material under condition of electrical breakdown. The data on dielectric strength for all the samples are shown in Table 3. When comparing them, the results are obvious: the values for dielectric strength vary insignificantly for the PP/TPO blend. The polypropylene and pure blends reach dielectric strength at approximately 17 kV/mm. This is in contrast with the polymer nanocomposites, where a small yet pronounced trend can be observed. These exhibit lower values of only 13.6 to 15.5 kV/mm. The decrease in breakdown strength is attributed to the significant inhomogeneity of the local electric field, caused by differences between the polarizability of the dielectric polymer matrix and the dielectric microparticles of the filler.

Table 3. Electrical properties of the compounds.

Sample	Resistivity, R		Dielectric Strength
	Surface Resistivity (Ω/sq)	Volume Resistivity ($\Omega \cdot$cm)	Mean $\pm$ SD [a] (kV/mm)
Ag-ZnO filler	-	6.6×10^{6}	-
Neat PP100	1.1×10^{13}	9.8×10^{15}	17.4 ± 0.5
PP100/Ag-ZnO1	3.6×10^{12}	5.5×10^{15}	15.6 ± 1.4
PP100/Ag-ZnO3	3.0×10^{12}	4.1×10^{15}	14.8 ± 1.2
PP100/Ag-ZnO5	3.5×10^{12}	4.2×10^{15}	13.6 ± 0.6
Neat PP75	7.5×10^{12}	8.7×10^{15}	16.8 ± 0.7
PP75/Ag-ZnO1	3.2×10^{12}	4.8×10^{15}	15.5 ± 1.2
PP75/Ag-ZnO3	4.0×10^{12}	5.5×10^{15}	15.2 ± 1.3
PP75/Ag-ZnO5	3.7×10^{12}	6.1×10^{15}	14.1 ± 0.6
Neat PP50	3.2×10^{12}	7.2×10^{15}	17.2 ± 1.1
PP50/Ag-ZnO1	2.7×10^{12}	5.5×10^{15}	14.4 ± 1.2
PP50/Ag-ZnO3	3.0×10^{12}	4.0×10^{15}	15.5 ± 2.6
PP50/Ag-ZnO5	3.2×10^{12}	3.9×10^{15}	14.6 ± 1.4

[a] Standard deviation.

3.6. Surface Antibacterial Activity

The relatively finite concentration of the filler utilized should not influence the mechanical or electrical properties of the neat PP or PP/TPO blend dramatically; indeed, it was observed. However, antibacterial activity has to be significantly evident if the filler can truly impart this desired property to prepared compounds even at a 1 wt % loading. Herein, the antibacterial activities of the polymer nanocomposite materials were tested according to the standard ISO 22196:2007 (E) [42] against *E. coli* and *S. aureus*.

The results are summarised in Table 4, where neat PP100, PP75 and PP50 comprise reference samples, giving the U_t value, while the R value indicates the surface antibacterial activity of all the prepared polymer nanocomposites. The earlier JIS Z 2801 standard [50], which preceded ISO 22196:2007 (E) [42], specified an R-value of 2.0 or more as demonstration of antibacterial activity. Therefore, most of the prepared polymer nanocomposite materials can be categorized as exhibiting antibacterial activity if they show values exceeding 2.0 on their surface; this categorization is brought about by compounding them with fillers under testing This critical value of 2.0 is adequate for hygienic and similar applications. Nonetheless, a value of 6.0, i.e., a 99.9999% reduction in cell count against controls, is considered applicable for advanced medical uses of plastics [26].

All the prepared polymer nanocomposites containing 5 wt % of nanostructured Ag-ZnO filler showed surface antibacterial activity against *E. coli* of greater than 6.2. The surface antibacterial activity of said polymer nanocomposites against *S. aureus* exceeds 2.5, although only the PP100/Ag-ZnO5 sample demonstrated an R-value that reached 4.8. With respect to the peculiarity of *S. aureus* inhibition, this result can also be evaluated as excellent, as the lower sensitivity of *S. aureus* to antibacterial agents is generally recognized [6,9,20,26,51]. The antibacterial activity against both bacteria is lower for 3 wt % content, but it is still sufficient for utilization in hygienic applications. It can be seen from Table 3, that the higher concentration of TPO is present in the compound the lower R-value is experienced. This effect is more pronounced for *S. aureus* and for lower concentrations of the filler.

Table 4. Evaluation of surface antibacterial activity of Ag-ZnO polymer nanocomposites according to ISO 22196:2007 (E) [42].

Sample	R-Value for S. aureus (-)	Efficiency against S. aureus (%)	R-Value for E. coli (-)	Efficiency against E. coli (%)
Neat PP100	U_t = 5.4		U_t = 6.2	
PP100/Ag-ZnO1	2.2	99.3690	4.2	99.9937
PP100/Ag-ZnO3	4.6	99.9975	5.5	99.9997
PP100/Ag-ZnO5	4.8	99.9984	>6.2	99.9999
Neat PP75	U_t = 5.7		U_t = 6.3	
PP75/Ag-ZnO1	0.7	80.0474	2.5	99.6838
PP75/Ag-ZnO3	2.2	99.3690	5.5	99.9997
PP75/Ag-ZnO5	2.8	99.8415	>6.3	99.9999
Neat PP50	U_t = 5.9		U_t = 6.3	
PP50/Ag-ZnO1	0.5	68.3772	1.5	96.8377
PP50/Ag-ZnO3	1.8	98.4151	2.5	99.6838
PP50/Ag-ZnO5	2.5	99.6838	>6.3	99.9999

The antimicrobial effects of materials coming from the mixture of an antimicrobial agent and a non-active polymer are similar to some extent as the mechanism of the agent itself. The possible mechanisms of killing microorganisms by Ag-ZnO filler were partly discussed elsewhere [6,8,20]. The effect of silver ions may be explained as follows: (1) uptake of free silver ions followed by disruption of ATP (Adenosine triphosphate) production and DNA replication; (2) silver nanoparticle and silver ion generation of reactive oxygen species (ROS); and (3) silver nanoparticle direct damage to cell membranes [52–55]. Similarly, the release of Zn^{2+} cations or its complex forms is one of the proposed mechanisms of ZnO's activity [56] while the generation of reactive oxygen species (ROS) is another relevant explanation [57], but it requires illumination of the material by light with at least some portion of energy transferred by photons in the ultraviolet A (UVA) region [26,58]. Since the particles are embedded in the polymer matrix, it can be expected that mechanical detachment of the nanoparticles from the composite surface and their attachment to the surface of bacterial cells is not the principal mechanism imparting antibacterial property to the material. Previously reported synergy between silver and zinc oxide nanoparticles is based on direct contact and processes at the interface between the metallic particle and its semiconductor counterpart which may enhance both release of ions as well as ROS generation but not the mechanical particle action [8,20,51].

The presence of filler particles in the surface-region of the composites seems to be the first approach to understand observed trends in Table 4 [3,59]. Such direct mechanism would explain dependence of antibacterial activity on the filler concentration. However, a strong effect of used polymer matrix is clearly manifested too. Without any doubt, PP can be assessed as much efficient matrix than TPO. Embedding of particles into polymer matrix in both surface and near subsurface-region invokes a three step mechanism. First, it requires diffusion of water into the matrix, then corrosion of the filler particles resulting into the release of ions or ROS generation and finally, replenishment of active species on the polymer composite surface by diffusion [3,52].

According to the summary of results in Table 4, *E. coli* was more responsive to the effect of used filler than *S. aureus*. It can be explained by the difference in the cell wall structure. The thick peptidoglycan cell wall of Gram-positive bacteria protects its cell from silver penetration while Gramm-negative bacteria lack this protection [60,61]. Similar effect of the cell wall may apply for the case of Zn^{2+} ions as well as for ROS action [56,57,62].

The synergy effect of metal semiconductor hybrid Ag-ZnO antibacterial filler was demonstrated in our previous work although on material prepared by other synthetic routes [9,20,51]. However, increase of antibacterial efficiency towards *S. aureus* remains still a highly challenging issue which can be further influenced by the effect of polymer matrix choice as discussed above.

4. Conclusions

This study has detailed a method with real potential for preparing an antibacterial polymer nanocomposite. A conventional synthesis technique is applied that is simple, and information is given on how to avoid issues when processing the nanomaterials into the polymer matrix.

This filler boasts excellent homogenous distribution, dispersion and adhesion to selected representatives of polypropylene and thermoplastic elastomer and their blends. Furthermore, it can support the antibacterial performance of the polymer nanocomposite. The surface antibacterial activity observed herein of the prepared materials is assessed as excellent against *E. coli* and very high against *S. aureus*; hence they compare favourably against other materials available currently. It seems, that addition of TPO into polymer blend decreases the antibacterial activity of the nanocomposite keeping filler concentration constant. On the other hand, the mechanical and other properties of PP can be modified significantly by addition of TPOs in small concentrations only, as can be seen from relatively small effect of doubling the TPO concentration in the compound from 25 to 50 wt %.

Moreover, the mechanical and electrical properties of the polymer resin and blends utilized are not affected by adding a small amount of filler. Indeed, the prepared polymer nanocomposites possess the same resistivity as the neat matrix while dielectric strength is lowered only a little.

These facts testify to the fact that antibacterial polymer systems comprising Ag-ZnO nanostructured microparticles could potentially be employed as additives in plastic medical devices, in addition to finding uses in industries that require antibacterial action by a material, e.g., sanitary, hygienic or other interior applications.

Acknowledgments: This work was supported by the Ministry of Education, Youth and Sports of the Czech Republic-Program NPU I (LO1504). This article was written with the support of Operational Program Research and Development for Innovations co-funded by the European Regional Development Fund (ERDF) and national budget of the Czech Republic, within the framework of the project CPS—strengthening research capacity (reg. number: CZ.1.05/2.1.00/19.0409) and an internal grant from TBU in Zlin no. IGA/CPS/2017/007.

Author Contributions: Pavel Bazant, Ivo Kuritka and David Podlipny conceived and designed the experiments; Pavel Bazant, David Podlipny and Pavlina Holcapkova performed the experiments; Tomas Sedlacek and Ivo Kuritka analyzed the data; Pavel Bazant, Ivo Kuritka and Tomas Sedlacek wrote the paper.

Conflicts of Interest: The authors declare no conflict of interest.

References

1. Shaviv, E.; Schubert, O.; Alves-Santos, M.; Goldoni, G.; Di Felice, R.; Vallée, F.; Del Fatti, N.; Banin, U.; Sönnichsen, C. Absorption Properties of Metal–Semiconductor Hybrid Nanoparticles. *ACS Nano* **2011**, *5*, 4712–4719. [CrossRef] [PubMed]
2. Paul, D.R.; Robeson, L.M. Polymer Nanotechnology: Nanocomposites. *Polymer* **2008**, *49*, 3187–3204. [CrossRef]
3. Palza, H. Antimicrobial Polymers with Metal Nanoparticles. *Int. J. Mol. Sci.* **2015**, *16*, 2099–2116. [CrossRef] [PubMed]
4. Aricò, A.S.; Bruce, P.; Scrosati, B.; Tarascon, J.-M.; van Schalkwijk, W. Nanostructured Materials for Advanced Energy Conversion and Storage Devices. *Nat. Mater.* **2005**, *4*, 366–377. [CrossRef] [PubMed]
5. Scholes, G.D. Book Review of Semiconductor Nanocrystal Quantum Dots: Synthesis, Assembly, Spectroscopy and Applications. *J. Am. Chem. Soc.* **2008**, *130*, 18028. [CrossRef]
6. Bazant, P.; Kuritka, I.; Hudecek, O.; Machovsky, M.; Mrlik, M.; Sedlacek, T. Microwave-Assisted Synthesis of Ag/ZnO Hybrid Filler, Preparation, and Characterization of Antibacterial Poly(Vinyl Chloride) Composites Made from the Same. *Polym. Compos.* **2014**, *35*, 19–26. [CrossRef]
7. Lu, W.; Liu, G.; Gao, S.; Xing, S.; Wang, J. Tyrosine-Assisted Preparation of Ag/ZnO Nanocomposites with Enhanced Photocatalytic Performance and Synergistic Antibacterial Activities. *Nanotechnology* **2008**, *19*, 445711. [CrossRef] [PubMed]
8. Ghosh, S.; Goudar, V.S.; Padmalekha, K.G.; Bhat, S.V.; Indi, S.S.; Vasan, H.N. ZnO/Ag Nanohybrid: Synthesis, Characterization, Synergistic Antibacterial Activity and Its Mechanism. *RSC Adv.* **2012**, *2*, 930–940. [CrossRef]
9. Bazant, P.; Munster, L.; Machovsky, M.; Sedlak, J.; Pastorek, M.; Kozakova, Z.; Kuritka, I. Wood Flour Modified by Hierarchical Ag/ZnO as Potential Filler for Wood–plastic Composites with Enhanced Surface Antibacterial Performance. *Ind. Crops Prod.* **2014**, *62*, 179–187. [CrossRef]
10. Dufour, D.; Leung, V.; Lévesque, C.M. Bacterial Biofilm: Structure, Function, and Antimicrobial Resistance. *Endod. Top.* **2010**, *22*, 2–16. [CrossRef]
11. Lindsay, D.; von Holy, A. Bacterial Biofilms within the Clinical Setting: What Healthcare Professionals Should Know. *J. Hosp. Infect.* **2006**, *64*, 313–325. [CrossRef] [PubMed]
12. Beuchat, L.R. Pathogenic Microorganisms Associated with Fresh Produce. *J. Food Prot.* **1996**, *59*, 204–216. [CrossRef]
13. Samuel, U.; Guggenbichler, J.P. Prevention of Catheter-Related Infections: The Potential of a New Nano-Silver Impregnated Catheter. *Int. J. Antimicrob. Agents* **2004**, *23*, 75–78. [CrossRef] [PubMed]
14. Okelo, P.O.; Wagner, D.D.; Carr, L.E.; Wheaton, F.W.; Douglass, L.W.; Joseph, S.W. Optimization of Extrusion Conditions for Elimination of Mesophilic Bacteria during Thermal Processing of Animal Feed Mash. *Anim. Feed Sci. Technol.* **2006**, *129*, 116–137. [CrossRef]
15. Flores, G.E.; Bates, S.T.; Knights, D.; Lauber, C.L.; Stombaugh, J.; Knight, R.; Fierer, N. Microbial Biogeography of Public Restroom Surfaces. *PLoS ONE* **2011**, *6*, e28132. [CrossRef] [PubMed]
16. Bartlett, K.H.; Kennedy, S.M.; Brauer, M.; van Netten, C.; Dill, B. Evaluation and Determinants of Airborne Bacterial Concentrations in School Classrooms. *J. Occup. Environ. Hyg.* **2004**, *1*, 639–647. [CrossRef] [PubMed]
17. Seaton, A.; Tran, L.; Aitken, R.; Donaldson, K. Nanoparticles, Human Health Hazard and Regulation. *J. R. Soc. Interface* **2009**. [CrossRef] [PubMed]
18. De Jong, W.H.; Borm, P.J. Drug Delivery and Nanoparticles: Applications and Hazards. *Int. J. Nanomed.* **2008**, *3*, 133–149. [CrossRef]
19. Wiesner, M.R.; Lowry, G.V.; Alvarez, P.; Dionysiou, D.; Biswas, P. Assessing the Risks of Manufactured Nanomaterials. *Environ. Sci. Technol.* **2006**, *40*, 4336–4345. [CrossRef] [PubMed]
20. Bazant, P.; Kuritka, I.; Munster, L.; Machovsky, M.; Kozakova, Z.; Saha, P. Hybrid Nanostructured Ag/ZnO Decorated Powder Cellulose Fillers for Medical Plastics with Enhanced Surface Antibacterial Activity. *J. Mater. Sci. Mater. Med.* **2014**, *25*, 2501–2512. [CrossRef] [PubMed]
21. Li, S.; Meng, L.M.; Toprak, M.S.; Kim, D.K.; Muhammed, M. Nanocomposites of Polymer and Inorganic Nanoparticles for Optical and Magnetic Applications. *Nano Rev.* **2010**, *1*, 5214. [CrossRef] [PubMed]
22. Clemons, C.M.; Caulfield, D.F. Natural Fibers. In *Functional Fillers for Plastics*; Xanthos, M., Ed.; Wiley-VCH Verlag GmbH & Co. KGaA: Weinheim, Germany, 2005; pp. 195–206. [CrossRef]

23. Müller, K.; Bugnicourt, E.; Latorre, M.; Jorda, M.; Echegoyen Sanz, Y.; Lagaron, J.M.; Miesbauer, O.; Bianchin, A.; Hankin, S.; Bölz, U.; et al. Review on the Processing and Properties of Polymer Nanocomposites and Nanocoatings and Their Applications in the Packaging, Automotive and Solar Energy Fields. *Nanomaterials* **2017**, *7*, 74. [CrossRef] [PubMed]
24. Tanahashi, M. Development of Fabrication Methods of Filler/Polymer Nanocomposites: With Focus on Simple Melt-Compounding-Based Approach without Surface Modification of Nanofillers. *Materials* **2010**, *3*, 1593–1619. [CrossRef]
25. Hornsby, P. Compounding of Particulate-Filled Thermoplastics. In *Polymers and Polymeric Composites: A Reference Series*; Palsule, S., Ed.; Springer: Berlin/Heidelberg, Germany, 2016; pp. 1–16.
26. Machovsky, M.; Kuritka, I.; Bazant, P.; Vesela, D.; Saha, P. Antibacterial Performance of ZnO-Based Fillers with Mesoscale Structured Morphology in Model Medical PVC Composites. *Mater. Sci. Eng. C Mater. Biol. Appl.* **2014**, *41*, 70–77. [CrossRef] [PubMed]
27. Jang, Y.H.; Kochuveedu, S.T.; Cha, M.-A.; Jang, Y.J.; Lee, J.Y.; Lee, J.; Lee, J.; Kim, J.; Ryu, D.Y.; Kim, D.H. Synthesis and Photocatalytic Properties of Hierarchical Metal Nanoparticles/ZnO Thin Films Hetero Nanostructures Assisted by Diblock Copolymer Inverse Micellar Nanotemplates. *J. Colloid Interface Sci.* **2010**, *345*, 125–130. [CrossRef] [PubMed]
28. Zheng, Y.; Zheng, L.; Zhan, Y.; Lin, X.; Zheng, Q.; Wei, K. Ag/ZnO Heterostructure Nanocrystals: Synthesis, Characterization, and Photocatalysis. *Inorg. Chem.* **2007**, *46*, 6980–6986. [CrossRef] [PubMed]
29. Dou, P.; Tan, F.; Wang, W.; Sarreshteh, A.; Qiao, X.; Qiu, X.; Chen, J. One-Step Microwave-Assisted Synthesis of Ag/ZnO/Graphene Nanocomposites with Enhanced Photocatalytic Activity. *J. Photochem. Photobiol. A Chem.* **2015**, *302*, 17–22. [CrossRef]
30. Motshekga, S.C.; Ray, S.S.; Onyango, M.S.; Momba, M.N.B. Microwave-Assisted Synthesis, Characterization and Antibacterial Activity of Ag/ZnO Nanoparticles Supported Bentonite Clay. *J. Hazard. Mater.* **2013**, *262*, 439–446. [CrossRef] [PubMed]
31. Ye, X.-Y.; Zhou, Y.-M.; Sun, Y.-Q.; Chen, J.; Wang, Z.-Q. Preparation and Characterization of Ag/ZnO Composites via a Simple Hydrothermal Route. *J. Nanopart. Res.* **2009**, *11*, 1159–1166. [CrossRef]
32. Kakhki, R.M.; Tayebee, R.; Ahsani, F. New and Highly Efficient Ag Doped ZnO Visible Nano Photocatalyst for Removing of Methylene Blue. *J. Mater. Sci. Mater. Electron.* **2017**, *28*, 5941–5952. [CrossRef]
33. Patil, S.S.; Mali, M.G.; Tamboli, M.S.; Patil, D.R.; Kulkarni, M.V.; Yoon, H.; Kim, H.; Al-Deyab, S.S.; Yoon, S.S.; Kolekar, S.S.; et al. Green Approach for Hierarchical Nanostructured Ag-ZnO and Their Photocatalytic Performance under Sunlight. *Catal. Today* **2016**, *260*, 126–134. [CrossRef]
34. Huang, Q.; Zhang, Q.; Yuan, S.; Zhang, Y.; Zhang, M. One-Pot Facile Synthesis of Branched Ag-ZnO Heterojunction Nanostructure as Highly Efficient Photocatalytic Catalyst. *Appl. Surf. Sci.* **2015**, *353*, 949–957. [CrossRef]
35. Liu, Y.; Wei, S.; Gao, W. Ag/ZnO Heterostructures and Their Photocatalytic Activity under Visible Light: Effect of Reducing Medium. *J. Hazard. Mater.* **2015**, *287*, 59–68. [CrossRef] [PubMed]
36. Drobny, J.G. 7-Polyolefin-Based Thermoplastic Elastomers. In *Handbook of Thermoplastic Elastomers*; Plastics Design Library; William Andrew Publishing: Norwich, NY, USA, 2007; pp. 191–199, ISBN 9780323221368.
37. O'Connor, K.S.; Watts, A.; Vaidya, T.; LaPointe, A.M.; Hillmyer, M.A.; Coates, G.W. Controlled Chain Walking for the Synthesis of Thermoplastic Polyolefin Elastomers: Synthesis, Structure, and Properties. *Macromolecules* **2016**, *49*, 6743–6751. [CrossRef]
38. Leone, G.; Mauri, M.; Pierro, I.; Ricci, G.; Canetti, M.; Bertini, F. Polyolefin Thermoplastic Elastomers from 1-Octene Chain-Walking Polymerization. *Polymer* **2016**, *100*, 37–44. [CrossRef]
39. Rouquerol, J.; Rouquerol, F.; Llewellyn, P.; Maurin, G.; Sing, K.S.W. *Adsorption by Powders and Porous Solids: Principles, Methodology and Applications*; Academic Press: Cambridge, MA, USA, 2013; ISBN 978-0-12-598920-6.
40. *Plastics—Determination of Flexural Properties*; ISO 178:2010; International Organization for Standardization: Geneva, Switzerland, 2010.
41. *Standard Test Methods for DC Resistance or Conductance of Insulating Materials*; ASTM D257-14; ASTM International Standard: West Conshohocken, PA, USA, 2014.
42. *Plastics—Measurement of Antibacterial Activity on Plastics Surfaces*; ISO 22196:2007; International Organization for Standardization: Geneva, Switzerland, 2007.

43. Zhou, X.; Liu, D.; Bu, H.; Deng, L.; Liu, H.; Yuan, P.; Du, P.; Song, H. XRD-based quantitative analysis of clay minerals using reference intensity ratios, mineral intensity factors, Rietveld, and full pattern summation methods: A critical review. *Solid Earth Sci.* **2018**, *3*, 16–29. [CrossRef]
44. Saoud, K.; Alsoubaihi, R.; Bensalah, N.; Bora, T.; Bertino, M.; Dutta, J. Synthesis of Supported Silver Nano-Spheres on Zinc Oxide Nanorods for Visible Light Photocatalytic Applications. *Mater. Res. Bull.* **2015**, *63*, 134–140. [CrossRef]
45. Machovský, M.; Mrlík, M.; Plachý, T.; Kuřitka, I.; Pavlínek, V.; Kožáková, Z.; Kitano, T. The Enhanced Magnetorheological Performance of Carbonyl Iron Suspensions Using Magnetic Fe3O4/ZHS Hybrid Composite Sheets. *RSC Adv.* **2015**, *5*, 19213–19219. [CrossRef]
46. *Plastics—Determination of Tensile Properties—Part 2: Test Condition for Moulding and Extrusion Plastics*; ISO 527-2:2012; International Organization for Standardization: Geneva, Switzerland, 2012.
47. Look, D.C. Progress in ZnO Materials and Devices. *J. Electron. Mater.* **2006**, *35*, 1299–1305. [CrossRef]
48. Matula, R.A. Electrical Resistivity of Copper, Gold, Palladium, and Silver. *J. Phys. Chem. Ref. Data* **1979**, *8*, 1147–1298. [CrossRef]
49. Gulrez, S.K.H.; Ali Mohsin, M.E.; Shaikh, H.; Anis, A.; Pulose, A.M.; Yadav, M.K.; Qua, E.H.P.; Al-Zahrani, S.M. A Review on Electrically Conductive Polypropylene and Polyethylene. *Polym. Compos.* **2014**, *35*, 900–914. [CrossRef]
50. *Antimicrobial Products—Test for Antimicrobial Activity and Efficacy*; JIS Z 2801:2010; Japanese Industrial Standard: Tokyo, Japan, 2000.
51. Bazant, P.; Kuritka, I.; Munster, L.; Kalina, L. Microwave Solvothermal Decoration of the Cellulose Surface by Nanostructured Hybrid Ag/ZnO Particles: A Joint XPS, XRD and SEM Study. *Cellulose* **2015**, *22*, 1275–1293. [CrossRef]
52. Marambio-Jones, C.; Hoek, E.M.V. A Review of the Antibacterial Effects of Silver Nanomaterials and Potential Implications for Human Health and the Environment. *J. Nanopart. Res.* **2010**, *12*, 1531–1551. [CrossRef]
53. Kong, H.; Jang, J. Antibacterial Properties of Novel Poly(Methyl Methacrylate) Nanofiber Containing Silver Nanoparticles. *Langmuir* **2008**, *24*, 2051–2056. [CrossRef] [PubMed]
54. AshaRani, P.; Hande, M.P.; Valiyaveettil, S. Anti-Proliferative Activity of Silver Nanoparticles. *BMC Cell Biol.* **2009**, *10*, 65. [CrossRef] [PubMed]
55. Klapiszewski, Ł.; Rzemieniecki, T.; Krawczyk, M.; Malina, D.; Norman, M.; Zdarta, J.; Majchrzak, I.; Dobrowolska, A.; Czaczyk, K.; Jesionowski, T. Kraft Lignin/Silica-AgNPs as a Functional Material with Antibacterial Activity. *Colloid Surf. B* **2015**, *134*, 220–228. [CrossRef] [PubMed]
56. Reddy, K.M.; Feris, K.; Bell, J.; Wingett, D.G.; Hanley, C.; Punnoose, A. Selective Toxicity of Zinc Oxide Nanoparticles to Prokaryotic and Eukaryotic Systems. *Appl. Phys. Lett.* **2007**, *90*, 213902. [CrossRef] [PubMed]
57. Padmavathy, N.; Vijayaraghavan, R. Enhanced Bioactivity of ZnO Nanoparticles—An Antimicrobial Study. *Sci. Technol. Adv. Mater.* **2008**, *9*, 035004.[CrossRef] [PubMed]
58. Kołodziejczak-Radzimska, A.; Jesionowski, T. Zinc Oxide-From Synthesis to Application: A Review. *Materials* **2014**, *7*, 2833–2881. [CrossRef] [PubMed]
59. Nowacka, M.; Modrzejewska-Sikorska, A.; Chrzanowski, Ł.; Ambrożewicz, D.; Rozmanowski, T.; Myszka, K.; Czaczyk, K.; Bula, K.; Jesionowski, T. Electrokinetic and Bioactive Properties of CuO·SiO$_2$ Oxide Composites. *Bioelectrochemistry* **2012**, *87*, 50–57. [CrossRef] [PubMed]
60. Fortunati, E.; Armentano, I.; Zhou, Q.; Iannoni, A.; Saino, E.; Visai, L.; Berglund, L.A.; Kenny, J.M. Multifunctional Bionanocomposite Films of Poly(Lactic Acid), Cellulose Nanocrystals and Silver Nanoparticles. *Carbohydr. Polym.* **2012**, *87*, 1596–1605. [CrossRef]
61. Tomacheski, D.; Pittol, M.; Ferreira Ribeiro, V.; Marlene Campomanes Santana, R. Efficiency of Silver-Based Antibacterial Additives and Its Influence in Thermoplastic Elastomers. *J. Appl. Polym. Sci.* **2016**, *133*. [CrossRef]
62. Pittol, M.; Tomacheski, D.; Simoes, D.N.; Ribeiro, V.F.; Campomanes Santana, R.M. Antimicrobial Performance of Thermoplastic Elastomers Containing Zinc Pyrithione and Silver Nanoparticles. *Mater. Res.* **2017**, *20*, 1266–1273. [CrossRef]

© 2018 by the authors. Licensee MDPI, Basel, Switzerland. This article is an open access article distributed under the terms and conditions of the Creative Commons Attribution (CC BY) license (http://creativecommons.org/licenses/by/4.0/).

Article

TiO$_2$-ZnO Binary Oxide Systems: Comprehensive Characterization and Tests of Photocatalytic Activity

Katarzyna Siwińska-Stefańska [1,*], Adam Kubiak [1], Adam Piasecki [2], Joanna Goscianska [3], Grzegorz Nowaczyk [4], Stefan Jurga [4] and Teofil Jesionowski [1]

1. Faculty of Chemical Technology, Institute of Chemical Technology and Engineering, Poznan University of Technology, Berdychowo 4, PL-60965 Poznan, Poland; adam.kubiak@outlook.com (A.K.); teofil.jesionowski@put.poznan.pl (T.J.)
2. Faculty of Mechanical Engineering and Management, Institute of Materials Science and Engineering, Poznan University of Technology, Jana Pawla II 24, PL-60965 Poznan, Poland; adam.piasecki@put.poznan.pl
3. Faculty of Chemistry, Laboratory of Applied Chemistry, Adam Mickiewicz University in Poznan, Umultowska 89b, PL-61614 Poznan, Poland; asiagosc@amu.edu.pl
4. NanoBioMedical Centre, Adam Mickiewicz University in Poznan, Umultowska 85, PL-61614 Poznan, Poland; nowag@amu.edu.pl (G.N.); stjurga@amu.edu.pl (S.J.)
* Correspondence: katarzyna.siwinska-stefanska@put.poznan.pl; Tel.: +48-61-665-3626

Received: 17 April 2018; Accepted: 14 May 2018; Published: 18 May 2018

Abstract: A series of TiO$_2$-ZnO binary oxide systems with various molar ratios of TiO$_2$ and ZnO were prepared using a sol-gel method. The influence of the molar ratio and temperature of calcination on the particle sizes, morphology, crystalline structure, surface composition, porous structure parameters, and thermal stability of the final hybrids was investigated. Additionally, to confirm the presence of characteristic surface groups of the material, Fourier transform infrared spectroscopy was applied. It was found that the crystalline structure, porous structure parameters, and thermal stability were determined by the molar ratio of TiO$_2$ to ZnO and the calcination process for the most part. A key element of the study was an evaluation of the photocatalytic activity of the TiO$_2$-ZnO hybrids with respect to the decomposition of C.I. Basic Blue 9, C.I. Basic Red 1, and C.I. Basic Violet 10 dyes. It was found that the TiO$_2$-ZnO material obtained with a molar ratio of TiO$_2$:ZnO = 9:1 and calcined at 600 °C demonstrates high photocatalytic activity in the degradation of the three organic dyes when compared with pristine TiO$_2$. Moreover, an attempt was made to describe equilibrium aspects by applying the Langmuir-Hinsherlwood equation.

Keywords: titanium dioxide; zinc oxide; binary oxide material; sol-gel method; organic dyes decomposition; photocatalysis

1. Introduction

Photocatalysis is an effective process for creating minerals out of pollutants in the air and water such as simple inorganic compounds in the presence of a catalyst [1]. The most common and widely described heterogeneous photocatalysts are transition metal oxides and semiconductors such as TiO$_2$, ZnO, SnO$_2$, and CeO$_2$ [2–5]. Titanium dioxide is the most active of the compounds that have been tested. It is relatively cheap, photochemically stable, non-toxic, easily UV-activated, and insoluble in most reaction environments [6,7]. However, its application is limited because of its narrow photocatalytic region (α < 400 nm) and its ability to absorb only a small fraction (5%) of incident solar irradiation, which results from its relatively large band gap (anatase, ~3.2 eV) [8]. Many recent studies have focused on modifying the morphology and crystalline structure of TiO$_2$ to improve its photocatalytic activity. Modification may be performed by adding transition metal ions (such as Cr, Zr, Mn, and Mo) [9–12], preparing a reduced form of TiO$_{2-x}$, sensitization using dyes [13,14], doping with non-metals (such as N, S, C) [15–17], and using hybrid semiconductors such as TiO$_2$-ZnO, TiO$_2$-SiO$_2$, etc. [18–20]. To increase the response of TiO$_2$ to solar radiation, it is modified with ZnO, ZrO$_2$, and SnO$_2$ [21–23]. It has been proven that the formation of oxide hybrids is an appropriate tool for improving the photocatalytic ability of TiO$_2$ materials. Selection of an appropriate modifier and its compatibility with the material are important for the hybrid's physicochemical and optical properties. Each of the modifiers substantially affect the surface charge of the material and, therefore, enhance or weaken its photocatalytic capacity [24–26].

Zinc oxide is an attractive material due to its unique properties such as high chemical stability, high electrochemical coupling coefficient, high refractive index, high thermal conductivity, binding, antibacterial, and UV-protection properties [27]. Because of these properties, ZnO is added to materials and products including plastics, rubber, ceramics, paints, pigments, glass, cement, lubricants, ointments, adhesives, sealants, concrete, foods, batteries, ferrites, and fire retardants. Generally, zinc oxide occurs in nature in two main forms, which are hexagonal wurtzite and cubic zinc blende [28,29].

Titanium dioxide and zinc oxide have very similar physicochemical properties including nontoxicity, biocompatibility, thermal and chemical stability, insolubility in water, resistance to chemical breakdown and photo corrosion, and mechanical strength [30]. Many methods for the production of TiO$_2$-ZnO oxide systems have been proposed using different precursors of titania and zinc oxide. These methods include an electrospinning technique [31], a chemical co-precipitation method [32], the sol-gel technique [33,34], or solvothermal and hydrothermal methods [35–37]. These methods enable precise control of the synthesis to obtain materials with useful properties. The process can be controlled by temperature changes, sequence and type of reagent dosage, rate of stirring, pH, the ratio of water to precursors, and calcination conditions. Depending on the process parameters, the products exhibit different physicochemical properties and structure. The physicochemical properties of TiO$_2$-ZnO oxide systems depend on their morphology, the size of crystallites, and the crystallographic structure [31–37]. Moreover, ZnO has a band gap of 3.37 eV, which is slightly more negative than that of TiO$_2$. Therefore, the synthesis of TiO$_2$-ZnO oxide systems can result in the injection of conduction band electrons from ZnO to TiO$_2$, which is favorable to electron-hole separation [38,39]. Therefore, the incorporation of these two materials into an integrated structure is of great significance because the resulting products may possess improved specific and well-defined physical and chemical properties, which were determined during their synthesis [31–37].

The aim of this work was to study the correlation between conditions of preparation (molar ratio of precursors, temperature of calcination) of TiO$_2$-ZnO oxide systems and their properties including particle size distribution, morphology, crystalline and porous structure, and thermal stability. For the first time, this type of oxide material was used as a photocatalyst in the decomposition process of selected organic dyes (C.I. Basic Blue 9, C.I. Basic Red 1, and C.I. Basic Violet 10). In addition, theoretical description of performed photocatalytic process was presented. The Langmuir-Hinsherlwood equation and an assumption that pollutant decomposition is of pseudo-first-order reaction (PFO) were tested for this purpose.

2. Materials and Methods

2.1. Materials

Titanium(IV) isopropoxide (TTIP, 97%), C.I. Basic Blue 9 (Methylene Blue—MB, 95%), C.I. Basic Red 1 (Rhodamine 6G—R6G, 95%) and C.I. Basic Violet 10 (Rhodamine B—RhB, 95%) were purchased from Sigma-Aldrich. Zinc acetate dihydrate (99.5%), propan-2-ol (IPA, 99.5%), and ammonia (25%) were purchased from Chempur (Piekary, Śląskie, Poland). All reagents were of analytical grade and used without any further purification. The water used in all experiments was deionized.

2.2. Preparation of TiO$_2$-ZnO Oxide Systems Using the Sol-Gel Method

The synthesis of TiO$_2$-ZnO oxide hybrids with TiO$_2$:ZnO molar ratios of 9:1, 5:2, and 1:3 was performed by the sol-gel method. First, a reactor equipped with a T25 Basic type high-speed stirrer (IKA Werke GmbH, Staufen im Breisgau, Germany) was filled with a mixture containing an appropriate amount of TTIP in IPA. The resulting mixture was stirred at 1000 rpm. Afterward, an appropriate amount of 15% zinc acetate solution (the precursor of ZnO was dissolved in a mixture of IPA:H$_2$O at a volume ratio of 1:3) was introduced at a constant rate of 5 cm^3/min using an ISM833A peristaltic pump (ISMATEC, Wertheim, Germany). The synthesis was performed at room temperature. The reaction system was additionally stirred for 10 min. After this time, the promoter of hydrolysis (a mixture of ammonia and deionized water at a volume ratio of 1:3) was added at a constant rate of 1 cm^3/min. The colloidal suspension was mixed for 1 h and the resulting alcogel was dried at 120 °C for 24 h. To remove impurities, the white precipitate was washed several times with deionized water. Lastly, the powder was dried at 80 °C for 3 h and additionally calcined at 600 °C for 2 h (Nabertherm P320 Controller, Lilienthal, Germany). The methodology of the synthesis of TiO$_2$-ZnO oxide materials is presented in Figure 1.

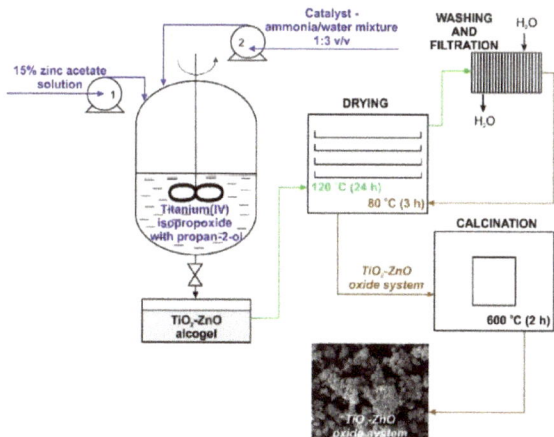

Figure 1. Synthesis of TiO$_2$-ZnO binary oxide powders via the sol-gel method.

2.3. Analysis of Materials

The particle sizes of the synthesized materials were measured by the non-invasive backscattering method (NIBS) using a Zetasizer Nano ZS (Malvern Instruments Ltd., Worcester, UK) instrument and enabling measurements in the diameter range 0.6–6000 nm. Each sample was prepared by dispersing 0.01 g of the tested product in 25 cm^3 of propan-2-ol. The resulting system was sonicated for 15 min and then placed in a cuvette and analyzed.

The surface microstructure and morphology of the TiO$_2$-ZnO binary oxide systems were examined on the basis of SEM images recorded from an EVO40 scanning electron microscope (Zeiss, Jena, Germany). Before testing, samples were coated with gold (Au) for 5 s using a Balzers PV205P (Oerlikon Balzers Coating SA,. Brügg, Switzerland) coater.

The crystalline structure of the synthesized binary oxide materials was analyzed by the X-ray diffraction method (XRD) using a D8 Advance diffractometer (Bruker, Karlsruhe, Germany) operating with Cu Kα radiation (α = 1.5418 Å), Ni filtered. The patterns were obtained in a step-scanning mode ($\Delta 2\theta$ = 0.05°) over an angular range of 10° to 80°.

High resolution transmission electron microscopy (HRTEM) images as well as dark field scanning TEM (DF STEM) selected area TEM diffractograms and EDS elemental maps were recorded by means of JEOL ARM 200F microscope (JEOL, Peabody, MA, USA) operating at an accelerating voltage of 200 kV. In order to prepare specimens, particular powders were dispersed in alcohol and then a few drops of such solution were placed on copper grids coated with carbon and formvar.

The surface composition of TiO_2-ZnO oxide hybrids (content of TiO_2 and ZnO) was analyzed by using energy dispersive X-ray spectroscopy (EDS) using a Princeton Gamma-Tech unit equipped with a prism digital spectrometer (Princeton Gamma-Tech, Princeton, NJ, USA). Representative parts of each sample (500 μm^2) were analyzed to determine their actual surface composition.

The parameters of the porous structure of the obtained oxide powders were measured using a physisorption analyzer (ASAP 2020, Micromeritics Instrument Co., Norcross, GA, USA) operating based on a low-temperature adsorption of nitrogen. Before measurement, all materials were degassed at 120 °C for 4 h. Surface area was determined by the multipoint BET method using adsorption data in a relative pressure (p/p_0) range of 0.05–0.30. The desorption isotherm was used to determine the pore size distribution based on the Barrett, Joyner, Halenda (BJH) model.

Characteristic functional groups present on the surface of the obtained materials were identified using Fourier transform infrared spectroscopy (FTIR). The measurements were performed using a Vertex 70 spectrophotometer (Bruker, Karlsruhe, Germany). Samples were prepared by mixing with KBr and pressing into small tablets. FTIR spectra were obtained in the transmission mode between 4000 cm^{-1} and 400 cm^{-1}.

A thermogravimetric analyzer (Jupiter STA 449F3, Netzsch, Selb, Germany) was used to investigate the thermal stability of the synthesized materials. Measurements were carried out under nitrogen flow (10 cm^3/min) at a heating rate of 10 °C/min over a temperature range of 30 °C to 1000 °C with an initial sample weight of approximately 5 mg.

2.4. Photocatalytic Tests

The photocatalytic activity of the obtained TiO_2-ZnO binary oxide systems was evaluated based on the decomposition of C.I. Basic Blue 9 (MB), C.I. Basic Red 1 (R6G), and C.I. Basic Violet 10 (RhB) dyes (see, Table 1) in an initial concentration of 5 mg/dm^3.

Table 1. Organic dyes used in photocatalytic tests.

Color Index Name	C.I. Basic Blue 9	C.I. Basic Red 1	C.I. Basic Violet 10
Chemical structure			
Molecular formula	$C_{16}H_{18}ClN_3S \cdot 3H_2O$	$C_{28}H_{31}ClN_2O_3$	$C_{28}H_{31}ClN_2O_3$
Molar mass (g/mol)	373.85	479.01	479.01
λ_{max} (nm)	664	526	553

Photocatalysis was carried out in a laboratory reactor of UV-RS2 type (Heraeus, Hanau, Germany) equipped with a 150 W medium-pressure mercury lamp as a UV light source surrounded by a water-cooling quartz jacket. First, an appropriate amount of photocatalyst (TiO_2-ZnO binary oxide material) was added to a glass tube reactor containing 100 cm^3 of the model organic impurity. The suspension was stirred using an R05 IKAMAG magnetic stirrer (IKA Werke GmbH, Staufen im Breisgau, Germany) for 30 min in darkness to determine the adsorption/desorption equilibrium. After this time, the radiation was turned on to initiate the photocatalytic reaction. The process was carried out for a maximum of 150 min. In the next step, the irradiated mixtures were collected from the reactor at regular intervals and centrifuged to separate the photocatalyst. The concentration of C.I. Basic Blue 9, C.I. Basic Red 1, or C.I. Basic Violet 10 (after adsorption and UV irradiation) was analyzed using a UV-Vis spectrophotometer (V-750, Jasco, Oklahoma City, OK, USA) at a wavelength of 664 nm (for MB), 526 nm (for R6G), or 553 nm (for RhB) using water as a reference. The photocatalytic activity of the TiO$_2$-ZnO binary oxide systems was determined by calculating the yield of dye degradation (W) using the formula below.

$$W(\%) = \left(1 - \frac{C_t}{C_0}\right) \cdot 100\% \tag{1}$$

where C_0 and C_t are the concentrations of the dye prior to and after irradiation, respectively.

2.5. Kinetic Study

Kinetic energy of the photocatalytic decomposition of selected organic dyes was described based on the Langmuir-Hinsherlwood equation [40] assuming that pollutant decomposition is of a pseudo-first-order reaction nature. The equation presents dependence between the dye concentration in the aqueous vs. time of UV irradiation.

$$r = \frac{dC}{dt} = k\left(\frac{KC}{1+KC}\right) \tag{2}$$

Assuming that the degradation process of the dye is of pseudo-first-order reaction nature, the constant reaction rate can be determined as the slope of the linear regression.

$$-\ln\left(\frac{C_t}{C_0}\right) = kt \tag{3}$$

where k is the degradation rate of organic dye, min^{-1}, K is the equilibrium constant of adsorption of the dye on the surface of the catalyst, C_0, C_t are concentrations of the dye compound in aqueous solution before irradiation ($t = 0$) and after define time t.

Estimation of constant reaction rate k enables determination of the half-life time of the model organic pollutant.

$$t_{\frac{1}{2}} = \frac{\ln 2}{k} \tag{4}$$

3. Results and Discussion

3.1. Dispersive and Morphological Characteristics

The results of dispersive analysis (see Table 2) show that both synthetic TiO$_2$ and ZnO (without thermal treatment) have monomodal particle size distributions. The TiO$_2$ and ZnO samples (denoted Ti and Zn) contain particles in the diameter ranges of 531–1720 nm and 220–615 nm, respectively. Dispersive analysis of the synthetic TiO$_2$-ZnO oxide systems showed that the molar ratio of the precursors significantly affects the particle sizes of the resulting materials. Samples obtained with TiO$_2$:ZnO molar ratios of 9:1; 5:2, and 1:3 denoted as Ti9Zn1, Ti5Zn2, and Ti1Zn3, which contain particles in the ranges 459 nm to 1110 nm, 459 nm to 1480 nm, and 396 nm to 825 nm, respectively. The results show that products with smaller particles were obtained when a higher content of ZnO was used.

It was also confirmed that increasing the temperature of calcination leads to the production of products with larger particles as a result of sintering and agglomerate formation. This situation was observed for all of the oxide materials except samples Ti5Zn2_600 and Ti1Zn3_600. All calcined TiO$_2$-ZnO oxide systems exhibit monomodal particle size distributions. Synthetic oxide systems obtained with different TiO$_2$:ZnO molar ratios (samples Ti9Zn1_600, Ti5Zn2_600, and Ti1Zn3_600) contain particles in the diameter ranges 531 nm to 1280 nm, 459 nm to 955 nm, and 255 nm to 615 nm, respectively. Among the calcined samples, those obtained with the highest molar contribution of ZnO were composed of the smallest particles.

The SEM microphotographs of TiO$_2$ and ZnO samples (uncalcined and calcined at 600 °C, Figure 2a,b,i,) show the presence of particles of almost spherical shape with high homogeneity.

Moreover, the SEM micrographs for all analyzed oxide samples confirm the presence of particles, which exhibit a high tendency towards agglomeration. SEM observations of the synthesized TiO_2-ZnO oxide systems (see Figure 2c–h) show that the molar ratio of the precursors does not have any significant effect on the morphology of the resulting systems. The SEM microphotographs of the studied samples confirm the presence of particles with precisely designed diameters, which corresponds to those indicated in the particle size distributions. Wang et al. [41] who synthesized a TiO_2-ZnO oxide system through a sol-gel method using ammonia as a catalyst obtained results analogous to those reported here. Tsai et al. [42] noted that the TiO_2-ZnO oxide system contains particles of a spherical shape, which show a high tendency towards agglomeration. Similarly, Ullah et al. [43] demonstrated that a TiO_2-ZnO oxide system synthesized via a sol-gel method using dimethylaminoethanol was composed of particles of spherical shape with a high tendency to agglomerate.

Table 2. Dispersive properties of TiO_2-ZnO oxide systems obtained via the sol-gel method.

Sample Name	Molar Ratio TiO_2:ZnO	Temperature of Calcination (°C)	Particle Diameter Range (nm)	Dominant Particles Diameter with Maximum Volume Contribution (%)	Polydispersity Index
Ti	10:0		531–1720	955 nm—21.6	0.183
Ti9Zn1	9:1		459–1110	712 nm—27.5	0.312
Ti5Zn2	5:2	-	459–1480	825 nm—25.9	0.361
Ti1Zn3	1:3		396–825	615 nm—31.7	0.150
Zn	0:10		220–615	396 nm—25.7	0.178
Ti_600	10:0		615–1990	1110 nm—23.8	0.220
Ti9Zn1_600	9:1		531–1280	825 nm—32.6	0.408
Ti5Zn2_600	5:2	600	459–955	712 nm—32.2	0.504
Ti1Zn3_600	1:3		255–615	396 nm—26.7	0.182
Zn_600	0:10		51–122 220–1110	79 nm—5.1 531 nm—14.2	0.434

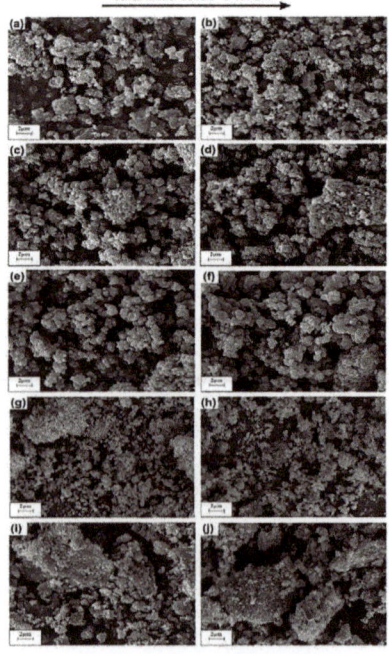

Figure 2. SEM images of obtained samples: (**a**) Ti; (**b**) Ti_600; (**c**) Ti9Zn1; (**d**) Ti9Zn1_600; (**e**) Ti5Zn2; (**f**) Ti5Zn2_600; (**g**) Ti1Zn3; (**h**) Ti1Zn3_600; (**i**) Zn and (**j**) Zn_600.

3.2. Structural Characteristics

The XRD pattern of titanium dioxide calcined at 600 °C (see Figure 3a) shows a strong diffraction peak at 2θ = 25.2, which corresponds to the anatase structure (JCPDS (Joint Committee on Powder Diffraction Standards), No. 21-1272). Less intense, characteristic diffraction peaks found at 36.95°, 37.8°, 38.58°, 48.05°, 53.89°, 55.06°, 62.12°, 62.69°, 68.76°, 70.31°, 75.03°, and 76.02° are also strictly related to the anatase phase. The rutile TiO_2 phase is not detected in this sample. Moreover, the XRD patterns of un-calcined zinc oxide and zinc oxide calcined at 600 °C (see Figure 3b,c) correspond to the wurtzite phase of ZnO with high intensity peaks located at 31.77°, 34.42°, and 36.25° (JCPDS No. 36-1451). These results prove that thermal treatment does not change the crystalline structure of the zinc oxide materials.

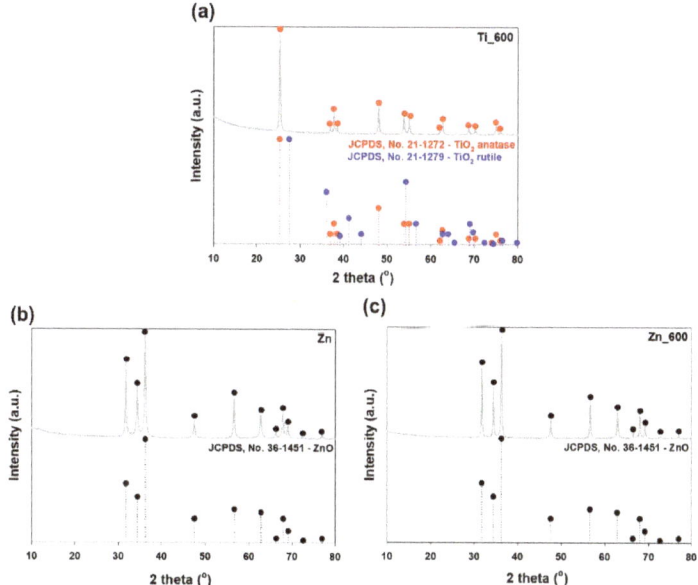

Figure 3. WAXS patterns of (**a**) calcined TiO_2; (**b**) uncalcined ZnO; and (**c**) calcined ZnO.

The XRD patterns of the synthetic, un-calcined TiO_2-ZnO oxide hybrids (see Figure 4a) do not show diffraction peaks of the TiO_2 and ZnO phases. The obtained samples are amorphous. It has been reported that the combination of titania with zinc oxide leads to inhibition of the phase formation of the ZnO crystalline structure. These obtained results suggest that some Zn^{2+} cations can incorporate into the titania network [44], which follows from the fact that the ionic radii of Zn^{2+} (ca. 60 pm) and Ti^{4+} (ca. 60.5 pm) are similar [45]. The XRD pattern of sample Ti9Zn1_600 (obtained with a TiO_2:ZnO molar ratio of 9:1 and calcined at 600 °C) confirms the formation of a crystalline material containing both titania and zinc oxide phases (see Figure 4b). Our results are in agreement with those of Stroyanova, Shalaby, and Moriadi [21,46,47]. Anatase was observed to be the dominant phase in sample Ti9Zn1_600. Characteristic diffraction peaks found at 25.28°, 36.95°, 37.8°, 38.58°, 48.05°, 53.89°, 55.06°, 70.31°, 75.03° and 76.02° were attributed to the anatase phase. For this sample, the peaks located at 27.45°, 39.19°, 41.23°, 44.05°, 54.32°, 56.64°, 65.48°, 69.01°, 69.79° and 79.82° correspond to the rutile phase. The XRD pattern of the obtained material also exhibited characteristic peaks with low intensity observed at 2θ = 36.25°, 56.6°, 62.86°, and 67.96°, which are characteristic for the ZnO structure. Additionally, the reflections observed at 2θ = 23.86°, 32.73°, 35.25°, 41.52°, 48.92°, 52.96°, 56.79°, 61.70°, 63.40°, 68.72°, and 70.91° can be identified with a cubic $ZnTiO_3$ phase (JCPDS

No. 14-0033). For sample Ti5Zn2_600 (obtained with a TiO$_2$:ZnO molar ratio of 5:2 and calcined at 600 °C; Figure 4c) with increasing content of ZnO, the characteristic peaks of anatase, and rutile TiO$_2$ gradually decreased [48]. Moreover, characteristic diffraction peaks found at 36.25°, 56.6°, 62.86°, and 67.96° are strictly related to the ZnO phase. In Ti9Zn1_600 and Ti5Zn2_600 samples crystallization of photoactive ZnTiO$_3$, which is the result of reaction between titania and zinc oxide was observed. For the analyzed materials, the intensity of the ZnTiO$_3$ peaks also increased. The XRD pattern of the TiO$_2$-ZnO oxide system obtained with a TiO$_2$:ZnO molar ratio of 1:3 and calcined at 600 °C (sample Ti1Zn3_600, Figure 4d) contained diffraction signals at 2θ values of 23.86°, 35.25° 43.10°, 48.92°, 56.79°, 61.70° and 70.91°, which is characteristic for the ZnTiO$_3$ structure. We also observed that increasing the molar ratio of ZnO until 3 leads to formation of photoinactive Zn$_2$TiO$_4$. The results obtained here are identical to those reported by other researchers [48–51].

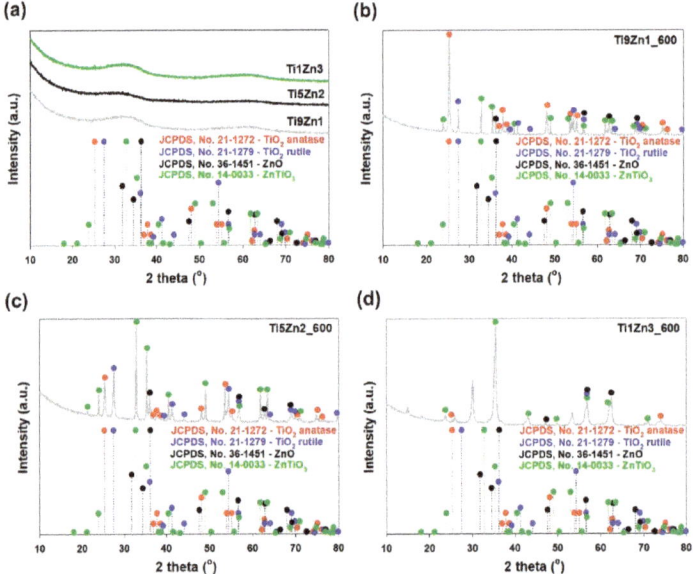

Figure 4. WAXS patterns of TiO$_2$-ZnO oxide systems: (**a**) synthesized with different molar ratios without thermal treatment and prepared with TiO$_2$:ZnO molar ratios of (**b**) 9:1; (**c**) 5:2; (**d**) 1:3, and subjected to thermal treatment.

HRTEM measurements confirmed that all prepared materials exhibit highly crystalline forms (see Figure 5a–c).

In order to confirm the crystalline structure of studied samples in the selected area, TEM diffraction experiments were conducted. The obtained results clearly confirmed high crystallinity of investigated samples and proved that crystallinity of TiO$_2$-ZnO is quite complex. However, the TEM diffractograms of the same high resolution as XRD results show well-distinguishing diffraction rings, which corresponds to data from XRD. The results of SATEM diffraction are presented in Figure 5d–f. The diffractograms were analyzed using CHT Diffraction Analysis [52]. The most distinctive rings of diffraction were indexed and compared to XRD data.

EDS (energy dispersive spectroscopy) experiments were carried out to verify the distribution of materials' components (elements) within the samples. The results (see Figure 6) indicate that distribution of Zn is not uniform.

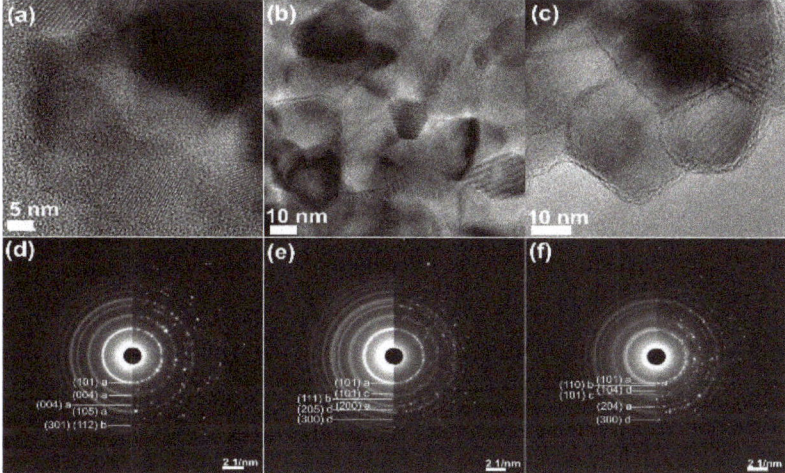

Figure 5. HRTEM images of samples: (**a**) Ti_600; (**b**) Ti9Zn1_600; and (**c**) Ti5Zn2_600, SATEM diffractograms of samples: (**d**) Ti_600; (**e**) Ti9Zn1_600; and (**f**) Ti5Zn2_600. Additionally, hkl planes of individual crystalline phases were denoted as: a—anatase, b—rutile, c—ZnO, d—ZnTiO$_3$.

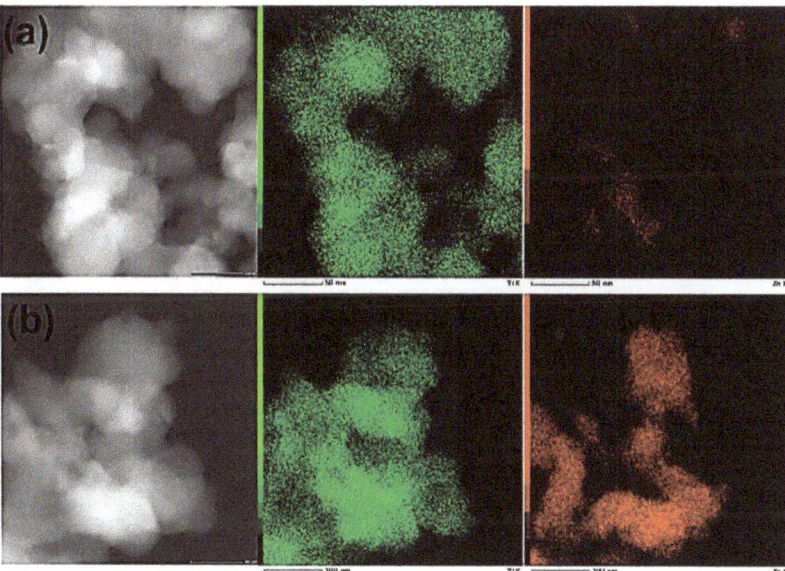

Figure 6. DFSTEM images and elemental maps of samples: (**a**) Ti9Zn1_600 and (**b**) Ti5Zn2_600. Ti and Zn are indicated as a green and a red color, respectively.

3.3. Surface Composition

Figure 7 presents the percentage content of titanium oxide and zinc oxide in the analyzed oxide systems. The results confirmed the efficiency of the sol-gel route of synthesis. Moreover, energy dispersive X-ray microanalysis showed that changing the molar ratio of the initial precursors affects the content of the corresponding oxides in the structure of the final materials. As was expected, the highest quantity of titania (84.0%) was observed in sample Ti9Zn1_600 (obtained with the molar ratio $TiO_2:ZnO$ = 9:1) and the highest quantity of zinc oxide (50.6%) in sample Ti1Zn3_600. It was concluded that the sol-gel method makes it possible to obtain materials with strictly defined properties whose composition is mainly determined by the molar ratio of the initial precursors.

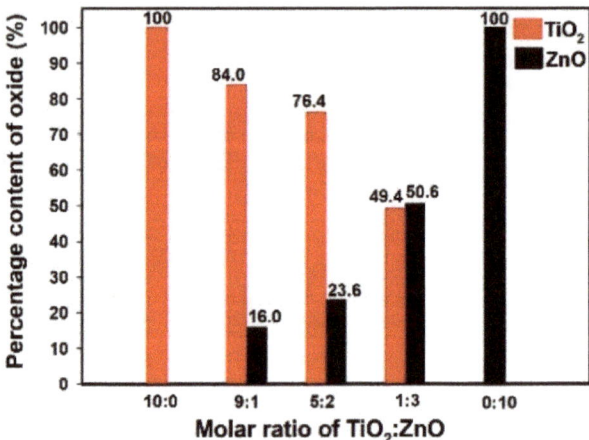

Figure 7. The percentage content of titania and zinc oxide in the calcined materials.

3.4. Porous Structure Parameters

The surface area of any material is the most important factor for influencing its catalytic activity. The results of textural characteristics of the obtained materials are summarized in Table 3. Samples of TiO_2-ZnO oxide systems that were not subjected to thermal treatment exhibit a relatively high surface area. The highest value, A_{BET} = 494.7 m^2/g, was observed for sample Ti9Zn1, which may be directly related to the dispersive nature of the analyzed material. This sample contained particles with smaller diameters than those of pure TiO_2, which is directly linked to the porous structure parameters of the products of synthesis. Slightly poorer porous structure parameters were observed for pure TiO_2 and sample Ti5Zn2 (with a molar ratio of TiO_2:ZnO = 5:2), which had surface areas (A_{BET}) of 488.6 m^2/g and 475.8 m^2/g. Moreover, an increase in the molar contribution of zinc oxide in the final product caused a significant decrease in the specific surface area, which was measured at 97.0 m^2/g for sample Ti1Zn3 (with a molar ratio of TiO_2:ZnO = 1:3) and 27.2 m^2/g for ZnO. Our observations align with those of Prasannalakshmi and Shanmugam [51].

Table 3. Porous structure parameters of TiO_2-ZnO oxide systems obtained by the sol-gel method.

Sample Name	Molar Ratio TiO_2:ZnO	Temperature of Calcination (°C)	Specific Surface Area A_{BET} (m^2/g)	Total Pore Volume V_p (cm^3/g)	Average Pore Size S_p (nm)
Ti	10:0		488.6	0.046	1.9
Ti9Zn1	9:1		494.7	0.079	1.9
Ti5Zn2	5:2	-	475.8	0.051	1.9
Ti1Zn3	1:3		97.0	0.030	2.0
Zn	0:10		27.2	0.008	2.1
Ti_600	10:0		26.5	0.010	2.2
Ti9Zn1_600	9:1		7.6	0.003	2.1
Ti5Zn2_600	5:2	600	7.2	0.005	2.2
Ti1Zn3_600	1:3		7.5	0.005	2.3
Zn_600	0:10		11.5	0.007	2.2

The samples that had undergone calcination exhibited a large decrease in the surface area. The highest surface area (7.6 m^2/g) for these TiO_2-ZnO oxide materials was recorded for sample

Ti9Zn1_600. Thermal treatment also led to a significant decrease in the pore volume and a slight increase in the pore diameters of the obtained materials. The calculated values also imply that the surface area decreases with increased ZnO content.

3.5. FTIR Analysis

The FTIR spectra of TiO$_2$-ZnO binary oxide materials (see Figure 8) show absorption peaks at 550 cm^{-1} and 650 cm^{-1} ascribed to symmetric stretching vibrations of ≡Ti–O–Ti≡ and the vibration mode of –Zn–O–Ti≡ groups [47,50,53]. The band at 1400 cm^{-1} indicates stretching vibrations of C–O bonds [54]. Moreover, the FTIR spectra of the synthetic TiO$_2$-ZnO oxide systems contain absorption peaks at 3440 cm^{-1} and 1630 cm^{-1}, which is attributed to physically adsorbed water (–OH) and N–H stretching vibrations [55–57].

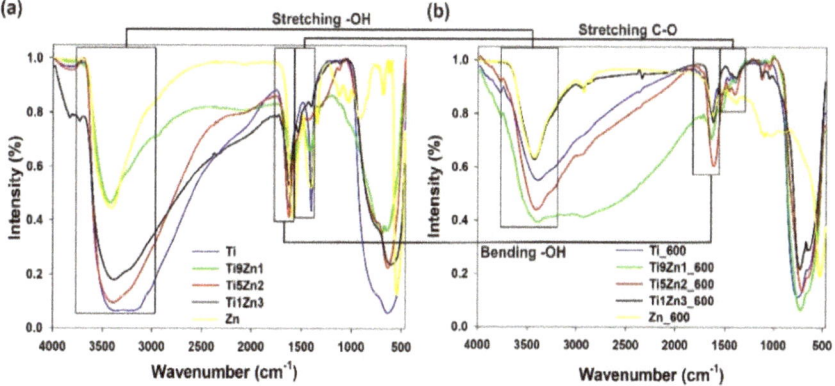

Figure 8. FTIR spectra of (**a**) un-calcined and (**b**) calcined TiO$_2$, ZnO, and TiO$_2$-ZnO samples.

The FTIR spectra of titanium dioxide (uncalcined and calcined) show three characteristic bands at 550 cm^{-1}, 1400 cm^{-1}, and 3400 cm^{-1}, which is associated respectively with stretching vibrations of ≡Ti–O, C–O, and –OH bonds. Analysis of the FTIR spectra of zinc oxide reveals a peak characteristic for zinc oxide (Zn–O) at 500 cm^{-1}. The broad absorption peak appearing at 700 cm^{-1} to 1100 cm^{-1} is characteristic for non-reacted products such as CH$_3$COO$^-$ and NH$_4^+$. Moreover, the peak at approximately 3400 cm^{-1} is ascribed to stretching vibrations of O–H bonds, which are indirectly related to water physically adsorbed on the surface. The FTIR results for synthetic TiO$_2$-ZnO oxide hybrids showed absorption peaks for ≡Ti–O–Ti≡ bonds at 550 cm^{-1}, Zn–Ti–O bonds at 650 cm^{-1}, and –OH groups at 3400 cm^{-1}. Analysis of the FTIR spectra for samples Ti9Zn1, Ti5Zn2, and Ti1Zn3 reveals significant changes in the intensities of the relevant bands, which depend on the molar ratio of the precursors. Moreover, for TiO$_2$-ZnO oxide systems calcined at 600 °C (Ti9Zn1_600, Ti5Zn2_600, Ti1Zn3_600), a decrease in the intensity of the –OH peak at 3400 cm^{-1} was observed. The spectra show that the intensity of the absorption bands around 650 cm^{-1}, which correspond to ≡Ti–O–Ti≡ bonds, increases with a growing molar ratio of the TiO$_2$ precursor. It was also observed that the peaks at 1630 cm^{-1} for O–H bending vibrations at 1400 cm^{-1} for C–O stretching vibrations decrease when the calcination temperature is increased.

3.6. Thermal Analysis

Analysis of the thermograms of samples Ti, Ti9Zn1, Ti5Zn2, and Zn (see Figure 9a) indicates a one-step degradation process. The degradation step in the temperature range 30 °C to 380 °C is associated with a significant decrease in mass by about 19%, 18.5%, 17.0%, and 2.5% for samples Ti, Ti5Zn2, Ti9Zn1, and Zn, respectively. The mass loss is mainly related to the local elimination of water bonded with the surface of the materials. When the temperature is above 380 °C, the samples stabilize and their mass remains almost unchanged. For sample Ti1Zn3, three mass losses were observed on the TGA curves. The first sample in the range of 30 °C to 300 °C corresponds to the loss of free water and amounts to about 7.5%. In the range 300°C to 470 °C, there is a second mass loss of about 5%, which is related to the thermal decomposition of the ZnO precursor. The total mass loss for sample Ti1Zn3 was 14.0%.

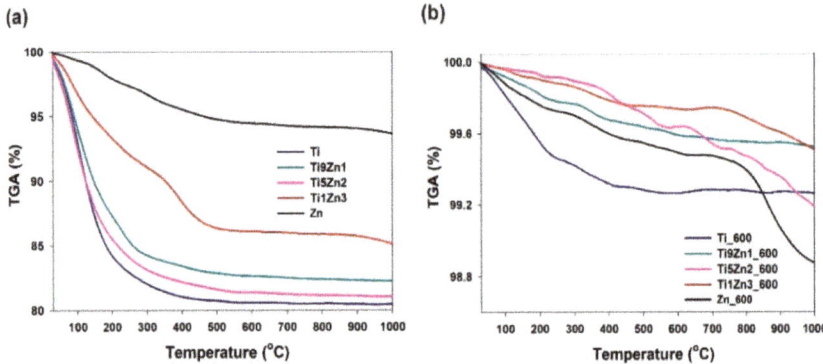

Figure 9. Thermal stability of TiO_2, ZnO, and TiO_2-ZnO oxide systems: (a) un-calcined and (b) calcined.

The thermograms of samples Ti_600 and Ti9Zn1_600 (see Figure 9b) show a minor mass loss in the temperature range of 30 °C to 350 °C by about 0.6% and 0.2%, respectively. This is related to the presence of a small amount of moisture in the systems [58,59]. In the range of 350 °C to 1000 °C, there is a second mass loss of about 0.7% and 0.4% for samples Ti_600 and Ti9Zn1_600, respectively. A slightly different thermogravimetric curve was observed for samples Ti5Zn2_600, Ti1Zn3_600, and Zn_600. In all three cases, the first degradation step in the temperature range of 30 °C to 300 °C with a mass loss of about 0.1% (for samples Ti5Zn2_600 and Ti1Zn3_600) and 0.2% (for sample Zn_600) is related to the local elimination of water bonded with the surface of the products. The next mass loss of about 0.2%, 0.4%, and 0.5% for samples Ti1Zn3_600, Ti5Zn2_600, and Zn_600, respectively, in the temperature range of 300 °C to 800 °C is related to the thermal decomposition of unreacted zinc acetate [60]. The third degradation step is probably associated with the phase transformation as a result of the applied high temperatures. The total mass loss for samples Ti5Zn2_600, Ti1Zn3_600, and Zn_600 is slightly above 0.8%, 0.3%, and 1.1%, respectively.

Wang et al. [61] who obtained nanoparticles of TiO_2-ZnO via a sol-gel process observed three steps of mass loss, which are associated with the evaporation of water, the dehydroxylation of precursors, and the polymorphic transformation of TiO_2. The results presented above indicate that the obtained TiO_2-ZnO oxide systems have greater thermal stability than materials obtained in other studies [41,61]. In addition, titanium dioxide and zinc oxide following thermal treatment have similar thermal stability to what was reported in the literature [58–61].

3.7. Photocatalytic Activity

Titanium dioxide is known as an effective photocatalyst for the photo-oxidation of different kinds of hazardous organic pollutants in waste water [15]. Zinc oxide is another attractive semiconductor oxide with similar photocatalytic properties [28]. For this reason, a key element of the present research was an evaluation of the photocatalytic activity of the obtained TiO_2-ZnO binary oxide systems. The evaluation was based on the decomposition of MB, R6G, and RhB dyes under UV irradiation. Titanium dioxide and samples obtained with TiO_2:ZnO molar ratios of 9:1 and 5:2, additionally calcined at 600 °C, were subjected to photocatalytic tests. The results are presented in Figure 10.

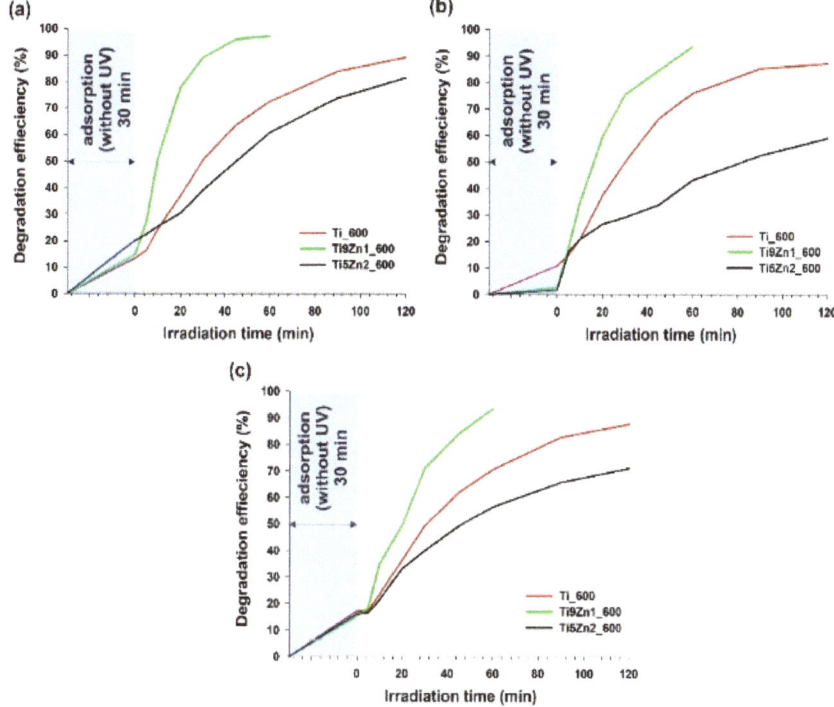

Figure 10. Efficiency of photocatalytic decomposition of (a) C.I. Basic Blue 9; (b) C.I. Basic Red 1; and (c) C.I. Basic Violet 10 in the presence of the synthesized hybrids.

The first stage of photocatalytic testing involved evaluating the photocatalytic activity of TiO_2-ZnO oxide systems in the removal of C.I. Basic Blue 9 (see Figure 10a). The TiO_2-ZnO sample obtained with a molar ratio of TiO_2:ZnO = 9:1 exhibited significantly better photocatalytic activity than pure titanium. Applying the Ti9Zn1_600 photocatalyst, the degree of decomposition of MB dye was 97.2% after 60 min of UV irradiation. The efficiency of C.I. Basic Blue 9 photodegradation in the presence of samples Ti_600 and Ti5Zn2_600 was 89.3% and 81.6% (after 120 min), respectively. Additionally, the photodecomposition of this organic dye increased with increasing irradiation time.

The decolorization of R6G under UV irradiation showed that sample Ti9Zn1_600 had good photo-oxidation activity (the efficiency of its degradation of C.I. Basic Red 1 was 93.6% after 60 min), which is shown in Figure 10b. Samples Ti_600 and Ti5Zn2_600 showed lower photocatalytic activity in the decomposition of C.I. Basic Red 1. The degradation efficiency was 87.2% (after 120 min) in the presence of sample Ti_600 and slightly lower (59.1%) in the case of photocatalysis using the sample Ti5Zn2_600.

Lastly, the photocatalytic experiments showed that a combination of titania with zinc oxide in a molar ratio of 9:1 exhibited significantly better photocatalytic activity than samples Ti_600 and Ti5Zn2_600 in the degradation of RhB (see Figure 10c). After 60 min of UV irradiation applying the Ti9Zn1_600 photocatalyst, the degree of decomposition of C.I. Basic Violet 10 reached 93.4%. The efficiency of degradation of RB dye in the presence of samples Ti_600 and Ti5Zn2_600 was 87.7% and 71.1%, respectively.

Our results imply that the photocatalytic activity of the synthesized samples depends not only on their BET surface area or crystallinity but can rather be attributed to dispersion and surface morphology. Moreover, based on research reports regarding heterogeneous photocatalysis [31,62,63], we propose a probable mechanism (see Figure 11) and reactions of the photodegradation of organic dyes using TiO$_2$-ZnO oxide materials.

$$TiO_2 - ZnO \quad photocatalysts \; + h\nu \rightarrow e^- + h^+ \quad (5)$$

$$h^+ + H_2O \rightarrow H^+ + OH^* \quad (6)$$

$$h^+ + OH^- \rightarrow OH^* \quad (7)$$

$$h^+ + dye \rightarrow oxidation \; products \quad (8)$$

$$e^- + O_2 \rightarrow {}^*O_2^- \quad (9)$$

$${}^*O_2^- + 2H^+ \rightarrow H_2O_2 \quad (10)$$

$$e^- + H_2O_2 \rightarrow OH^* + OH^- \quad (11)$$

$$e^- + dye \rightarrow reduction \; products \quad (12)$$

$$day + OH^+ + O_2 \rightarrow CO_2 + H_2O + other \; decomposition \; products \quad (13)$$

Prasannalakshmi and Shanmugam [51] reported that TiO$_2$-ZnO oxide hybrids obtained using a sol-gel method produce almost complete degradation of C.I. Basic Blue 9 within 25 min of irradiation. Pérez-González et al. [45] obtained (TiO$_2$)$_{1-x}$-(ZnO)$_x$ thin films, with x = 0.00, 0.25, 0.50, 0.75, and 1.00, by the sol-gel process, which were deposited on glass. The synthesized films were evaluated for their ability to degrade MB. The authors found that the photocatalytic performance was improved by decreasing the value of x with the TiO$_2$ thin films displaying the highest response. Araújo et al. [64] produced TiO$_2$-ZnO hierarchical hetero nanostructures following a two-step procedure in which the hydrothermal growth of nanorods took place on the surface of decorated electrospun fibers. The resulting material was applied as a photocatalyst in the photodegradation of Rhodamine B. Photocatalytic tests showed the TiO$_2$-ZnO composite to have good photocatalytic activity. Agrawal et al. [65] obtained hierarchically nanostructured hollow spheres composed of ZnO-TiO$_2$ mixed oxides as a potential candidate for photocatalytic application. Pei and Leung [62] prepared TiO$_2$-ZnO nanofibers from a nozzle-less electrospinning solution system. The authors evaluated the photocatalytic activities of different TiO$_2$-ZnO composites in the photodegradation of Rhodamine B (RhB) under irradiation with 420 nm visible light. ZnO/TiO$_2$ hybrid nanofibers were prepared via electrospinning by Chen et al. [63]. Based on the photodegradation of RhB, it was shown that the synthesized products exhibited high degradation efficiency. The ZnO/TiO$_2$ (1 wt %) nanofibers degraded 90% of the dye in about 15 min.

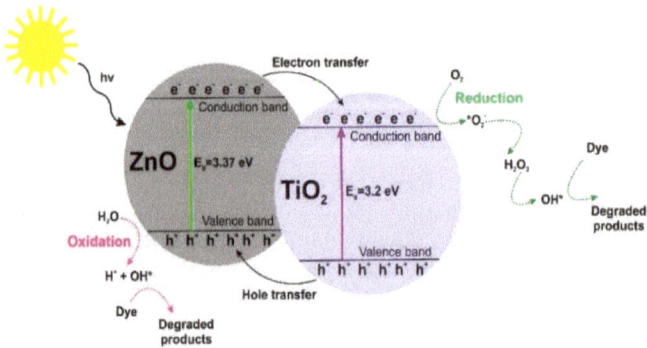

Figure 11. Mechanism of photodegradation of organic dyes using TiO$_2$-ZnO oxide materials.

Table 4 presents results from the literature concerning the efficiency of decomposition of C.I. Basic Blue 9, C.I. Basic Red 1, and C.I. Basic Violet 10 dyes when different photocatalysts were used.

Analysis of the kinetics of photochemical decomposition of organic dyes shows significant differences in the rate of degradation of the analyzed impurities in the presence of catalysts (see Table 5). Regardless of the type of organic dye, the highest values of the degradation reaction rate k (0.0596 min^{-1}—C.I. Basic Blue 9, 0.0459 min^{-1}—C.I. Basic Red 1 and 0.0453 min^{-1}—C.I. Basic Violet 10) were recorded when the Ti9Zn1_600 oxide system was used as a photocatalyst. Furthermore, in the presence of Ti9Zn1_600 material, the highest values of the half-life time ($t_{1/2}$ = 11.632 min—C.I. Basic Blue 9, 15.086 min—C.I. Basic Red 1 and 15.301 min—C.I. Basic Violet 10) of tested organic dyes were noted.

Table 4. Efficiency of decomposition of selected organic dyes.

Sample Name	Concentration of Dye Solution (mg/dm^3)	Efficiency of Decomposition (%)			Ref.
		C.I. Basic Blue 9	C.I. Basic Red 1	C.I. Basic Violet 10	
Ti9Zn1_600	5	97.2	93.6	93.4	this work
(TiO$_2$)$_{1-x}$-(ZnO)$_x$	10	45.0–62.0	-	-	[45]
TZO1-TZO4	1	~100	-	-	[51]
TiO$_2$	5	98.3	-	-	[59]
TiO$_2$/ZnO	0.5	-	-	~100	[62]
ZnO/TiO$_2$	2	-	-	90.0	[63]
TiO$_2$/ZnO	4.8	-	-	90.0	[64]
TiO$_2$/ZnO	4.8	-	~100	-	[65]
ZnO-TiO$_2$	4.8	-	-	~100	[66]
TiO$_2$/ZnO	20	96.0	-	83.0	[67]

Table 5. The reaction rate constant (k), correlation coefficient (R^2), and half-life time ($t_{1/2}$) of tested organic dyes during the photocatalytic process.

Sample Name	k (1/min)	R^2	$t_{1/2}$ (min)
C.I. Basic Blue 9			
Ti_600	0.0186	0.9937	37.269
Ti9Zn1_600	0.0596	0.9731	11.632
Ti5Zn2_600	0.0141	0.9957	49.178
C.I. Basic Red 1			
Ti_600	0.0172	0.9584	40.398
Ti9Zn1_600	0.0459	0.9930	15.086
Ti5Zn2_600	0.0074	0.9922	93.225
C.I. Basic Violet 10			
Ti_600	0.0174	0.9918	39.727
Ti9Zn1_600	0.0453	0.9897	15.301
Ti5Zn2_600	0.0103	0.9765	67.007

4. Conclusions

The proposed methodology of synthesis of the TiO$_2$-ZnO binary oxide materials using the sol-gel method proved to be very effective.

We studied how the TiO$_2$:ZnO molar ratio and calcination temperature affects the physicochemical and photocatalytic properties of synthetic TiO$_2$-ZnO oxide hybrids. It was found that the particle sizes, crystalline phase, surface area, pore structures, and photocatalytic activity of the TiO$_2$-ZnO oxide systems are strongly dependent on the amount of zinc oxide in the product as well as on the calcination temperature.

The results of XRD analysis show that the quantity of zinc oxide in the product and the calcination temperature have significant effects on crystallizing the resulting materials. The porous structure parameters of the TiO$_2$-ZnO oxide systems decreased with an increasing quantity of zinc oxide and temperature of calcination.

The TiO$_2$-ZnO oxide hybrid obtained in a molar ratio of TiO$_2$:ZnO = 9:1 and calcined at 600 °C (sample T9Zn1_600) showed the highest photocatalytic activity. This is attributed to the fact that this sample is composed with titanium, zinc oxide, and ZnTiO$_3$ phases as well as with anatase as the dominant phase. Moreover, analysis of the kinetics of the photocatalytic process performed based on the Langmuir-Hinshelwood equation confirmed that degradation of the model organic dyes occurred most intensely in the presence of the Ti9Zn1_600 catalyst.

Author Contributions: K.S.-S. supervised the whole project. A.K. was responsible for the synthesis and characterization of the materials. A.P. was responsible for EDS analysis. J.G. was responsible for XRD analysis. G.N. and S.J. took part in HRTEM and DF STEM analysis and interpreted the results. The manuscript was written by K.S.-S. while T.J. critically revised the manuscript.

Funding: This research was funded by the Polish National Centre of Science (Research Grant No. 2017/01/X/ST5/01050).

Acknowledgments: This work was supported by the Polish National Centre of Science (Research Grant No. 2017/01/X/ST5/01050).

Conflicts of Interest: The authors declare no conflicts of interest.

References

1. Gross, S.; Muller, K. Sol-gel derived silica-based organic-inorganic hybrid materials as composite precursors for the synthesis of highly homogeneous nanostructured mixed oxides: An overview. *J. Sol-Gel Sci. Technol.* **2011**, *60*, 283–298. [CrossRef]
2. Sanchez, C.; Belleville, P.; Popall, M.; Nicole, L. Applications of advanced hybrid organic-inorganic nanomaterials: From laboratory to market. *Chem. Soc. Rev.* **2011**, *40*, 696–753. [CrossRef] [PubMed]
3. Yuan, C.; Wu, H.B.; Xie, Y.; Lou, X.W. Mixed transition-metal oxides: Design, synthesis, and energy-related applications. *Angew. Chem. Int. Ed.* **2014**, *53*, 1488–1504. [CrossRef] [PubMed]
4. Chauhan, I.; Aggrawal, S.; Mohanty, C.; Mohanty, P. Metal oxide nanostructures incorporated/immobilized paper matrices and their applications: A review. *RSC Adv.* **2015**, *5*, 83036–83055. [CrossRef]
5. Debecker, D.P.; Hulea, V.; Mutin, P.H. Mesoporous mixed oxide catalysts via non-hydrolytic sol-gel: A review. *Appl. Catal. A* **2013**, *451*, 192–206. [CrossRef]
6. Pirzada, B.M.; Mir, N.A.; Qutub, N.; Mehraj, O.; Sabir, S.; Muneer, M. Synthesis, characterization and optimization of photocatalytic activity of TiO$_2$/ZrO$_2$ nanocomposite heterostructures. *Mater. Sci. Eng. B* **2015**, *193*, 137–145. [CrossRef]
7. Du, X.; Men, K.; Xu, Y.; Li, B.; Yang, Z.; Liu, Z.; Li, L.; Li, L.; Feng, T.; ur Rehman, W.; et al. Enhanced capacitance perfomance of Al$_2$O$_3$-TiO$_2$ composite thin film via sol-gel using double chelators. *J. Colloid Interface Sci.* **2015**, *443*, 170–176. [CrossRef] [PubMed]
8. Wu, H.; Yang, Y.; Suo, H.; Qing, M.; Yan, L.; Wu, B.; Xu, J.; Xiang, H.; Li, Y. Effects of ZrO$_2$ promoter on physic-chemical properties and activity of Co/TiO$_2$-SiO$_2$ Fischer-Tropsch catalysts. *J. Mol. Catal. A Chem.* **2015**, *396*, 108–119. [CrossRef]
9. Michalow, K.A.; Otal, E.H.; Burnat, D.; Fortunato, G.; Emerich, H.; Ferri, D.; Heel, A.; Graule, T. Flame-made visible light active TiO$_2$:Cr photocatalysts: Correlation between structural, optical and photocatalytic properties. *Catal. Today* **2013**, *209*, 47–53. [CrossRef]
10. Pouretedal, H.R. Visible photocatalytic activity of co-doped TiO$_2$/Zr,N nanoparticles in wastewater treatment of nitrotoluene samples. *J. Alloys Compd.* **2018**, *735*, 2507–2511. [CrossRef]
11. Park, J.-H.; Jang, I.; Song, K.; Oh, S.-G. Surfactants-assisted preparation of TiO$_2$-Mn oxide composites and their catalytic activities for degradation of organic pollutant. *J. Phys. Chem. Solids* **2013**, *74*, 1056–1062. [CrossRef]
12. Cui, M.; Pan, S.; Tang, Z.; Chen, X.; Qiao, X.; Xu, Q. Physiochemical properties of n-n heterostructured TiO$_2$/Mo-TiO$_2$ composites and their photocatalytic degradation of gaseous toluene. *Chem. Speciat. Bioavailab.* **2017**, *29*, 60–69. [CrossRef]

13. Garmaroudi, Z.A.; Mohammadi, M.R. Design of TiO_2 dye-sensitized solar cell photoanode electrodes with different microstructures and arrangement modes of the layers. *J. Sol-Gel Sci. Technol.* **2015**, *76*, 666–678. [CrossRef]
14. Barea, E.M.; Bisquert, J. Properties of chromophores determining recombination at the TiO_2-dye-electrolyte interface. *Langmuir* **2013**, *29*, 8773–8781. [CrossRef] [PubMed]
15. Wang, S.Q.; Liu, W.B.; Fu, P.; Cheng, W.L. Enhanced photoactivity of N-doped TiO_2 for Cr(VI) removal: Influencing factors and mechanism. *Korean J. Chem. Eng.* **2017**, *34*, 1584–1590. [CrossRef]
16. He, Y.; Fu, Z.; Zhou, Q.; Zhong, M.; Yuan, L.; Wei, J.; Yang, X.; Wang, C.; Zeng, Y. Fabrication and electrochemical behavior of a lithium-sulfur cell with a TiO_2-sulfur-carbon aerogel-based cathode. *Ionics* **2015**, *21*, 3065–3073. [CrossRef]
17. Orge, C.A.; Faria, J.L.; Pereira, M.F.R. Photocatalytic ozonation of aniline with TiO_2-carbon composite materials. *J. Environ. Manag.* **2017**, *195*, 208–215. [CrossRef] [PubMed]
18. Delsouz Khaki, M.R.; Shafeeyan, M.S.; Raman, A.A.A.; Daud, W.M.A.W. Evaluating the efficiency of nano-sized Cu doped TiO_2/ZnO photocatalyst under visible light irradiation. *J. Mol. Liq.* **2018**, *258*, 354–365. [CrossRef]
19. Soltan, S.; Jafari, H.; Afshar, S.; Zabihi, O. Enhancement of photocatalytic degradation of furfural and acetophenone in water media using nano-TiO_2-SiO_2 deposited on cementitious materials. *Water Sci. Technol.* **2016**, *74*, 1689–1697. [CrossRef] [PubMed]
20. Sobhanardakani, S.; Zandipak, R. Synthesis and application of TiO_2/SiO_2/Fe_3O_4 nanoparticles as novel adsorbent for removal of Cd(II), Hg(II) and Ni(II) ions from water samples. *Clean Technol. Environ.* **2017**, *19*, 1913–1925. [CrossRef]
21. Stoyanova, A.; Hitkova, H.; Bachvarova-Nedelcheva, A.; Iordanova, R.; Ivanova, M.; Sredkova, N. Synthesis and antibacterial activity of TiO_2/ZnO nanocomposites prepared via nonhydrolytic route. *J. Chem. Technol. Metall.* **2013**, *48*, 154–161.
22. Fan, M.; Hu, S.; Ren, B.; Wang, J.; Jing, X. Synthesis of nanocomposite TiO_2/ZrO_2 prepared by different templates and photocatalytic properties for the photodegradation of Rhodamine B. *Powder Technol.* **2013**, *235*, 27–32. [CrossRef]
23. Duraisamy, N.; Thangavelu, R.R. Synthesis, characterization and photocatalytic properties of TiO_2-SnO_2 composite nanoparticles. *Adv. Mater. Res.* **2013**, *678*, 373–377. [CrossRef]
24. Kumar, S.G.; Rao, K.S.R.K. Comparison of modification strategies towards enhanced charge carrier separation and photocatalytic degradation activity of metal oxide semiconductors (TiO_2, WO_3 and ZnO). *Appl. Surf. Sci.* **2017**, *391*, 124–148. [CrossRef]
25. Nolan, M.; Iwaszuk, A.; Lucid, A.K.; Carey, J.J.; Fronzi, M. Design of novel visible light active photocatalyst materials: Surface modified TiO_2. *Adv. Mater.* **2016**, *28*, 5425–5446. [CrossRef] [PubMed]
26. Fujishima, A.; Zhang, X.; Tryk, D.A. TiO_2 photocatalysis and related surface phenomena. *Surf. Sci. Rep.* **2008**, *63*, 515–582. [CrossRef]
27. Kolodziejczak-Radzimska, A.; Jesionowski, T. Zinc oxide—From synthesis to application: A review. *Materials* **2014**, *7*, 2833–2881. [CrossRef] [PubMed]
28. Gao, L.; Li, Q.; Luan, W.L. Preparation and electric properties of dense nanocrystalline zinc oxide ceramics. *J. Am. Ceram. Soc.* **2002**, *85*, 1016–1018. [CrossRef]
29. Ulyankina, A.; Leontyev, I.; Avramenko, M.; Zhigunov, D.; Smirnova, N. Large-scale synthesis of ZnO nanostructures by pulse electrochemical method and their photocatalytic properties. *Mater. Sci. Semicond. Process.* **2018**, *76*, 7–13. [CrossRef]
30. Tian, J.; Chen, L.; Dai, J. Preparation and characterization of TiO_2, ZnO, and TiO_2/ZnO nanofilms via sol-gel process. *Ceram. Int.* **2009**, *35*, 2261–2270. [CrossRef]
31. Li, J.; Yan, L.; Wang, Y.; Kang, Y.; Wang, C. Fabrication of TiO_2/ZnO composite nanofibers with enhanced photocatalytic activity. *J. Mater. Sci. Mater. Electron.* **2016**, *27*, 7834–7838. [CrossRef]
32. Arabnezhad, M.; Afarani, M.S.; Jafari, A. Co-precipitation synthesis of ZnO-TiO_2 nanostructure composites for arsenic photodegradation from industrial wastewater. *Int. J. Environ. Sci. Technol.* **2017**, 1–6. [CrossRef]
33. Kwiatkowski, M.; Bezverkhyy, I.; Skompskab, M. ZnO nanorods covered with a TiO_2 layer: Simple sol-gel preparation, and optical, photocatalytic and photoelectrochemical properties. *J. Mater. Chem. A* **2015**, *3*, 12748–12760. [CrossRef]

34. Li, W.; Wu, D.; Yu, Y.; Zhang, P.; Yuan, J.; Cao, Y.; Cao, Y.; Xu, J. Investigation on a novel ZnO/TiO$_2$-B photocatalyst with enhanced visible photocatalytic activity. *Physica E* **2014**, *58*, 118–123. [CrossRef]
35. Vlazan, P.; Ursu, D.H.; Irina-Moisescu, C.; Mirona, I.; Sfirloaga, P.; Rusu, E. Structural and electrical properties of TiO$_2$/ZnO core-shell nanoparticles synthesized by hydrothermal method. *Mater. Charact.* **2015**, *101*, 153–158. [CrossRef]
36. Lin, L.; Yang, Y.; Men, L.; Wang, X.; He, D.; Chai, Y.; Zhao, B.; Ghoshroy, S.; Tang, Q. A highly efficient TiO$_2$@ZnO n–p–n heterojunction nanorod photocatalyst. *Nanoscale* **2013**, *5*, 588–593. [CrossRef] [PubMed]
37. Zha, R.; Nadimicherla, R.; Guo, X. Ultraviolet photocatalytic degradation of methyl orange by nanostructured TiO$_2$/ZnO heterojunctions. *J. Mater. Chem. A* **2015**, *3*, 6565–6574. [CrossRef]
38. Hu, Z.; Chen, G. Novel nanocomposite hydrogels consisting of layered double hydroxide with ultrahigh tensibility and hierarchical porous structure at low inorganic content. *Adv. Mater.* **2014**, *26*, 5950–5956. [CrossRef] [PubMed]
39. Hassan, S.A.; El-Salamony, R.A. Photocatalytic disc-shaped composite systems for removal of hazardous dyes in aqueous solutions. *Can. Chem. Trans.* **2014**, *2*, 56–70. [CrossRef]
40. Behnajady, M.A.; Modirshahla, N.; Hamzavi, R. Kinetic study on photocatalytic degradation of C.I. Acid Yellow 23 by ZnO photocatalyst. *J. Hazard. Mater.* **2006**, *133*, 226–232. [CrossRef] [PubMed]
41. Wang, J.; Mi, W.; Tian, J.; Dai, J.; Wang, X.; Liu, X. Effect of calcinations of TiO$_2$/ZnO composite powder at high temperature on photodegradation of methyl orange. *Compos. Part B Eng.* **2013**, *45*, 758–767. [CrossRef]
42. Tsai, M.T.; Chang, Y.Y.; Huang, H.L.; Hsu, J.-T.; Chen, Y.-C.; Wu, A.Y.-J. Characterization and antibacterial performance of bioactive Ti-Zn-O coatings deposited on titanium implants. *Thin Solid Films* **2013**, *528*, 143–150. [CrossRef]
43. Ullah, H.; Khan, K.A.; Khan, W.U. ZnO/TiO$_2$ nanocomposite synthesized by sol gel from highly soluble single source molecular precursor. *Chin. J. Chem. Phys.* **2014**, *27*, 548–554. [CrossRef]
44. Perez-Larios, A.; Lopez, R.; Hernandez-Gordillo, A.; Tzompantzi, F.; Gomez, R.; Torres-Guerra, L.M. Improved hydrogen production from water splitting using TiO$_2$-ZnO mixed oxides photocatalysts. *Fuel* **2012**, *100*, 139–143. [CrossRef]
45. Pérez-González, M.; Tomás, S.A.; Morales-Luna, M.; Arvizua, M.A.; Tellez-Cruz, M.M. Optical, structural, and morphological properties of photocatalytic TiO$_2$-ZnO thin films synthesized by the sol-gel process. *Thin Solid Films* **2015**, *594*, 304–309. [CrossRef]
46. Shalaby, A.; Dimitriev, Y.; Iordanova, R.; Bachvarova-Nedelcheva, A.; Iliev, T. Modified sol-gel synthesis of submicron powders in the system ZnO-TiO$_2$. *J. Univ. Chem. Technol. Mater.* **2011**, *46*, 137–142.
47. Moradi, S.; Azar, P.A.; Farshid, S.R.; Khorrami, S.A.; Givianrad, M.H. Effect of additives on characterization and photocatalytic activity of TiO$_2$/ZnO nanocomposite prepared via sol-gel process. *Int. J. Chem. Eng.* **2012**, *2012*, 215373. [CrossRef]
48. Chen, Y.; Zhang, C.; Huang, W.; Yang, C.; Huang, T.; Situ, Y.; Huang, H. Synthesis of porous ZnO/TiO$_2$ thin films with superhydrophilicity and photocatalytic activity via a template-free sol-gel method. *Surf. Coat. Technol.* **2014**, *258*, 531–538. [CrossRef]
49. Tian, J.; Chen, L.; Yin, Y.; Wang, X.; Dai, J.; Zhu, Z.; Liu, X.; Wu, P. Photocatalyst of TiO$_2$/ZnO nano composite film: Preparation, characterization, and photodegradation activity of methyl orange. *Surf. Coat. Technol.* **2009**, *204*, 205–214. [CrossRef]
50. Naseri, N.; Yousefi, M.; Moshfegh, A.Z. A comparative study on photoelectrochemical activity of ZnO/TiO$_2$ and TiO$_2$/ZnO nanolayer systems under visible irradiation. *Solar Energy* **2011**, *85*, 1972–1978. [CrossRef]
51. Prasannalakshmi, P.; Shanmugam, N. Fabrication of TiO$_2$/ZnO nanocomposites for solar energy driven photocatalysis. *Mater. Sci. Semicond. Process.* **2017**, *61*, 114–124. [CrossRef]
52. Mitchell, D.R.G. Circular Hough transform diffraction analysis: A software tool for automated measurement of selected area electron diffraction patterns within Digital Micrograph™. *Ultramicroscopy* **2008**, *108*, 367–374. [CrossRef] [PubMed]
53. Wang, J.; Li, J.; Xie, Y.; Li, C.; Han, G.; Zhang, L.; Xu, R.; Zhang, X. Investigation on solar photocatalytic degradation of various dyes in the presence of Er^{3+}:YAlO$_3$/ZnO-TiO$_2$ composite. *J. Environ. Manag.* **2010**, *91*, 677–684. [CrossRef] [PubMed]
54. Du, X.-W.; Fu, Y.-S.; Sun, J.; Han, X.; Liu, J. Complete UV emission of ZnO nanoparticles in a PMMA matrix. *Semicond. Sci. Technol.* **2006**, *21*, 1202–1206. [CrossRef]

55. Morsi, R.E.; Elsalamony, R.A. Superabsorbent enhanced-catalytic core/shell nanocomposites hydrogels for efficient water decolorization. *New J. Chem.* **2016**, *1542*, 33–36. [CrossRef]
56. Wu, L.; Yu, J.C.; Zhang, L.; Wang, X.; Ho, W. Preparation of a highly active nanocrystalline TiO_2 photocatalyst from titanium oxo cluster precursor. *J. Solid State Chem.* **2004**, *177*, 2584–2590. [CrossRef]
57. Lotus, A.F.; Tacastacas, S.N.; Pinti, M.J.; Britton, L.A.; Stojilovic, N.; Ramsier, R.D.; Chase, G.G. Fabrication and characterization of TiO_2-ZnO composite nanofibers. *Physica E* **2011**, *43*, 857–861. [CrossRef]
58. Suzuki, Y.; Yoshikawa, S. Synthesis and thermal analyses of TiO_2-derived nanotubes prepared by the hydrothermal method. *J. Mater. Res.* **2004**, *19*, 982–985. [CrossRef]
59. Siwińska-Stefańska, K.; Zdarta, J.; Paukszta, D.; Jesionowski, T. The influence of addition of a catalyst and chelating agent on the properties of titanium dioxide synthesized via the sol-gel method. *J. Sol-Gel Sci. Technol.* **2015**, *75*, 264–278. [CrossRef]
60. Das, S.; Meena, S.S.; Pramanik, A. Zinc oxide functionalized human hair: A potential water decontaminating agent. *J. Colloid Interface Sci.* **2016**, *462*, 307–314. [CrossRef] [PubMed]
61. Wang, L.; Fu, X.; Han, Y.; Chang, E.; Wu, H.; Wang, H.; Li, K.; Qi, X. Preparation, characterization, and photocatalytic activity of TiO_2/ZnO nanocomposites. *J. Nanomater.* **2013**, *2013*, 321459. [CrossRef]
62. Pei, C.C.; Leung, W.W.-F. Photocatalytic degradation of Rhodamine B by TiO_2/ZnO nanofibers under visible-light irradiation. *Sep. Purif. Technol.* **2013**, *114*, 108–116. [CrossRef]
63. Chen, J.D.; Liao, W.S.; Jiang, Y.; Yu, D.N.; Zou, M.L.; Zhu, H.; Zhang, M.; Du, M.L. Facile fabrication of ZnO/TiO_2 heterogeneous nanofibres and their photocatalytic behaviour and mechanism towards Rhodamine B. *Nanomater. Nanotechnol.* **2016**, *6*, 9. [CrossRef]
64. Araújo, E.S.; da Costa, B.P.; Oliveira, R.A.P.; Libardi, J.; Faia, P.M.; de Oliveira, H.P. TiO_2/ZnO hierarchical heteronanostructures: Synthesis, characterization and application as photocatalysts. *J. Environ. Chem. Eng.* **2016**, *4*, 2820–2829. [CrossRef]
65. Agrawal, M.; Gupta, S.; Pich, A.; Zafeiropoulos, N.E.; Stamm, M. A facile approach to fabrication of $ZnO-TiO_2$ hollow spheres. *Chem. Mater.* **2009**, *21*, 5343–5348. [CrossRef]
66. Zhang, D. Effectiveness of photodecomposition of Rhodamine B and Malachite Green upon coupled tricomponent TiO_2(Anatase-Rutile)/ZnO nanocomposite. *Acta Chim. Slovaca* **2013**, *2*, 245–255. [CrossRef]
67. Rahman, M.M.; Roy, D.; Mukit, M.S.H. Investigation on the relative degradation of Methylene Blue (MB) and Rhodamine-B (RB) dyes under UV-Visible light using thermally treated commercial and doped TiO_2/ZnO photocatalysts. *Int. J. Integr. Sci. Technol.* **2016**, *2*, 14–18.

© 2018 by the authors. Licensee MDPI, Basel, Switzerland. This article is an open access article distributed under the terms and conditions of the Creative Commons Attribution (CC BY) license (http://creativecommons.org/licenses/by/4.0/).

Article

Thermal and Mechanical Properties of Silica–Lignin/Polylactide Composites Subjected to Biodegradation

Aleksandra Grząbka-Zasadzińska, Łukasz Klapiszewski, Sławomir Borysiak * and Teofil Jesionowski *

Institute of Chemical Technology and Engineering, Faculty of Chemical Technology, Poznan University of Technology, Berdychowo 4, PL-60965 Poznan, Poland;
aleksandra.grzabka-zasadzinska@put.poznan.pl (A.G.-Z.); lukasz.klapiszewski@put.poznan.pl (Ł.K.)
* Correspondence: slawomir.borysiak@put.poznan.pl (S.B.); teofil.jesionowski@put.poznan.pl (T.J.); Tel.: +48-61-665-3549 (S.B.); +48-61-665-3720 (T.J.)

Received: 19 October 2018; Accepted: 10 November 2018; Published: 13 November 2018

Abstract: In this paper, silica–lignin hybrid materials were used as fillers for a polylactide (PLA) matrix. In order to simulate biodegradation, PLA/hybrid filler composite films were kept in soil of neutral pH for six months. Differential scanning calorimetry (DSC) allowed analysis of nonisothermal crystallization behavior of composites, thermal analysis provided information about their thermal stability, and scanning electron microscopy (SEM) was applied to define morphology of films. The influence of biodegradation was also investigated in terms of changes in mechanical properties and color of samples. It was found that application of silica–lignin hybrids as fillers for PLA matrix may be interesting not only in terms of increasing thermal stability, but also controlled biodegradation. To the best knowledge of the authors, this is the first publication regarding biodegradation of PLA composites loaded with silica–lignin hybrid fillers.

Keywords: silica–lignin hybrid materials; polylactide; physicochemical and morphological properties; mechanical properties; biodegradation

1. Introduction

During the last few years, due to the growing environmental concerns regarding waste disposal and increasing prices of fossil fuels, demand for green products obtained from renewable resources has significantly increased. Great deal of interest was generated especially by biodegradable polymers. Among many biodegradable polymers, poly(lactic acid) (PLA) appears to be one of the most attractive materials. PLA is so popular because of renewability of it resources (it can be produced from 100% renewable resources like corn, sugar beets or rice), good mechanical properties, and biodegradability [1]. Apart from these indisputable advantages PLA has also some serious drawbacks that limit its wider application—poor thermal stability, low crystallization ability, and low barrier properties [2]. Nevertheless, it has been found that the above problems can be overcome by addition of fillers.

Silica–lignin hybrid filler combines highly available, low cost kraft lignin and silica with good mechanical and thermal properties [3,4]. The kraft lignin has in its structure numerous hydroxyl groups which are often found problematic and has to be masked or modified [5,6]. These hydroxyl groups are also responsible for biodegradation of lignin [7] and its degradation in temperature over 170 °C [8]. Hence, combining of lignin with a thermostable, inorganic filler such as silica is thought to provide an increased thermal stability. Although intensive research on silica–lignin hybrid fillers is being carried out [9–11], there are still not many publications dealing with polymer/hybrid fillers composites [3,12–

14]. Silica–lignin hybrid coupled with ammonium polyphosphate was tested as a novel intumescent flame-retardant system to improve the fire retardancy of PLA [13]. Bula et al. [3] confirmed that silica–lignin hybrid filler enhances thermal stability of polypropylene (PP) composites. It was also stated that as the percentage content of the hybrid filler in the PP matrix increases, thermal stability of composite increases, too. What is more, based on DMTA analysis, Strzemiecka et al. [12] concluded that the thermo-mechanical properties of the composite containing silica–lignin filler depended on the lignin-to-silica ratio in the hybrid filler. Also, composites consisting of phenolic resin and silica–lignin hybrid filler were characterized with better thermo-mechanical properties than systems with lignin or silica alone [14]. Therefore, application of such hybrid fillers in polymer composites is well-grounded and should be further investigated.

Polymers that biodegrade are already very desirable in agriculture or packaging because they eliminate the waste disposal issue. In case of PLA biodegradation, the most important process is hydrolysis which is catalyzed by end groups of carboxylic acids (autocatalytic reaction). The speed of this process is mainly determined by the molecular weight and stereochemical composition of sample [15,16]. However, it turns out that in case of PLA composites fillers may also have an effect on the biodegradation process of PLA. There are information that addition of starch blends or wood flour to PLA accelerates the thermal decomposition of composites [17]. Blending organically modified montmorillonites with PLA was found to increase the degradation rate [18], while PLA/PEG/nano-silica composites showed similar degradation behavior as PLA/PEG films [19]. Although detailed literature research, no study on biodegradation of PLA/hybrid fillers was found.

In view of the above considerations, it would appear that one of the directions that ought to be taken in order to further develop PLA-based composites, e.g., for packaging applications, is to define how biodegradation process of such materials affects their properties.

For that reason, the aim of this work was to determine changes taking place during biodegradation of PLA composites filled with silica–lignin fillers of different composition. For this purpose, thermal stability, crystallization behavior, change of color, morphology, and mechanical properties of composites before and after simulated biodegradation process were defined. This work is a continuation of our study on PLA composites with silica–lignin hybrid fillers [20].

2. Materials and Methods

2.1. Materials

Silica–lignin hybrid filler was produced using commercial Syloid 244 silica (W.R. Grace & Co., Columbia, MD, USA) and kraft lignin (Sigma-Aldrich, Steinheim am Albuch, Germany). Polylactide, type Ingeo 2500 HP was purchased at Nature Works, Minnetonka, MN, USA.

2.2. Preparation of Silica–Lignin Hybrid Materials

Four silica–lignin hybrid materials containing silica and lignin in weight ratios 1:1, 2:1, 5:1, and 20:1 were prepared in the same way as in previous publications [3,20–22]. Later in this paper, silica–lignin hybrid materials will be described as "hybrid", e.g., 20:1 hybrid stands for silica–lignin in ratio 20:1.

2.3. Preparation of Polylactide/Silica–Lignin Hybrid Composites

Composite pellets of polylactide and 7.5% (w/w) of each hybrid filler were prepared in a co-rotating twin screw extruder (ø = 16 mm, L/D = 40, EHP 1614, Zamak Mercator Sp. z o.o., Skawina, Poland) and then cut with knife mill (25-16/TC-SL, TRIA, Novi, MI, USA). The process parameters for this first extrusion were following: a barrel temperature of 180–200 °C and a screw rotation speed of 145 rpm. Films for further characterization were produced with single screw extruder (ø = 25 mm, L/D = 30, Metalchem, Warszawa, Poland) with slit die and chill-roll puller (pull speed 2.5 m/s). In this second extrusion the process parameters were subsequently 17–220 °C and 80 rpm.

The samples were named according to the following convention: "d" stands for sample after biodegradation, PLA stands for polylactide, and numbers stand for filler type. For example, name "dPLA/20:1" means biodegraded sample of polylactide filled with 20:1 hybrid filler.

2.4. Biodegradation of Composite Films

Films were cut into pieces of 10 mm width and 100 mm length and subsequently put into stainless steel mesh envelops. They were buried in the conditioned soil of neutral pH, at 8 cm depth. The initial temperature and relative humidity were 23 ± 2 °C and 60 ± 5%, respectively, and soil was also aerated. All these parameters, temperature, humidity and aeration, were periodically assessed throughout the entire period of the biodegradation process which lasted 6 month. After that time film samples were carefully removed from soil and analyzed following the experimental protocol.

2.5. Characterization of Materials

2.5.1. Particle Size Distribution and Porous Properties of Fillers

Zetasizer Nano ZS (0.6–6000 nm) (Malvern Instruments Ltd., Malvern, UK) using the non-invasive backscattering technique was applied to determine particle size and the dispersive properties of the silica–lignin samples. Size of pores were determined using an ASAP 2020 instrument (Micromeritics Instrument Co., Norcross, GA, USA).

2.5.2. Differential Scanning Calorimetry

Thermal properties of materials in form of films were evaluated using DSC (DSC 1, Mettler Toledo, Greifensee, Switzerland) under argon atmosphere. For nonisothermal crystallization investigations, the samples were first heated from 40 °C to 210 °C at the rate 20 °C/min and kept at this temperature for 4 min to eliminate the previous thermal and/or mechanical history. Then the samples were quenched to 40 °C at the rate 5 °C/min. This procedure was repeated two times and the second tour was used in the calculations. Based on the determined values for the enthalpy of crystallization (H), the extent of crystallization (crystal conversion), α was calculated (Equation (1)).

$$\alpha = \frac{\int_0^t \left(\frac{dH}{dt}\right) \times dt}{\int_0^1 \left(\frac{dH}{dt}\right) \times dt} \tag{1}$$

From the curves of $\alpha = f(t)$, the half-time of crystallization ($t_{0.5}$) was determined as time when crystal conversion was 50%. The crystallinity degree (X_c) of materials was evaluated according to Equation (2).

$$X_c = \left(\frac{\Delta H_m}{\Delta H_m° \times \left(1 - \frac{\%wt\ filler}{100}\right)}\right) \times 100 \tag{2}$$

where: ΔH_m is the melting enthalpy (from second heating scan), $\Delta H_m°$ is the melting enthalpy of a 100% crystalline polymer matrix (93.0 J/g for PLA [23]) and %wt filler is the filler weight percentage. Furthermore, characteristic temperatures—melting (T_m) and crystallization (T_c) temperatures—were defined.

2.5.3. Thermogravimetric Analysis

A Jupiter STA analyzer (Jupiter STA 449F3, Netzsch, Selb, Germany) was used to investigate the influence of filler type on thermal stability of the composites. Measurements were conducted in the atmosphere of nitrogen (flow rate 20 cm³/min) at a heating rate of 10 °C/min over a temperature range of 30–800 °C, with an initial sample weight of approximately 5 mg.

2.5.4. Colorimetric Analysis

Colorimeter (Testan DT-145, Anticorr, Gdańsk, Poland) was used to measure differences in colors of films before and after biodegradation. Method CIE76 was applied and the differences were calculated using Equation (3).

$$\Delta E_{ab}* = \sqrt{(L_2*-L_1*)^2 + (a_2*-a_1*)^2 + (b_2*-b_1*)^2} \tag{3}$$

where L_1, a_1, and b_1 stand for the measured parameters of the standard color, whereas L_2, a_2, and b_2 stand for the parameters of the sample. Value $\Delta E_{ab}*$~2.3 corresponds to a just noticeable difference [24].

2.5.5. Tensile Properties

Tensile properties of produced composite films were defined using Zwick and Roell Allround-Line Z020 TEW testing machine (Wrocław, Poland). Samples of 10 mm width and thickness ca. 100 µm were tested with speed 5 mm/min and initial force 0.2 N in accordance to standard ISO 527-3.

2.5.6. Scanning Electron Microscopy

The morphology of samples was observed using a scanning electron microscopy (EVO40, Zeiss, Jena, Germany), at acceleration voltage of 18 kV. Before testing, all the specimens were sputter-coated with gold for 5 s using a Balzers coater (PV205P, Oerlikon Balzers Coating SA, Brügg, Switzerland).

3. Results and Discussion

3.1. Characterization of Hybrid Fillers

In Table 1, particle size distributions, as well as average size of pores of hybrid fillers and its precursors are presented. The effectiveness of hybrid filler formation was already proved and discussed in our earlier paper [20]. Sizes of silica particles were in two ranges: 39–71 nm and 1440–4800 nm. In case of lignin particles no particles smaller than 100 nm were observed, lignin sample consisted of particles with 1720–5560 nm diameters. For each hybrid filler particle diameters were found to have two ranges, which were a resultant of presence of both precursors, lignin and silica. Fillers containing high amount of silica, e.g., filler 5:1, were characterized with smaller particles than fillers with high content of lignin (e.g., filler 1:1). Similar relationship can be found while considering the average size of pores. Silica had pores three times smaller than lignin (3.9 nm versus 12.3 nm). Diameter of pores of hybrid filler 20:1, containing definitely more silica than lignin, were comparable to those of silica itself. With the increasing amount of lignin in hybrid filler, the mean size of pores also increased.

Table 1. Particle size distribution and mean size of pores for hybrid fillers and their precursors.

Sample	Particle Size Distribution Range (nm)	Mean Size of Pores (nm)
Silica	39–71; 1440–4800	3.9
Lignin	1720–5560	12.3
Filler 20:1	39–79; 1720–4800	4.1
Filler 5:1	68–122; 1990–4800	4.8
Filler 2:1	79–220; 1720–4800	5.5
Filler 1:1	91–220; 1990–5560	8.0

3.2. Differential Scanning Analysis

It is known that introduction of filler into a polymer matrix may affect its crystallization behavior. Defining the nucleating abilities of PLA matrix in presence of hybrid fillers after biodegradation process is believed to be important in terms of defining the biodegradation mechanism of such composites. Therefore, DSC technique was applied to investigate kinetic parameters of PLA crystallization in composites. Figure 1a shows the DSC curves of the PLA and composite materials before biodegradation. Figure 1b presents analogical curves of samples subjected to simulated biodegradation. Characteristic temperatures, X_c and $t_{0.5}$ presented in Table 2 were calculated based on data obtained from DSC measurements.

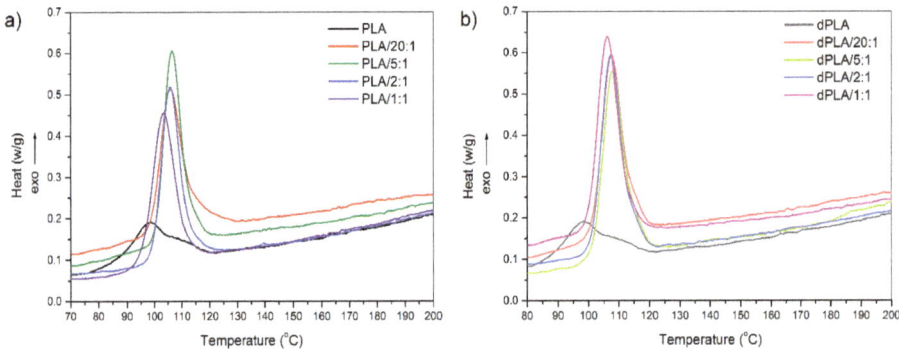

Figure 1. Differential scanning calorimetry (DSC) curves measured during the second cooling of samples: (**a**) before biodegradation and (**b**) after biodegradation.

Table 2. Tabulated values of melting temperatures (T_m), crystallization temperatures (T_c), crystallinity degree (X_c), and half-time of crystallization ($t_{0.5}$) of samples before and after biodegradation.

Sample	T_m (°C)	T_c (°C)	X_c (%)	$t_{0.5}$ (min)
PLA	177.0	98.0	60	2.9
PLA/20:1	177.4	105.9	47	2.8
PLA/5:1	176.9	106.6	48	2.5
PLA/2:1	175.9	105.8	51	2.4
PLA/1:1	174.5	103.5	52	2.2
dPLA	176.8	98.1	66	2.5
dPLA/20:1	177.0	107.6	49	2.2
dPLA/5:1	177.0	107.9	51	2.2
dPLA/2:1	176.9	107.2	55	2.1
dPLA/1:1	175.9	106.1	57	2.0

For samples that were not subjected to simulated biodegradation the T_m was in range 174.5–177.4 °C, while for the biodegraded samples T_m was in range 175.9–177.0 °C. It is known that small changes in T_m of composites are caused by the perfection of spherulite structure of PLA. T_m values obtained in this study were very comparable and consistent with the literature [2,25].

The peaks at 98–108 °C are attributed to crystallization of PLA matrix. The crystallization temperature provides information about the nucleation ability of PLA in presence of hybrid fillers. In case of all, non- and biodegraded, samples addition of hybrid filler was responsible for shifting of DSC exothermic peak towards higher temperature. In comparison with unfilled PLA, T_c of composite samples before biodegradation markedly increased (by 5.5–8.5 °C). For biodegraded samples that difference in T_c was even slightly higher, and reached 8–10 °C. Obtained results are compliant with changes in T_c reported for PLA/lignin [26] and PLA/graphene composites [27]. Given the above, it can be stated that hybrid fillers act as heterogeneous nucleation agents of PLA matrix.

The calculated X_c was in range from 47% to 60% for pristine films and from 49% to 66% for degraded samples. Among both, non- and biodegraded samples, the highest value of X_c parameter was noted for the unfilled PLA. However, samples after biodegradation were characterized with slightly higher values of X_c. Similar tendency was observed by Zimmermann et al. [28]. According to literature this is probably a result of biodegradation process that in the first stage takes place in amorphous regions and thus enhances the crystallinity degree [29].

Nucleation effect and polymer chain mobility—these are two competing processes that take place during crystallization process. It is likely that introduction of hybrid filler into PLA matrix intensifies the first stage of crystallization, formation of nuclei, but at the same impairs the chain mobility.

The analysis of crystal conversion curves (Figure 2) and crystallization half-times (Table 2) reveals that incorporation of each type of hybrid filler caused a drop of $t_{0.5}$. In case of non-degraded samples, the lowest noted value of $t_{0.5}$ was 2.2 min (2.9 min for pristine PLA). Silica, the main constituent of our hybrid fillers, is a well-known nucleating agent for PLA [30,31]. Generally, it is believed that as the particle size of filler decreases, its surface area increases providing more nucleating sites. However, our findings do not support the thesis that decrease of particle size (the higher was the amount of silica in hybrid filler, the smaller were particles of filler) enhances the crystallization of PLA.

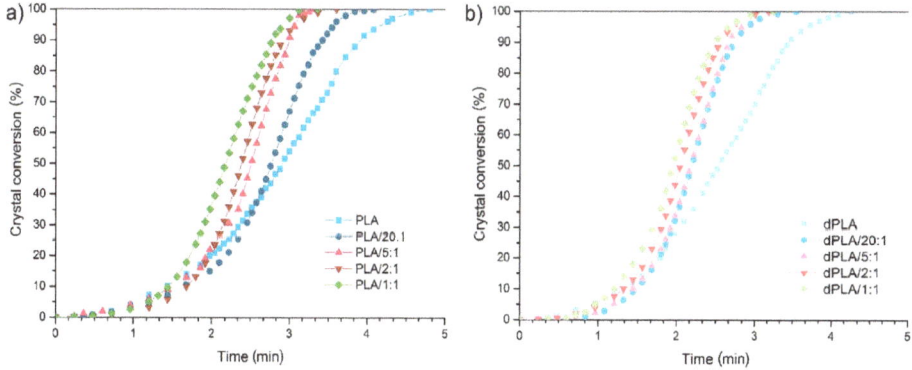

Figure 2. Crystal conversion curves of samples: (**a**) before biodegradation and (**b**) after biodegradation.

It seems rather that the composition of the hybrid filler was a major factor influencing the $t_{0.5}$ parameter, especially in non-degraded composites. It was shown that crystallization half-times of PLA matrix in presence of hybrid fillers depend mainly on mobility of polymer chains. Size and surface area of fillers are crucial in terms of nucleating processes. Nonetheless, hybrid fillers with highly active surface may induce segmental limitation of polymer chains mobility, and in result, formation of spherulitic structures. It is likely that in case of PLA/hybrid filler composites the amount of small nuclei formed in matrix by silica particles was so high that it restricted the mobility of polymeric chains and consequently limited the formation of spherulites. Another factor possibly restricting the mobility of PLA chains was size of pores of the filler. Pores of 20:1 filler, containing high amounts of silica, were definitely smaller than in case of 1:1 hybrid filler (4.1 nm and 8.0 nm, respectively). Even though presence of these small pores is responsible for formation of transcrystalline layers in PLA/silica composites [20], it also hindered the movements of polymer chains. That explains why the $t_{0.5}$ parameter was lower for PLA/1:1 than for PLA/20:1 films. Surprisingly, not only silica was an agent hindering the formation of PLA spherulites. As it can be seen in Table 2, crystallization half-times of all biodegraded composites were rather similar and lower than those for non-degraded films (minimal value 2.1 min for dPLA/2:1, 2.5 min for dPLA). These comparable results provide information about the course of the biodegradation process of hybrid filler. Most likely products of decomposition of lignin, constituent of hybrid filler, act as a nucleating agent for PLA matrix. It is confirmed by differences in X_c of biodegraded samples, which for composite with 1:1 hybrid filler reached 57% and for films with 20:1 hybrid filler only 49%. Here, removal of the lignin from the PLA matrix (during the biodegradation) was also responsible for the fact that PLA polymer chains regained some of their mobility.

3.3. Thermal Stability of Composites

The curves of PLA-based composites with hybrid fillers are given in Figure 3.

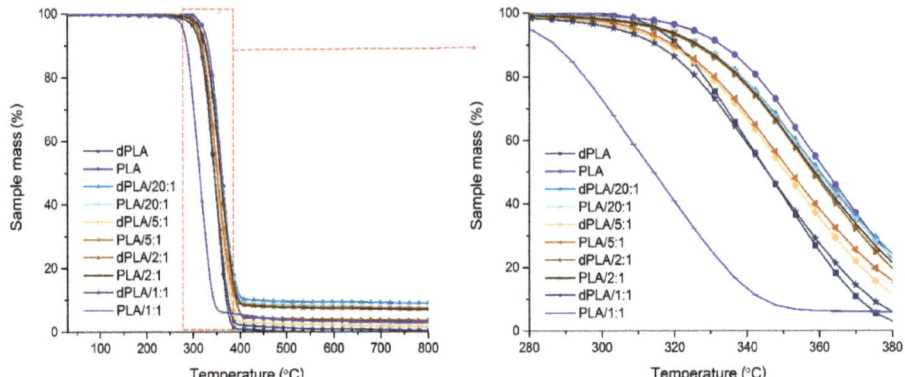

Figure 3. Curves for PLA-based films.

Curves in Figure 3 indicate that the thermal degradation of PLA and PLA/hybrid filler is a rather simple process. According to Kopinke et al. [32] who performed analysis and pyrolysis-MS of PLA, at temperature 295 °C lactide is released, whereas at 350 °C also higher cyclic oligomers are removed. The curves of silica–lignin hybrid filler showed that decomposition of such material takes places in three stages—first one is associated with loss of water, during second stage (210–600 °C) intensive sample mass loss (~40%) occurs, and in the last stage fragmentation and final degradation takes place [3]. The main weight loss step for both, pristine PLA and hybrid filler, occurs in similar temperature range (280–380 °C). It is possible that because of overlapping for PLA/hybrid filler composites only one stage weight loss was observed. The tabulated values obtained from these measurements are provided in Table 3. The results indicate that biodegradation of pristine PLA causes a decrease in its thermal stability—50% wt. loss for PLA and dPLA was observed at 361 °C and 345 °C, respectively. Decrease of thermal stability (from T_d = 351.5 °C to T_d = 339.6 °C) of PLA biodegraded for 6 months was also reported by Pinto et al. [33].

Table 3. Tabulated thermal analyses values of tested films.

Sample	5% wt. Loss	15% wt. Loss	50% wt. Loss
	Temperature (°C)		
PLA	326	337	361
PLA/20:1	314	331	359
PLA/5:1	303	326	350
PLA/2:1	314	332	358
PLA/1:1	281	291	315
dPLA	314	326	345
dPLA/20:1	314	331	360
dPLA/5:1	303	325	350
dPLA/2:1	314	332	356
dPLA/1:1	280	321	345

Gordobil et al. reported that incorporation in PLA acetylated kraft lignin, as well as unmodified one, improves the thermal stability of composites [34]. However, the presence of untreated lignin in PLA composites was also shown to have a negative impact on their thermal stability [35].

Analysis of thermal values reveals no substantial differences in characteristic temperatures of non- and biodegraded samples containing 20:1, 5:1, and 2:1 hybrid filler. These values were also comparable with those noted for pristine PLA. However, it turned out that the difference in thermal stability of dPLA/1:1 and PLA/1:1 was particularly important. The starting PLA/1:1 sample contained relatively high amount of lignin that was probably decomposed during biodegradation. In such case silica became the main constituent of the filler and thus contributed to increase of thermal stability of composites. Given that it can be assumed that in silica–lignin hybrid fillers the presence of silica is responsible for maintaining thermal stability of composites and the lignin is a constituent.

3.4. Colorimetric Analysis

In view of potential applications of PLA/hybrid fillers composites not only thermal and mechanical properties are important. Appearance of such films is also an essential factor. Therefore, in Figure 4 values of ΔE_{ab}^* parameter obtained by comparison of samples before and after biodegradation process are presented.

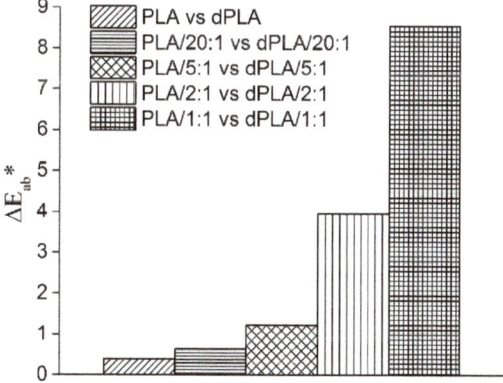

Figure 4. Values of ΔE_{ab}^* parameter for tested films.

Since value $\Delta E_{ab}^* \sim 2.3$ corresponds to a just noticeable difference it can be stated that only for samples of PLA/1:1 and PLA/2:1 a clearly visible change in color of films could be observed. In these samples the amount of lignin was the highest. These relationships are understandable—the lignin is known to change color not only during its isolation from biomass but also during biodegradation because of degradation of aromatic structures toward oligomeric chromophores [36,37]. This change of color, browning, is thought to be the main obstacle for high value-added use of lignin in areas such as sunscreen or dyestuff dispersants [37].

3.5. Mechanical Properties

Table 4 presents parameters obtained during tensile testing of primary PLA films and samples subjected to simulated biodegradation.

Table 4. Young's modulus (YM), tensile strength (TS), and elongation at break (EB) parameters calculated for PLA films before and after biodegradation.

Sample	YM (GPa)	TS (MPa)	EB (%)
PLA	2.08 ± 0.12	52.28 ± 5.59	46.33 ± 17.17
PLA/20:1	1.15 ± 0.18	27.94 ± 1.84	10.23 ± 1.20
PLA/5:1	1.33 ± 0.13	30.42 ± 2.56	20.37 ± 7.20
PLA/2:1	1.34 ± 0.21	38.17 ± 1.08	5.26 ± 1.38
PLA/1:1	1.12 ± 0.18	30.07 ± 3.14	3.67 ± 0.79
dPLA	2.11 ± 0.55	52.93 ± 2.53	12.30 ± 3.75
dPLA/20:1	0.67 ± 0.11	21.30 ± 0.25	1.75 ± 0.04
dPLA/5:1	0.37 ± 0.15	13.39 ± 6.12	2.01 ± 0.68
dPLA/2:1	0.29 *	7.58 *	1.67 *
dPLA/1:1	0.14 *	3.17 *	1.41 *

For samples marked with '*' standard deviation was not calculated.

As it should be expected, biodegradation of composites caused a decrease of TS and YM of films. Especially dPLA/2:1 and dPLA/1:1 films were so degraded that it was impossible to test more than one sample. Incorporation of hybrid filler of any type caused an impairment of all the tested parameters. The highest values of YM, TS, and EB were obtained for neat PLA. Maximal values of TS and YM for pristine PLA were definitely higher than those for some composites. What is more, TS and YM of PLA matrix were unaffected by the biodegradation process. Similar relationship for TS was described by Karamanlioglu et al. [38] who studied influence of biotic and abiotic factors on the rate of degradation of PLA samples buried in compost and soil. TS and YM of non-degraded composite samples were similar (approximately 28–38 MPa for TS and 1.12–1.34 GPa for YM) but lower than for pristine PLA. Obtaining a good interaction between components of composites is needed for effective stress transfer between matrix and filler. Accordingly, it is crucial in terms of enhancing mechanical properties of composites [39]. In general it is believed that the increase of particle size of reinforcement causes an enhanced debonding of the filler from polymer matrix [40]. However, in this research the influence of particle size on mechanical properties of films turned out to be rather negligible. Here, the decrease in tensile properties should be rather ascribed to presence of lignin, what was also reported elsewhere [41]. Even though lignin and PLA, due to their ability to form hydrogen bonds, offer quite good compatibility in order to observe significant increase of mechanical properties fractionation or modification of lignin is often needed [42,43].

Values of TS and YM calculated for biodegraded samples were divergent but there is some relationship between type of filler used and tested parameters. It can be noticed that the higher was the amount of lignin in filler, the lower values of TS and YM were obtained. In order to fully understand this relationship, one has to know that two main mechanisms of biodegradation are distinguished: *(i)* bulk erosion, when water diffuses rapidly between polymer chains, causing hydrolysis and *(ii)* surface erosion during which the polymer resorbs water from its outer surface toward its center [44]. It was found that PLA composites are first being hydrolyzed in whole bulk of the material [45] and then they may undergo surface erosion [46]. It was also proved that presence of hydroxyl groups in composite fillers enables penetration of water into ester groups of PLA matrix and facilitates biodegradation process [18]. Also hydroxyl groups present in kraft lignin can contribute to its biodegradation. It is most likely that as the amount of lignin in composite increases, the water adsorption increases and thus biodegradation accelerates. That is also consistent with other research [47].

The values of the EB parameter of the evaluated samples were significantly different. PLA generally suffers from a low deformation at break thus requires the use of plasticizers [48]. However, in this case the highest value of this parameter (ca. 46%) was noted for neat PLA. Incorporation of filler in PLA matrix resulted in decrease of EB. This is a typical behavior for some composites, especially nanocomposites. Again, biodegraded composites were found to have more uniform values of EB (ca. 1–2%) than non-biodegraded samples (ca. 4–20%). EB parameter for pristine PLA was the parameter affected by the biodegradation the most. While in terms of YM and TS of both PLA samples there was almost no difference, for EB for PLA reached 46%, whereas for degraded PLA it was only 12%.

3.6. Morphology of Films

SEM micrographs for non- and biodegraded samples (see Figure 5) were taken in order to define the morphological structure of composite films and thus provide better understanding of results of mechanical tests.

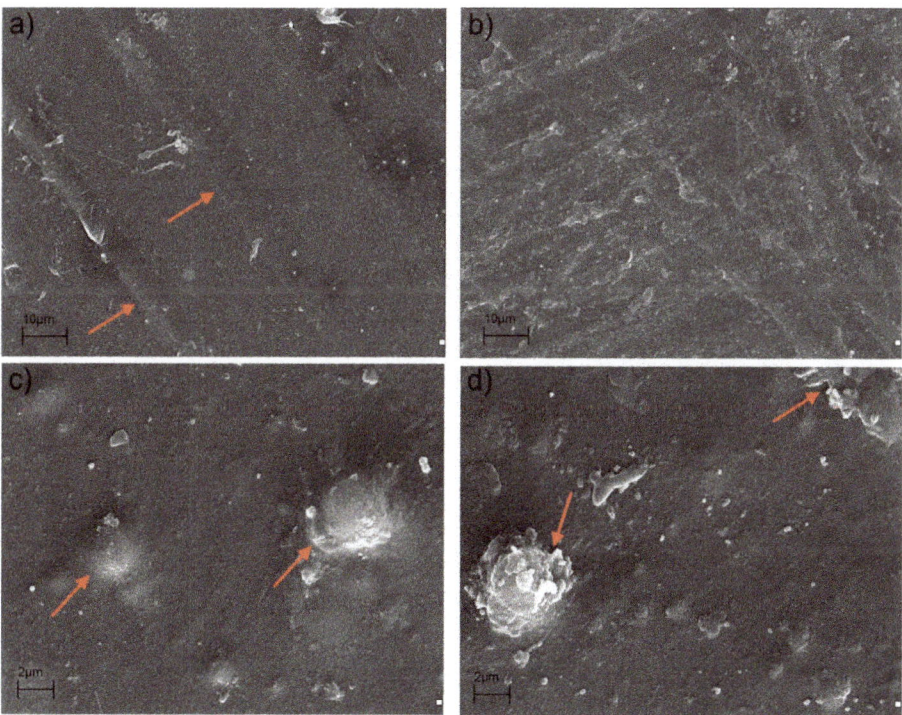

Figure 5. SEM images of composite films: (**a**) PLA, (**b**) dPLA, (**c**) PLA/5:1, and (**d**) dPLA/5:1.

Comparison of SEM pictures of non- and biodegraded PLA samples shows some interesting differences. On the contrary to dPLA sample (Figure 5b), the micrograph of non-degraded PLA (Figure 5a) reveals the presence of well visible parallel lines (marked with red arrows) which are the remains of orientation process occurring during extrusion of films. Orientation and polymer chain relaxation, these are two important effects that can have an impact on mechanical properties of materials. Yu et al. [49] have shown that during orientation of films under the different drawing speeds modulus and tensile strength remained almost the same, but an elongation decreased. Increased chain relaxation along with lower degree of orientation occurred during biodegradation of PLA films seem to be a reason for such significant decrease of EB parameter (from ca 46% for PLA to ca. 12% for dPLA).

In presented SEM microphotographs both, lignin and silica particles can be observed. Figure 5c,d reveal the presence of some aggregates of silica (seen as light points) with particles sizes below 1 μm as well as some bigger silica aggregates (ca. 2 μm) covered with polymer matrix. On the other hand, lignins are known to have spherical shape which is a result of their three-dimensional structure [50]. An example of such big, well-defined, spherical lignin particle, fully covered with PLA matrix, can be seen in Figure 5c. That sample was also characterized with good interphase adhesion between polymer matrix and particles of the filler. On the contrary, as shown in Figure 5d, the surface of the degraded sample was less uniform when compared to non-degraded sample. The PLA matrix itself remained intact, but the lignin particles were already partially decomposed, showing some delamination. What is crucial, in the biodegraded composite film the lignin particles and products of their degradation were almost entirely uncovered with polymer. Also some cracks at the matrix/filler interphase were present. These cracks were responsible for lowering the mechanical properties of composites subjected to biodegradation. According to literature, hydroxyl groups present in lignin are partially responsible for its biodegradation [7]. Moreover, weight loss studies have shown that samples containing high amount of lignin underwent the highest weight loss—7.7% and 6.5% for composites with 1:1 and 2:1 hybrid filler, respectively. Weight loss for other samples was as follows: 0.9% for unfilled PLA, 2.8% for PLA with 20:1 hybrid, and 4.3% for PLA filled with 5:1 hybrid. Therefore, it is believed that in PLA/hybrid filler films lignin was an ingredient that underwent biodegradation as first, causing a discontinuity of PLA matrix.

4. Conclusions

The subject of this study was to prepare PLA/hybrid fillers composites and determine changes taking place during biodegradation of PLA. Measurements were taken so to define thermal stability and crystallization behavior, while morphology of samples was investigated by SEM technique. Changes in color and mechanical properties of composites were examined as well. All of these methods were used to investigate both, non- and biodegraded samples. To the best knowledge of the authors this is the first publication regarding biodegradation of PLA composites with silica–lignin hybrids as fillers.

It was found that thermal studies, including phase transitions observations, can provide information about the course of biodegradation process of PLA/hybrid filler composites. The results show that during the biodegradation of hybrid fillers lignin is a component that undergoes degradation as first. Products of its decomposition turned out to act as heterogeneous nucleation agent for PLA. However, the major agent determining thermal stability of composites was the presence of silica. Susceptibility to biodegradation and thermo-mechanical properties of prepared composites are strongly related to composition of hybrid filler. Composites with hybrid filler containing high amount of lignin were the most active in terms of nucleating abilities. However, such composites were characterized with the lowest thermal stability. These findings suggest that composition of hybrid fillers should be further optimized. Obtained results indicate that composite with 2:1 hybrid filler was the most thermally stable (comparable to unfilled PLA) and underwent biodegradation process the most effectively.

Author Contributions: A.G.-Z. Preparation and characterization of thermal as well as mechanical properties of silica–lignin/polylactide composites. Results development. Manuscript preparation. Ł.K. Planning studies. Preparation and characterization of silica–lignin hybrid materials. Results development. Manuscript preparation. S.B. Research discussion. Elaboration of the obtained results. Coordination of all tasks in the paper. T.J. Planning studies. Results development. Coordination of all tasks in the paper.

Funding: This research was supported by the grants of Poznan University of Technology no. 03/32/DSPB/0803 and 03/32/DSPB/0806.

Conflicts of Interest: The authors declare no conflict of interest.

References

1. Vink, E.T.H.; Rábago, K.R.; Glassner, D.A.; Gruber, P.R. Applications of life cycle assessment to NatureWorks™ polylactide (PLA) production. *Polym. Degrad. Stabil.* **2003**, *80*, 403–419. [CrossRef]
2. Wen, X.; Lin, Y.; Han, C.; Zhang, K.; Ran, X.; Li, Y.; Dong, L. Thermomechanical and optical properties of biodegradable poly(L-lactide)/silica nanocomposites by melt compounding. *J. Appl. Polym. Sci.* **2009**, *114*, 3379–3388. [CrossRef]
3. Bula, K.; Klapiszewski, Ł.; Jesionowski, T. A novel functional silica/lignin hybrid material as a potential bio-based polypropylene filler. *Polym. Compos.* **2015**, *36*, 913–922. [CrossRef]
4. Jesionowski, T.; Zdarta, J.; Krajewska, B. Enzyme immobilization by adsorption: A review. *Adsorption* **2014**, *20*, 801–821. [CrossRef]

5. Sadeghifar, H.; Cui, C.; Argyropoulos, D.S. Toward thermoplastic lignin polymers. Part 1. Selective masking of phenolic hydroxyl groups in kraft lignins via methylation and oxypropylation chemistries. *Ind. Eng. Chem. Res.* **2012**, *51*, 16713–16720. [CrossRef]
6. Cui, C.; Sadeghifar, H.; Sen, S.; Argyropoulos, D.S. Toward thermoplastic lignin polymers. Part II: Thermal & polymer characteristics of kraft lignin & derivatives. *BioResources* **2013**, *8*, 864–886.
7. Wyman, C.E. *Aqueous Pretreatment of Plant Biomass for Biological and Chemical Conversion to Fuels and Chemicals*; John Wiley & Sons: Hoboken, NJ, USA, 2013; ISBN 9780470972021.
8. Toriz, G.; Denes, F.; Young, R.A. Lignin-polypropylene composites. Part 1: Composites from unmodified lignin and polypropylene. *Polym. Compos.* **2002**, *23*, 806–813. [CrossRef]
9. Hasegawa, I.; Fujii, Y.; Yamada, K.; Kariya, C.; Takayama, T. Lignin-silica hybrids as precursors for silicon carbide. *J. Appl. Polym. Sci.* **1999**, *73*, 1321–1328. [CrossRef]
10. Klapiszewski, Ł.; Rzemieniecki, T.; Krawczyk, M.; Malina, D.; Norman, M.; Zdarta, J.; Majchrzak, I.; Dobrowolska, A.; Czaczyk, K.; Jesionowski, T. Kraft lignin/silica-AgNPs as a functional material with antibacterial activity. *Colloids Surf. B* **2015**, *134*, 220–228. [CrossRef] [PubMed]
11. Qu, Y.; Tian, Y.; Zou, B.; Zhang, J.; Zheng, Y.; Wang, L.; Li, Y.; Rong, C.; Wang, Z. A novel mesoporous lignin/silica hybrid from rice husk produced by a sol-gel method. *Bioresour. Technol.* **2010**, *101*, 8402–8405. [CrossRef] [PubMed]
12. Strzemiecka, B.; Klapiszewski, Ł.; Jamrozik, A.; Szalaty, T.J.; Matykiewicz, D.; Sterzyński, T.; Voelkel, A.; Jesionowski, T. Physicochemical characterization of functional lignin-silica hybrid fillers for potential application in abrasive tools. *Materials* **2016**, *9*, 517. [CrossRef] [PubMed]
13. Zhang, R.; Xiao, X.; Tai, Q.; Huang, H.; Yang, J.; Hu, Y. Preparation of lignin-silica hybrids and its application in intumescent flame-retardant poly(lactic acid) system. *High Perform. Polym.* **2012**, *24*, 738–746. [CrossRef]
14. Strzemiecka, B.; Klapiszewski, Ł.; Matykiewicz, D.; Voelkel, A.; Jesionowski, T. Functional lignin-SiO$_2$ hybrids as potential fillers for phenolic binders. *J. Adhes. Sci. Technol.* **2016**, *30*, 1031–1048. [CrossRef]
15. Mohanty, A.K.; Misra, M.; Hinrichsen, G. Biofibres, biodegradable polymers and biocomposites: An overview. *Macromol. Mater. Eng.* **2000**, *276–277*, 1–24. [CrossRef]
16. Jain, J.P.; Yenet Ayen, W.; Domb, A.J.; Kumar, N. Biodegradable Polymers in Drug Delivery. In *Biodegradable Polymers in Clinical Use and Clinical Development*; Domb, A.J., Kumar, N., Ezra, A., Eds.; John Wiley & Sons: Hoboken, NJ, USA, 2011; pp. 1–58. ISBN 9780470424759.
17. Petinakis, E.; Liu, X.; Yu, L.; Way, C.; Sangwan, P.; Dean, K.; Bateman, S.; Edward, G. Biodegradation and thermal decomposition of poly(lactic acid)-based materials reinforced by hydrophilic fillers. *Polym. Degrad. Stabil.* **2010**, *95*, 1704–1707. [CrossRef]
18. Fukushima, K.; Abbate, C.; Tabuani, D.; Gennari, M.; Camino, G. Biodegradation of poly(lactic acid) and its nanocomposites. *Polym. Degrad. Stabil.* **2009**, *94*, 1646–1655. [CrossRef]
19. Liu, Y.; Chen, W.; Kim, H.-I. Synthesis, characterization, and hydrolytic degradation of polylactide/poly(ethylene glycol)/nano-silica composite films. *J. Macromol. Sci. Part A* **2012**, *49*, 348–354. [CrossRef]
20. Grząbka-Zasadzińska, A.; Klapiszewski, Ł.; Bula, K.; Jesionowski, T.; Borysiak, S. Supermolecular structure and nucleation ability of polylactide-based composites with silica/lignin hybrid fillers. *J. Therm. Anal. Calorim.* **2016**, *126*, 263–275. [CrossRef]
21. Borysiak, S.; Klapiszewski, Ł.; Bula, K.; Jesionowski, T. Nucleation ability of advanced functional silica/lignin hybrid fillers in polypropylene composites. *J. Therm. Anal. Calorim.* **2016**, *126*, 251–262. [CrossRef]
22. Klapiszewski, Ł.; Bula, K.; Sobczak, M.; Jesionowski, T. Influence of processing conditions on the thermal stability and mechanical properties of PP/silica-lignin composites. *Int. J. Polym. Sci.* **2016**, *1627258*, 1–9. [CrossRef]
23. Garlotta, D. A literature review of poly(lactic acid). *J. Polym. Environ.* **2001**, *9*, 63–84. [CrossRef]
24. Sharma, G.; Bala, R. *Digital Color Imaging Handbook*; CRC Press: Boca Raton, FL, USA, 2002; ISBN 9780849309007.
25. Rahman, M.M.; Afrin, S.; Haque, P.; Islam, M.M.; Islam, M.S.; Gafur, M.A. Preparation and characterization of jute cellulose crystals-reinforced poly(L-lactic acid) biocomposite for biomedical applications. *Int. J. Chem. Eng.* **2014**, *842147*, 1–7. [CrossRef]

26. Anwer, M.A.S.; Naguib, H.E.; Celzard, A.; Fierro, V. Comparison of the thermal, dynamic mechanical and morphological properties of PLA-Lignin & PLA-Tannin particulate green composites. *Compos. Part B Eng.* **2015**, *82*, 92–99. [CrossRef]
27. Valapa, R.; Hussain, S.; Iyer, P.; Pugazhenthi, G.; Katiyar, V. Influence of graphene on thermal degradation and crystallization kinetics behaviour of poly(lactic acid). *J. Polym. Res.* **2015**, *22*, 1–14. [CrossRef]
28. Zimmermann, M.V.G.; Brambilla, V.C.; Brandalise, R.N.; Zattera, A.J. Observations of the effects of different chemical blowing agents on the degradation of poly(lactic acid) foams in simulated soil. *Mater. Res.* **2013**, *16*, 1266–1273. [CrossRef]
29. Hakkarainen, M.; Albertsson, A.-C.; Karlsson, S. Weight losses and molecular weight changes correlated with the evolution of hydroxyacids in simulated in vivo degradation of homo- and copolymers of PLA and PGA. *Polym. Degrad. Stabil.* **1996**, *52*, 283–291. [CrossRef]
30. Murariu, M.; Dechief, A.-L.; Ramy-Ratiarison, R.; Paint, Y.; Raquez, J.-M.; Dubois, P. Recent advances in production of poly(lactic acid) (PLA) nanocomposites: A versatile method to tune crystallization properties of PLA. *Nanocomposites* **2015**, *1*, 71–82. [CrossRef]
31. Niaounakis, M. *Biopolymers: Applications and Trends*; Elsevier Inc.: Amsterdam, The Netherlands, 2015; ISBN 978-0-323-35399-1.
32. Kopinke, F.D.; Remmler, M.; Mackenzie, K.; Möder, M.; Wachsen, O. Thermal decomposition of biodegradable polyesters—II. Poly(lactic acid). *Polym. Degrad. Stabil.* **1996**, *53*, 329–342. [CrossRef]
33. Pinto, A.M.; Gonçalves, C.; Gonçalves, I.C.; Magalhães, F.D. Effect of biodegradation on thermo-mechanical properties and biocompatibility of poly(lactic acid)/graphene nanoplatelets composites. *Eur. Polym. J.* **2016**, *85*, 431–444. [CrossRef]
34. Gordobil, O.; Delucis, R.; Egüés, I.; Labidi, J. Kraft lignin as filler in PLA to improve ductility and thermal properties. *Ind. Crop. Prod.* **2015**, *72*, 46–53. [CrossRef]
35. Costes, L.; Laoutid, F.; Aguedo, M.; Richel, A.; Brohez, S.; Delvosalle, C.; Dubois, P. Phosphorus and nitrogen derivatization as efficient route for improvement of lignin flame retardant action in PLA. *Eur. Polym. J.* **2016**, *84*, 652–667. [CrossRef]
36. Hiltunen, E.; Alvila, L.; Pakkanen, T.T. Characterization of Brauns' lignin from fresh and vacuum-dried birch (Betula pendula) wood. *Wood Sci. Technol.* **2006**, *40*, 575. [CrossRef]
37. Wang, J.; Deng, Y.; Qian, Y.; Qiu, X.; Ren, Y.; Yang, D. Reduction of lignin color via one-step UV irradiation. *Green Chem.* **2016**, *18*, 695–699. [CrossRef]
38. Karamanlioglu, M.; Robson, G.D. The influence of biotic and abiotic factors on the rate of degradation of poly(lactic) acid (PLA) coupons buried in compost and soil. *Polym. Degrad. Stabil.* **2013**, *98*, 2063–2071. [CrossRef]
39. Fu, S.Y.; Feng, X.Q.; Lauke, B.; Mai, Y.W. Effects of particle size, particle/matrix interface adhesion and particle loading on mechanical properties of particulate-polymer composites. *Compos. Part B Eng.* **2008**, *39*, 933–961. [CrossRef]
40. Omar, M.F.; Akil, H.M.; Ahmad, Z.A. Particle size—Dependent on the static and dynamic compression properties of polypropylene/silica composites. *Mater. Des.* **2013**, *45*, 539–547. [CrossRef]
41. Li, J.; He, Y.; Inoue, Y. Thermal and mechanical properties of biodegradable blends of poly(L-lactic acid) and lignin. *Polym. Int.* **2003**, *52*, 949–955. [CrossRef]
42. Chung, Y.-L.; Olsson, J.V.; Li, R.J.; Frank, C.W.; Waymouth, R.M.; Billington, S.L.; Sattely, E.S. A renewable lignin–lactide copolymer and application in biobased composites. *ACS Sustain. Chem. Eng.* **2013**, *1*, 1231–1238. [CrossRef]
43. Zhu, J.; Xue, L.; Wei, W.; Mu, C.; Jiang, M.; Zhou, Z. Modification of lignin with silane coupling agent to improve the interface of poly (L-lactic) acid/lignin composites. *BioResources* **2015**, *10*, 4315–4325. [CrossRef]
44. Gajjar, C.R.; King, M.W. *Resorbable Fiber-Forming Polymers for Biotextile Applications*; Springer International Publishing: New York, NY, USA, 2014; ISBN 978-3-319-08304-9.
45. Fukushima, K.; Tabuani, D.; Abbate, C.; Arena, M.; Ferreri, L. Effect of sepiolite on the biodegradation of poly(lactic acid) and polycaprolactone. *Polym. Degrad. Stabil.* **2010**, *95*, 2049–2056. [CrossRef]
46. Lyu, S.; Untereker, D. Degradability of polymers for implantable biomedical devices. *Int. J. Mol. Sci.* **2009**, *10*, 4033–4065. [CrossRef] [PubMed]
47. Spiridon, I.; Leluk, K.; Resmerita, A.M.; Darie, R.N. Evaluation of PLA-lignin bioplastics properties before and after accelerated weathering. *Compos. Part B Eng.* **2015**, *69*, 342–349. [CrossRef]

48. Cicogna, F.; Coiai, S.; Monte, C.D.; Spiniello, R.; Fiori, S.; Franceschi, M.; Braca, F.; Cinelli, P.; Fehri, S.M.K.; Lazzeri, A.; et al. Poly(lactic acid) plasticized with low-molecular-weight polyesters: Structural, thermal and biodegradability features. *Polym. Int.* **2017**, *66*, 761–769. [CrossRef]
49. Yu, L.; Liu, H.; Xie, F.; Chen, L.; Li, X. Effect of annealing and orientation on microstructures and mechanical properties of polylactic acid. *Polym. Eng. Sci.* **2008**, *48*, 634–641. [CrossRef]
50. Fengel, D.; Wegener, G. *Wood: Chemistry, Ultrastructure, Reactions*; De Gruyter: Berlin, Germany, 1983; ISBN 10.

© 2018 by the authors. Licensee MDPI, Basel, Switzerland. This article is an open access article distributed under the terms and conditions of the Creative Commons Attribution (CC BY) license (http://creativecommons.org/licenses/by/4.0/).

Review

Titania-Based Hybrid Materials with ZnO, ZrO$_2$ and MoS$_2$: A Review

Adam Kubiak, Katarzyna Siwińska-Ciesielczyk * and Teofil Jesionowski *

Institute of Chemical Technology and Engineering, Faculty of Chemical Technology, Poznan University of Technology, Berdychowo 4, PL-60965 Poznan, Poland; adam.l.kubiak@doctorate.put.poznan.pl
* Correspondence: katarzyna.siwinska-ciesielczyk@put.poznan.pl (K.S.-C.);
 teofil.jesionowski@put.poznan.pl (T.J.); Tel.: +48-6-1665-3626 (K.S.-C.); +48-665-3720 (T.J.)

Received: 16 October 2018; Accepted: 12 November 2018; Published: 15 November 2018

Abstract: Titania has properties that enable it to be used in a variety of applications, including self-cleaning surfaces, air and water purification systems, hydrogen evolution, and photoelectrochemical conversion. In order to improve the properties of titanium dioxide, modifications are made to obtain oxide/hybrid systems that are intended to have the properties of both components. In particular, zinc oxide, zirconia and molybdenum disulfide have been proposed as the second component of binary systems due to their antibacterial, electrochemical and photocatalytic properties. This paper presents a review of the current state of knowledge on the synthesis and practical utility of TiO$_2$-ZnO and TiO$_2$-ZrO$_2$ oxide systems and TiO$_2$-MoS$_2$ hybrid materials. The first part focuses on the hydrothermal method; then a review is made of the literature on the synthesis of the aforementioned materials using the sol-gel method. In the last section, the literature on the electrospinning method of synthesis is reviewed. The most significant physico-chemical, structural and dispersive-morphological properties of binary hybrid systems based on TiO$_2$ are described. A key aim of this review is to indicate the properties of TiO$_2$-ZnO, TiO$_2$-ZrO$_2$ and TiO$_2$-MoS$_2$ hybrid systems that have the greatest importance for practical applications. The variety of utilities of titania-based hybrid materials is emphasized.

Keywords: titanium dioxide; zinc oxide; zirconia; molybdenum disulfide; binary systems; hybrid materials

1. Introduction

Constant scientific and technological progress, as well as the desire to create environment-friendly technologies, is leading to intensive work on obtaining new-generation, functional products with strictly designed physico-chemical and dispersive-morphological properties, dedicated to specific applications. These studies focus on changing the physico-chemical features of many well-known and widely used materials. This group of substances includes oxide and hybrid materials based on titanium dioxide, which are distinguished by specific, precisely defined physico-chemical and structural properties, determined chiefly at the stage of their synthesis. Research on the synthesis of advanced materials based on titanium dioxide is becoming oriented towards the conscious and skilled modification of those materials' properties, achieved by selecting appropriate methods and process conditions. For every method, the selection of conditions such as the precursors used, the implementation of the process, the final processing temperature and the pH of the reaction system has a decisive impact on properties of the product such as its dispersive character, morphology, thermal and colloidal stability, parameters of porous structure, and hydrophilicity or hydrophobicity. The wide range of possible methods of synthesis enables the design of materials of this type with diverse physico-chemical and structural parameters, and is crucial in view of the constant demand for such hybrids. Furthermore, technological progress is accompanied by increased interest in and development of methods enabling better control of the process, and thus also of the properties of the synthesized materials.

Metal oxide systems have attracted a great deal of attention in recent years on account of their special electronic and chemical properties. Among the metal oxide semiconductors, compounds such as TiO_2, ZrO_2 and ZnO have been investigated extensively due to their chemical stability and good photocatalytic properties [1]. Titanium dioxide is the most widely used metal oxide for environmental applications, paints, electronic devices [2], gas sensors [3] and solar cells [4,5]. It is a well-known semiconductor with excellent photocatalytic properties, which has been widely used in environmental pollutant elimination [6,7], antibacterial dopes, self-cleaning surface, etc. [8]. Its unique antibacterial properties make it a candidate for applications in medical devices and sanitary ware surfaces [9]. Moreover, titania incorporated into polyester fabric [10,11] and related materials can be used as an absorber of harmful UV irradiation. Zinc oxide, due to its antibacterial [12] and photocatalytic activity [13], as well as its wide bandgap, has found applications in various areas based on its optical [14], piezoelectric and gas sensing properties [15,16], and in addition zinc compounds have generally been regarded as safe [1,17]. Zirconia is one of the most intensively studied materials owing to its technologically important applications in gas sensors [18,19], fuel cell electrolytes [20], catalysts [21] and catalytic supports [22], metal oxide-semiconductor devices [23], and in view of its superior thermal and chemical stability [24] and other properties [25].

Among synthetic hybrid oxide systems, particular attention is given to TiO_2-ZrO_2 materials, which thanks to the addition of zirconium dioxide have much greater surface area and mechanical strength than pure TiO_2. Conjugation of zirconia together with titania leads to obtain products which are characterized with higher specific surface in comparison to pristine TiO_2. Moreover, the addition of ZrO_2 to TiO_2 inhibits phase transformation of anatase-to-rutile, and creates more active site groups on titania surfaces [22,24]. There are many publications concerning the application of TiO_2-ZrO_2 hybrids in photocatalysis [26,27]. The use of TiO_2-ZrO_2 hybrid materials in the photo-oxidation of organic compounds or degradation of dyes originating from various industrial plants is well known. They also have applications in the photo-reduction of atmospherically harmful oxides, like CO_2 and NOx, resulting for example from the combustion of fossil fuels. The advantages of the TiO_2-ZrO_2 hybrid are its mechanical strength, non-toxicity and corrosion resistance, and the ability to conduct photocatalytic processes using sunlight. These may be causes of increasing demand for this material in the near future [28–30].

Combining Titania with zinc oxide can also lead to a hybrid oxide system with good photocatalytic properties [31]. The resulting material can be used, for instance, in the degradation of organic impurities such as detergents, dyes and pesticides present in various types of wastewater. A TiO_2-ZnO hybrid material can be synthesized by both physical and chemical processes, which enables enhancement of its properties, for example by widening its spectrum of light absorption. Additionally, the photocatalytic activity of oxides may help reduce the tendency of pollutants to form aggregate structures [32].

The combination of TiO_2 with a semiconductor, such as molybdenum disulfide, allows the creation of a hybrid system that not only exhibits activity under the influence of visible radiation, but also allows the separation of photogenerated electron-hole pairs, which increases the catalytic capacity of the material [33,34].

The presented review focuses on recent stage on knowledge about sol-gel, hydrothermal and electrospinning synthesis of advanced, multifunctional materials based on titanium dioxide with strictly defined physico-chemical and structural properties, which significantly determine their multidirectional application. It is known that different properties of hybrid systems, such as TiO_2-ZnO, TiO_2-ZrO_2 and TiO_2-MoS_2, depend on their morphology, crystallites size, and crystalline structure, which can be modified by selecting the appropriate method for their synthesis, as well as selecting the right components which together with TiO_2 will create advanced hybrid materials with unique properties.

2. Method of Synthesis of Titania-Based Materials

2.1. The Hydrothermal Method

One of the most frequently used methods for obtaining powder materials with specific physico-chemical and dispersive-morphological properties is hydrothermal synthesis. This can be used to obtain oxides such as TiO_2, ZrO_2, Al_2O_3, etc. According to the definition proposed by Roy [35] and Rubenau [36], hydrothermal synthesis is a reaction that takes place in a water environment at elevated temperature (>100 °C) and pressure, carried out in an autoclave [37,38].

The physico-chemical properties of products formed using the hydrothermal method depend on many process parameters, including the temperature, pressure, time of reaction, reactor volume, type of solvent used, and ratio of reagents (Figure 1). Literature data demonstrate clearly that the temperature and time of hydrothermal treatment [39,40], as well as the relative quantities of reagents used, have an effect on the crystal structure [41,42] and crystallite size of the synthesized material. Another important parameter is the reactor volume; this conditions the creation of (hydrostatic) pressure, which has a direct impact on the size of the synthesized particles [43–45].

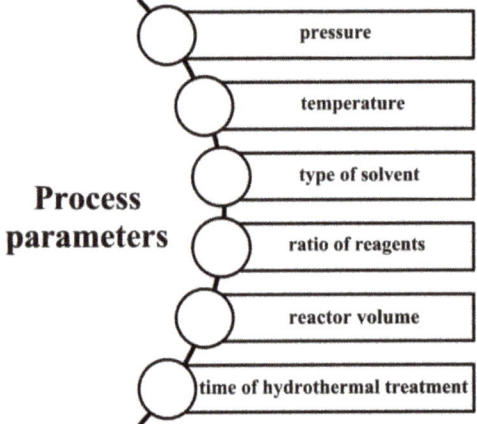

Figure 1. Process parameters determining the physico-chemical and structural properties of products obtained by the hydrothermal method.

The hydrothermal method offers many advantages, including a fast reaction rate, the quality and purity of the synthesized products, and the obtaining of materials with a crystalline structure and a smaller particle size. The hydrothermal method can also be used to obtain products in intermediate oxidation states, such as chromium(IV) oxide, as well as metastable compounds such as tellurium iodide (Te_2I) [39]. Furthermore, it is extremely advantageous both environmentally and economically, because of its relatively low energy costs. Compared with other well-known and widely used techniques in which the growth of crystals requires significant energy inputs (high temperature and pressure), the hydrothermal method offers conditions of crystallite growth similar to natural conditions. This is consistent with the principles of green chemistry [46–48].

Some unquestionable disadvantages of the hydrothermal method include the complexity and cost of the equipment (resulting from the need for the autoclaves to be resistant to high pressure). Moreover, this method does not permit direct observation of the process, and it requires the reagents used to be soluble in water, which serves as the process medium [49–51].

Despite these drawbacks, the hydrothermal method is often combined with other techniques (combined methods) to improve the physico-chemical properties of the synthesized materials. One such technique was described by Komarneni et al. [52] and Tompsett et al. [53], who introduced microwaves to the hydrothermal reactor, enabling it to be heated more rapidly to the required temperature. This enhancement of the hydrothermal process is made possible by the special properties of Teflon. That material has an optical gate which enables the passage of waves characteristic for the microwave range. The microwave-assisted hydrothermal method offers a high speed of crystallization (faster reaction kinetics) and leads to products with a very narrow particle size distribution and controlled morphology.

Due to the advantages of both conventional and microwave-assisted hydrothermal methods, in recent years these techniques have been used more frequently for synthesizing oxides and hybrid systems. To obtain oxide systems with diverse physico-chemical and structural properties, many researchers have used titanium dioxide (TiO_2)—a compound commonly applied in many branches of industry—as a base for combination with zinc oxide (ZnO), zirconium dioxide (ZrO_2) and molybdenum disulfide (MoS_2) [54,55]. These materials offer good photocatalytic and electrochemical properties. For this reason, it is a problem of interest to synthesize oxide systems such as TiO_2-ZnO, TiO_2-ZrO_2 as well as TiO_2-MoS_2 hybrid system using the principles of the conventional and microwave-assisted hydrothermal methods. Table 1 summarizes the most significant literature reports concerning the synthesis of TiO_2-ZnO binary systems using the hydrothermal method, with or without the action of microwaves.

Many literature reports indicate that the most photo-active crystalline form of titanium dioxide is anatase [4,5]. However, its use limited by relatively large band gap 3.2 eV [6]. Therefore, the scientists have focused on modifying the physico-chemical properties TiO_2 to improve its photocatalytic activity. Modification may be performed by formation hybrid materials [26,28,33]. Xu et al. [8] developed the synthesis of a TiO_2-ZnO oxide system based on the principles of the sol-gel and hydrothermal methods. A key part of their work was the evaluation of the photocatalytic properties of the resulting materials in the decomposition of methyl orange (C.I. Basic Orange 10). In the first step, titanium tetrabutoxide, acetic acid and distilled water were mixed with zinc acetate, diethylamine (DEA) and ethanol, after which the solution underwent aging until it became a gel. In the second step the product was hydrothermally treated at different temperatures and for different times. X-ray diffractograms revealed the presence of characteristic diffraction reflections for anatase. Increasing the temperature of the hydrothermal treatment was shown not to affect the crystal structure of the final product. Scanning electron microscope (SEM) images showed that the TiO_2-ZnO oxide systems contained particles with near-spherical shape and with a marked tendency to agglomerate. It was also found that an increase in the temperature of hydrothermal treatment, and the calcination process, did not significantly alter the morphology of the TiO_2-ZnO oxide system. In photocatalytic tests, all of the synthesized materials demonstrated high photoactivity. The material subjected to hydrothermal treatment at 150 °C for 24 h and additional calcination at 350 °C for 2 h produced a 55% yield of methyl orange degradation after 3 h.

Table 1. Resources, synthesis conditions and physico-chemical properties of TiO$_2$-ZnO oxide systems obtained by the hydrothermal method.

Resources	Synthesis Conditions	Physico-Chemical Properties of the Final Product	References
Ti(OC$_4$H$_9$)$_4$, Zn(CH$_3$COO)$_2$·2H$_2$O, ethanol, deionized water	hydrothermal treatment: 120, 180, 200 °C, reaction time: 1, 12, 24, 48 h, drying: 60 °C, calcination: 350 °C for 1 h	anatase crystal structure, spherical shaped particles with a tendency to agglomerate and particle sizes in the range 413–527 nm	[8]
TiO$_2$ microspheres (P25 Degussa), Zn(NO$_3$)$_2$·6H$_2$O, hexamethylenetetramine, distilled water	hydrothermal treatment: 100 °C, reaction time: 1–5 h, drying: 60 °C	crystal structure of anatase and wurtzite, spherical shaped particles coated with nanospindles, particle diameter 1.5–2 μm, BET surface area 122.2–34.8 m^2/g (reduction of specific surface area with increase of ZnO fraction)	[56,57]
Ti(SO$_4$)$_2$, ZnCl$_2$, NH$_3$·H$_2$O, cetyltrimethylammonium bromide, distilled water	hydrothermal treatment: 100 °C, reaction time: 4 h, drying: 60 °C	crystal structure of TiO, Ti$_3$O$_5$ and ZnO, particles with a smooth surface with a tendency to agglomerate	[58]
Ti[OCH(CH$_3$)$_2$]$_4$, Zn(NO$_3$)$_2$·6H$_2$O, NaOH, distilled water, poly(vinyl alcohol)	hydrothermal treatment: 220 °C, reaction time: 5 h, drying: 80 °C for 2 h	crystal structure of anatase and wurtzite, particles with spherical and cubic shapes with a tendency to agglomerate, particle diameter 30 nm	[59]
Ti(OC$_4$H$_9$)$_4$, Zn(CH$_3$COO)$_2$·2H$_2$O, acetic acid, ethylene glycol, deionized water	hydrothermal treatment: 180, 190, 200 °C, reaction time: 8, 12, 15, 24 h, drying: 60 °C	crystal structure of anatase and wurtzite, particle size distribution in the range 25–100 nm with average particle diameter 64 nm, BET surface area 97 m^2/g and 206 m^2/g for systems obtained at 180 and 200 °C respectively	[60]
Ti(SO$_4$)$_2$, Zn(NO$_3$)$_2$·6H$_2$O, NH$_4$F, CO(NH$_2$)$_2$, ethanol, distilled water	hydrothermal treatment: 120–180 °C, reaction time: 2–24 h, drying: 80 °C for 24 h	crystal structure anatase and wurtzite, spherical particles with a tendency to agglomerate, nanoparticles of diameter 0.5–2 μm	[61]
Ti[OCH(CH$_3$)$_2$]$_4$, Zn(CH$_3$COO)$_2$·2H$_2$O, NH$_3$·H$_2$O, ethylene glycol	hydrothermal treatment: 220 °C, reaction time: 5 h, drying: 60 °C	crystal structure of anatase and wurtzite, particles with a tendency to agglomerate, porous microstructure	[62]
TiCl$_4$, ZnCl$_2$, urea, ethanol, distilled water	hydrothermal treatment: 180 °C, reaction time: 16 h, calcination: 450 °C for 2 h	crystal structure of anatase and wurtzite, particles taking the shape of nanorods with a tendency to agglomerate	[63]
TiCl$_4$, Zn(NO$_3$)$_2$·6H$_2$O, NH$_3$·H$_2$O, ethanol, distilled water	hydrothermal treatment: 150 °C, reaction time: 1 h, freeze-drying: −55 °C in vacuum, calcination: 600, 700, 900 °C for 1 h	crystal structure of rutile, wurtzite, Zn$_2$TiO$_2$O$_8$, spherical particles with a tendency to agglomerate, particle diameters in the range 140–270 μm	[64]
Ti[OCH(CH$_3$)$_2$]$_4$, Zn(CH$_3$COO)$_2$·2H$_2$O, sodium hydroxide, deionized water	microwave treatment: 180 °C for 5 min, frequency: 2.45 GHz, calcination: 500, 600 °C	crystal structure of anatase, wurtzite and zinc titanates (Zn TiO$_3$, Zn$_2$Ti$_3$O$_4$), spherical and hexagonal particles with a tendency to agglomerate, mean particle size 36 and 31 nm for samples calcined at 500 and 600 °C	[65]
Ti(OC$_4$H$_9$)$_4$, Zn(CH$_3$COO)$_2$, deionized water	microwave treatment: 80 °C for 30 min, drying: 80 °C	crystal structure of anatase and wurtzite, band gap energy 3.15 eV, spherical shaped particles with a tendency to agglomerate, BET surface area 290 m^2/g, pore diameter 3.4 nm, pore volume 0.32 cm^3/g	[66]

Cheng et al. [56] obtained TiO_2-ZnO oxide systems which were used for energy conversion in dye-sensitized solar cells (DSSCs). The synthesis process used previously obtained TiO_2 nanospheres, which were hydrothermally treated with zinc nitrate. The products were found to contain the crystal structure of anatase and wurtzite, and had the surface areas (which was determined by the multipoint Brunauere Emmette Teller method (BET)) of 348.8, 187.3 and 122.2 m^2/g. It was shown that an increase in the molar fraction of ZnO in the synthesized materials led to increased intensity of the diffraction peaks corresponding to wurtzite, and deterioration of the porous structure parameters. The synthesized systems were used as photoanodes in DSSCs. The tests showed that the TiO_2-ZnO oxide systems had better optical properties and higher energy conversion efficiency (8.78%) than pure titanium dioxide (6.79%).

Similar results to the aforementioned reported by Xu et al. [8] were obtained by Li [58], who observed the presence of characteristic diffraction peaks for ZnO, Ti_3O_5 and Ti, and found the materials to contain particles with a fusiform shape. They also demonstrated that the crystallinity of ZnO declined with an increase in the fraction of TiO_2 in the final product. Tests of the photocatalytic activity of the TiO_2-ZnO binary oxide system showed that it yielded complete degradation of methyl orange (C.I. Basic Orange 10) after 30 min. With pristine ZnO, by contrast, complete degradation occurred after 50 min.

With regard to the extensively described good electrochemical properties of TiO_2 [67–69] and ZnO [70–72], Vlazan et al. [59] undertook work on the synthesis of a TiO_2-ZnO oxide system in the form of core-shell nanoparticles, and went on to determine the electrochemical properties of the synthesized material. In the first stage of the synthesis, ZnO was obtained as a result of hydrothermal treatment of a mixture of zinc nitrate and a sodium hydroxide solution. The ZnO was then added to a suspension of poly(vinyl alcohol) and titanium tetraisopropoxide, and the reaction mixture was treated hydrothermally at 220 °C for 5 h. Analysis by techniques including X-ray diffraction (XRD) and scanning electron microscopy (SEM) showed that the obtained binary oxide system had the crystal structure of anatase and wurtzite, and that its particles were of spherical shape and measured 30 nm in diameter. Compared with the materials described by Cheng et al. [56], the obtained TiO_2-ZnO oxide system had a smaller BET surface area (44.4 m^2/g) and total pore volume (0.1132 cm^3/g). The study also included an investigation of the electrochemical properties of the synthesized binary material. It was found that the electrical resistance of the TiO_2-ZnO oxide system decreased with increasing temperature, from $\varrho = 1.2 \cdot \times 10^8\ \Omega$ (-128 °C) to $\varrho = 7.52 \cdot \times 10^6\ \Omega$ (29 °C). At the same time the electrical conductivity increased from $\sigma = 3.78 \cdot \times 10^{-8}$ S/cm (-128 °C) to $\sigma = 6.04 \cdot \times 10^{-7}$ S/cm (29 °C). The results demonstrate that the synthesized material offers good semiconducting properties.

The common use of titania in photocatalysis [54] motivated Zhang et al. [60] to test an oxide system based on TiO_2 in a degradation process of C.I. (Colour Index) Basic Orange 10. Titanium tetrabutoxide and zinc acetate, dissolved in a poly(ethylene glycol) solution, were used as precursors. The obtained solutions were mixed and then treated hydrothermally at 120 °C for 6 h. To improve its crystallinity, the resulting oxide system was subjected to further hydrothermal treatment at 180–200 °C for a specified time. Based on X-ray diffraction results it was shown that the materials contained the crystal structure of anatase and wurtzite. It was observed that increasing the temperature and time of hydrothermal treatment improved the crystal structure of the analyzed materials. The TiO_2-ZnO oxide systems contained particles of spherical shape with a tendency to agglomerate, and a range of sizes between 25 nm and 100 nm. The BET surface area was 206 m^2/g and 97 m^2/g respectively for materials treated at 180 °C and 200 °C. The synthesized materials were tested in the photocatalytic degradation of methyl orange (C.I. Basic Orange 10). The TiO_2-ZnO oxide systems exhibited high activity in the degradation of that dye. The synthesized hybrids were also shown to have superior properties to those of pure TiO_2 and ZnO.

Spherical hollow structures are receiving a great deal of attention, because they offer the lowest surface/volume ratio for aggregated products. Recent literature reports indicate the high photocatalytic activity of this type of TiO_2 [73,74] and ZnO [75] structures due to their large surface area, low density and highly efficient light-harvesting abilities. Wang et al. [61] showed that the use of titanium(IV) sulphate(VI) and zinc nitrate(V) as precursors in the hydrothermal synthesis of TiO_2-ZnO binary materials enables the obtaining of products with a defined morphology, spherical hollow structures (Figure 2). The synthesized oxide systems were found to contain the crystal structure of anatase and wurtzite. It was confirmed that an increase in the time of hydrothermal treatment above 6 h did not significantly affect the crystal structure. Analysis of porous structure parameters showed the BET surface area to be equal to 55.9 m^2/g. To confirm the high photocatalytic activity of the products, a test was conducted using the degradation of methyl orange (C.I. Basic Orange 10). The TiO_2-ZnO oxide systems produced very high rates of dye degradation (total decoloration after 25 min in UV light and after 180 min in visible light).

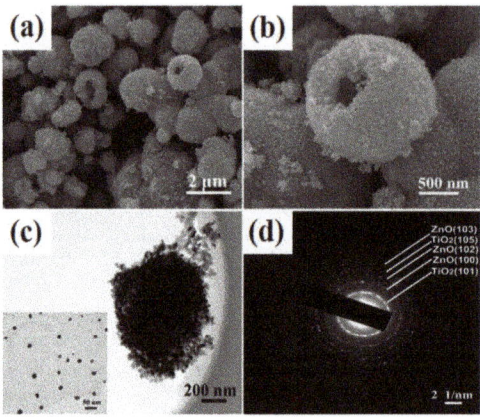

Figure 2. (**a**,**b**) SEM; (**c**) high-resolution transmission electron microscopy (HRTEM) and (**d**) selected area electron diffraction (SAED) images of the TiO_2-ZnO system (created based on [61] with permission from Elsevier Publisher).

Core-shell [76,77] nanoparticles have a core made of one material, coated with another material on top of it. In biological applications, core-shell nanoparticles have major advantages over simple nanoparticles, which can lead to improved antibacterial properties. Rusu et al. [62] developed a simple synthesis of a TiO_2-ZnO oxide system in the form of core-shell structures, with an application in a process of bacterial degradation. In the first stage of the synthesis, ammonia water was added to a zinc acetate solution. Then titanium tetraisopropoxide was added dropwise to the solution, and the resulting reaction mixture was hydrothermally treated at 220 °C for 5 h. X-ray diffraction results showed the product to contain the crystal structure of TiO_2 and ZnO. SEM images revealed particles of spherical shape with a tendency to agglomerate; the particle diameters were measured at 10–20 nm for TiO_2 and 30–80 nm for ZnO. Based on the results of antibacterial and antifungal tests, it was found that the material had good biological properties. The TiO_2-ZnO oxide system exhibited a biostimulating effect on the biosynthesis of fungi proteases, which is very important from the point of view of biotechnological applications.

Methylene blue (C.I. Basic Blue 9) is the most commonly used substance for dyeing cotton, timber and silk. Therefore, it is of great importance to deal effectively with the pollution resulting from such processes [78,79]. Chen, Zhang, Hu and Li [63] proposed a simple synthesis of TiO_2-ZnO binary oxide materials for use in methylene blue degradation. In the first stage of the synthesis, $TiCl_4$ and $ZnCl_2$ were dissolved in a water-ethanol mixture in the molar ratios Ti:Zn = 2:1, 1:1 and 1:2. Next, a urea solution was added with intense mixing for several hours, and then the mixture was subjected to hydrothermal treatment. The synthesized photocatalysts had similar crystal structure and morphology to the materials described in earlier studies. It was also observed that with an increase in the molar fraction of a given monoxide, there was an increase in the intensity of the diffraction peaks corresponding to its crystal structure. In the degradation of a model methylene blue solution, the synthesized TiO_2-ZnO oxide systems demonstrated high photoactivity in the decomposition of the tested organic pollutant. Complete decoloration of the dye solution was observed after 180 min in the presence of the material obtained in the molar ratio TiO_2:ZnO = 1:2.

The precisely defined crystal structure of this group of binary materials determines the range of their possible applications. In an investigation of the formation of specific mixed crystal structures, Wang et al. [64] carried out a synthesis of TiO_2-ZnO oxide systems using the hydrothermal method, supported by an additional calcination process. Diffractograms showed that the materials obtained after the hydrothermal treatment had an amorphous structure, whereas the TiO_2-ZnO oxide systems that had been calcined at 600, 700 and 900 °C showed the presence of diffraction reflections characteristic for the crystal structure of anatase (600, 700 °C), rutile (700, 900 °C), wurtzite, Zn_2TiO_8, Zn_2TiO_4 and $ZnTiO_3$. To determine the temperature of transformation of specific crystal structures, differential thermal analysis (DTA) was conducted. It was proved that the endothermic peaks at 140, 250, 800 and 940 °C were associated with, respectively, ammonia decomposition, the transformation of Zn_2TiO_8 to $ZnTiO_3$ and rutile, and the transformation of $ZnTiO_3$ to Zn_2TiO_4 and rutile. Furthermore, two exothermic peaks were observed at 560 and 690 °C, indicating the creation of a Zn_2TiO_8 form and the formation of $ZnTiO_3$ and rutile.

The microwave hydrothermal method is a recently developed technique to prepare binary materials in very short times. The advantages of this process over the conventional hydrothermal method include extremely rapid kinetics of crystallization, very rapid heating to the treatment temperature, and the possible formation of new metastable phases [80,81]. Ashok, Venkateswara and Rao [65] developed a methodology for producing a TiO_2-ZnO oxide system with the use of the hydrothermal method supported by the action of microwaves. In the first stage of the synthesis, a zinc acetate solution was added to titanium tetraisopropoxide, and then the mixture was alkalized with NaOH. The product was subjected to the action of microwaves for 5 min at 180 °C, using a frequency of 2.45 GHz. In the final step, the resulting oxide system underwent a calcination process in air at temperatures of 500 and 600 °C. X-ray analysis showed the presence of diffraction peaks corresponding to the crystal structure of anatase, wurtzite, and the zinc titanates ($ZnTiO_3$ and $Zn_2Ti_3O_4$). Images obtained by the SEM technique see (Figure 3) showed the TiO_2-ZnO oxide systems to contain spherical particles with a tendency to agglomerate, irrespective of the temperature of thermal treatment.

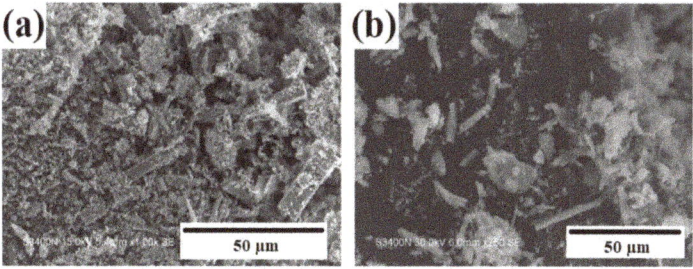

Figure 3. SEM images of a TiO_2-ZnO oxide system subjected to calcination at temperatures of (**a**) 500 °C and (**b**) 600 °C (created based on [65] with permission from Elsevier Publisher).

The synthesis of multi-component systems, despite its many advantages, such as the possibility of obtaining products with very good functional properties, is associated with certain problems, such as the need to optimize the process to obtain the best possible properties with substrates in given proportions. For this purpose, Divy et al. [66] used a combined method (a microwave-assisted hydrothermal process) to produce a TiO_2-ZnO oxide system, which was then used as a matrix for a TiO_2-ZnO-GO-Ag multi-component system. The matrix must have strictly defined parameters to produce a multi-component system with high photoactivity. For this reason, the synthesized TiO$_2$-ZnO oxide system was analyzed with the use of techniques including XRD and BET. Diffractograms showed the presence of characteristic reflections for anatase and wurtzite. From low-temperature nitrogen sorption isotherms, the BET surface area was calculated to be 290 m^2/g, the pore diameter 3.4 nm, and the pore volume 0.32 cm^3/g. Because of the theoretically high photocatalytic activity of each component of the TiO$_2$-ZnO-GO-Ag hybrid system, it was tested in the degradation of a model solution of rhodamine B (C.I. Basic Violet 10). The multi-component system exhibited high photoactivity (complete degradation of the dye after 60 min). It also demonstrated high stability, and could be used repeatedly over at least 5 catalytic cycles.

In the next part of this review a survey will be made of work on TiO$_2$-ZrO$_2$ binary systems, synthesized—like the materials discussed in this section—using a hydrothermal process (for review see Table 2).

In recent years, due to the increased emphasis on environmental protection, interest has grown in the use of biomass as a renewable source of fuel and organic chemicals, where fast or flash pyrolysis is used to produce bio-oils. In a study by Leahy [82], a TiO$_2$-ZrO$_2$ oxide material was obtained with the characteristic crystal structures of TiO$_2$ and ZrO$_2$. A scanning and transmission electron microscope (SEM and TEM) images (Figure 4) showed the deposition on nanorods of commercial TiO$_2$ of ZrO$_2$ particles with a diameter of approximately 20 nm. To enable the use of the synthesized oxide systems in the esterification of levulinic acid, surface modification with sulfate groups (SO$_4^{2-}$) was carried out. In tests of the extraction process, the TiO$_2$-ZrO$_2$ system attained a high yield of 70% (for a reaction taking place at 80 °C).

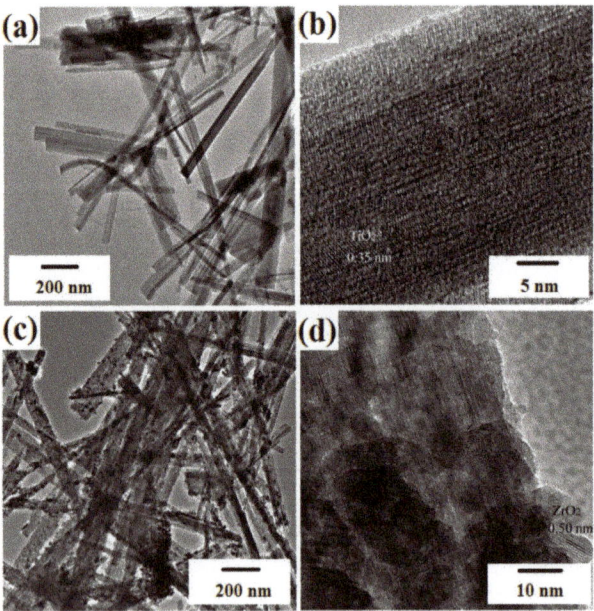

Figure 4. TEM images of (a) TiO$_2$; (b) TiO$_2$-ZrO$_2$; HRTEM images of (c) TiO$_2$ and (d) TiO$_2$-ZrO$_2$ (created based on [82] with permission from Elsevier Publisher).

Table 2. Resources, synthesis conditions and physico-chemical properties of TiO_2-ZrO_2 oxide systems obtained by the hydrothermal method.

Resources	Synthesis Conditions	Physico-Chemical Properties of the Final Product	References
TiO_2, $ZrOCl_2·8H_2O$, NaOH, distilled water	hydrothermal treatment: 180 °C, reaction time: 48 h, drying: 60 °C, calcination: 500 °C for 3 h	crystal structure of TiO_2 and tetragonal ZrO_2, nanorod particles up to 5 µm in length	[82]
$Ti[OCH(CH_3)_2]_4$, $Zr[OCH(CH_3)_2]_4$, propan-2-ol	hydrothermal treatment: 240 °C, reaction time: 24 h, drying: 100 °C for 2 h, calcination: 450 °C for 4 h	crystal structure of anatase, rutile and tetragonal ZrO_2, spherical particles with a tendency to agglomerate	[83]
$Ti[OCH(CH_3)_2]_4$, $Zr[OCH(CH_3)_2]_4$, 1,3-butanediol, ethylene glycol	hydrothermal treatment: 300 °C, reaction time: 2 h, drying: 60 °C, calcination: 500 and 800 °C	crystal structure of tetragonal ZrO_2, spherical particles with diameter up to 1 µm	[84]
$TiOSO_4$, $Zr(SO_4)_2$, distilled water	hydrothermal treatment: 200, 240 °C, reaction time: 24 h, drying: 60 °C, calcination: 400–1000 °C for 1 h	anatase crystal structure (for systems obtained at a molar ratio of TiO_2 > 50%) and monocrystalline ZrO_2 (for all synthesized systems), spherical particles with diameter 13 nm	[85]
$Ti[OCH(CH_3)_2]_4$, $Zr[OCH(CH_3)_2]_4$, $Zn(NO_3)_2·6H_2O$, propan-2-ol, nitric acid	hydrothermal treatment: 240 °C, reaction time: 24 h, drying: 100 °C for 2 h, calcination: 450 °C for 4 h	crystal structure of anatase, rutile and monolithic ZrO_2, average crystallite size 17.4 nm, band gap for TiO_2-ZrO_2 = 2.15 eV	[86]
$Ti[OCH(CH_3)_2]_4$, $ZrOCl_2·8H_2O$, poly(methyl methacrylate), ethanol	hydrothermal treatment: 160 °C, reaction time: 10 h, drying: 100 °C, calcination: 800 °C	crystal structure of anatase and monolithic ZrO_2 (for a calcined system), particles with a structure of hollow microspheres, diameters in the range 0.7–2 µm, BET surface area 224.6 m^2/g	[87]
$TiCl_4$, $ZrOCl_2·8H_2O$, $NH_3·H_2O$, distilled water	hydrothermal treatment: 220 °C, reaction time: 4 h, drying: 120 °C, calcination: 500 °C for 10 h	amorphous structure, high thermal stability (10% mass loss), BET surface area 209 m^2/g	[88]
commercial Ti, $ZrOCl_2·8H_2O$, $NH_3·H_2O$, H_2O_2, HNO_3, lactic acid	hydrothermal treatment: 180 °C, reaction time: 12 h, drying: 50 °C for 12 h	crystal structure of anatase and ZrO_2, spherical and spindle shaped particles	[89]
$Ti[OCH(CH_3)_2]_4$, $ZrCl_2$, HNO_3, methanol	hydrothermal treatment: 260 °C, reaction time: 12 h, drying: 80 °C for 24 h	crystal structure of anatase, BET surface area 40.6 m^2/g, pore volume 0.183 cm^3/g, pore diameter 5.7 nm	[90]

Semiconductors such as titania are very important materials with potential uses in diverse applications. A significant property of semiconductors is their electron gap [91], because a prerequisite for an efficient photocatalyst is that the redox potential for the evolution of hydrogen should lie within the band gap of the semiconductor. Tomar and Chakrabarty [83] described the synthesis of TiO_2-ZrO_2 oxide systems and determined their band gap energies. To obtain the oxide material, titanium tetraisopropoxide was mixed with zirconium tetraisopropoxide in the molar ratios TiO_2:ZrO_2 = 10:90, 30:70, 40:60, 60:40, 70:30 and 90:10, and the resulting reaction mixture was subjected to hydrothermal treatment at 240 °C for 24 h. The obtained materials were further enhanced by calcination at 450 °C for 4 h. Diffractograms revealed the crystal structures of anatase, rutile, and monoclinic and tetragonal ZrO_2. The molar ratio used was found to affect the crystal structure: with an increase in the zirconia content, the diffractions peaks corresponding to ZrO_2 became more intense. SEM images showed the particles to have a spherical shape. The band gaps, determined by the technique of UV-Vis spectroscopy, were measured at 2.48, 1.72, 1.34, 1.7, 1.49 and 1.49 eV respectively for the oxide systems synthesized in the molar ratios TiO_2:ZrO_2 = 10:90, 30:70, 40:60, 60:40, 70:30 and 90:10.

Kubo et al. [84] described the effect of a calcination process on the physico-chemical properties of a TiO_2-ZrO_2 oxide system, using zirconium tetraisopropoxide and titanium tetraisopropoxide. In the first stage, the precursors of TiO_2 and ZrO_2 were mixed in the molar ratios Zr:Ti = 1:0, 9:1, 6:1 and 3:1, and then underwent hydrothermal treatment at 300 °C for 2 h in a nitrogen atmosphere. After this process, the materials were subjected to additional calcination at 500 and 800 °C. X-ray diffraction analysis showed the presence of characteristic reflections for the crystal structure of tetragonal ZrO_2. SEM images showed the synthesized materials to contain particles of spherical shape with a tendency to agglomerate. The effect of fusion of particles was also observed for the materials that had undergone thermal treatment (calcination). Low-temperature nitrogen sorption measurements gave BET surface areas of 200 and 80 m^2/g respectively for the systems calcined at 500 and 800 °C. The conclusion of morphological and pore structure analysis was that an increase in calcination temperature caused a decrease in BET surface area.

Many literature reports indicate the high photocatalytic activity of zirconia in the degradation of dyes such as methylene blue [92–94]. Hirano et al. [85] developed a simple hydrothermal synthesis of TiO_2-ZrO_2 binary oxide systems using titanium(IV) sulphate(VI) and zirconium sulphate(VI). Wide-ranging physico-chemical analysis confirmed the hydrothermal method as an effective process for the synthesis of binary oxide systems with defined crystallinity and morphology. The reported results indicate the presence of the crystal structure of anatase and monoclinic ZrO_2. Transmission electron microscope images (Figure 5) showed that the particles of the TiO_2-ZrO_2 system had an almost spherical shape. In view of the presence of anatase and tetragonal ZrO_2 in the synthesized material, an attempt was made to use the binary system in the process of degradation of methylene blue (C.I. Basic Blue 9). The TiO_2-ZrO_2 systems exhibited greater photocatalytic activity than the pure mono-oxides. The highest photoactivity in the degradation of the dye was achieved by the material obtained in the molar ratio TiO_2:ZrO_2 = 90:10.

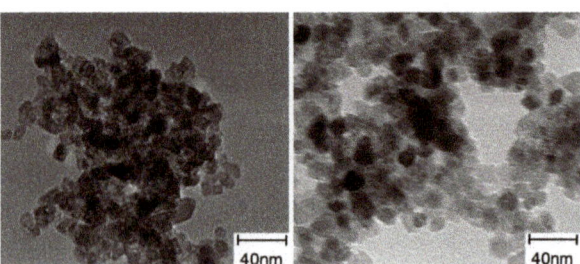

Figure 5. TEM images of the oxide system 90%TiO_2-10%ZrO_2 (created based on [85] with permission from Elsevier Publisher).

At present the main focus in the field of DSSCs is the development of a photoanode that can efficiently harvest light [95]. Increased dye pickup, light scattering ability, reduced recombination reaction and improved charge transport ability are parameters that need to be engineered for high conversion efficiency. The properties of titania make it one of the most important DSSC photoanode materials. Tomar et al. [86] worked on a DSSC using a TiO_2-ZrO_2 system as the anode. In testing of the cell, the binary oxide material demonstrated a higher energy conversion efficiency (η = 1.97%) and fill factor (FF = 0.256) than pure TiO_2 (η = 0.71%, FF = 0.061).

Yao et al. [87]—similarly to Hirano et al. [85]—described the use of a TiO_2-ZrO_2 oxide system in the process of photodegradation of methylene blue (C.I. Basic Blue 9). The material obtained in their study had a similar crystal structure to that described in [85]. SEM images revealed hollow microspheres of TiO_2-ZrO_2—this is closely linked to the high BET surface area of the material, which was determined from nitrogen adsorption/desorption isotherms to be 224 m^2/g. Finally, the photocatalytic properties of the obtained binary material were tested in the degradation of methylene blue (C.I. Basic Blue 9). Use of the material led to a high level of degradation of the dye, reaching 99% after exposure to light for 60 min.

Caillot et al. [88] described the use of a synthesized TiO_2-ZrO_2 binary oxide system as an effective adsorbent of NH_3 and SO_2. The use of titanium(IV) and zirconium(IV) chlorides, and hydrothermal treatment at 220 °C, led to an amorphous structure. The absence of a defined crystal structure is linked to the high BET surface area (209 m^2/g) and the unspecific particle shape. Results of energy-dispersive X-ray spectroscopy (EDX) indicated that the synthesized material contained 35% of zirconia and 65% of titania. Adsorption tests showed the material to have high adsorption capacities with respect to both NH_3 and SO_2; the values obtained were 573 µmol/g for NH_3 and 414 µmol/g for SO_2.

Ueda et al. [89] described the use of a titanium dioxide zirconium dioxide system to form bioinert foils. The synthesis of the material included hydrothermal treatment at 180 °C for 12 h. To enable use of the obtained materials in biological processes, the surface was modified using lactic acid. It was observed that while pure TiO_2 had a high propensity to proliferate MC3T3-E1 cells (MC3T3 is an osteoblast precursor cell line derived from *Mus musculus* (mouse) calvaria), the synthesized oxide material showed a good ability to suppress proliferation of the cells, caused by a change in the surface morphology (increased roughness), which has a direct impact on the cell reproduction process. The experiment shows that the synthesized TiO_2-ZrO_2 material may be used for the purpose of slowing down osteointegration. The researchers also carried out a wide-ranging physico-chemical analysis to determine the morphological and crystal structure of the material. The synthesized binary oxide materials were found to contain the crystal structures of TiO_2 and ZrO_2, with the formation of a thin layer of ZrO_2 on the surface of the TiO_2.

Kim et al. [90] synthesized a TiO_2-ZrO_2 system for use in a DSSC. In the first stage, the material was synthesized with the use of hydrothermal treatment at 260 °C for 12 h. X-ray diffractograms showed the presence of characteristic peaks for the anatase structure. The synthesized material was then used to prepare paste electrodes. Analysis of the porous structure of the electrode material showed it to have a BET surface area of 40.6 m^2/g, a pore volume of 0.18 cm^3/g and a pore diameter of 5.7 nm. In electrochemical tests, the material exhibited a high energy conversion efficiency of 2.88%, compared with 0.09% for commercial TiO_2 (P25).

The next part of this review will contain a survey of the literature on the synthesis and potential applications of TiO_2-MoS_2 hybrid systems (Table 3).

Table 3. Resources, synthesis conditions and physico-chemical properties of TiO_2–MoS_2 hybrid systems obtained by the hydrothermal method.

Resources	Synthesis Conditions	Physico-Chemical Properties of the Final Product	References
$TiCl_4$, $Na_2MoO_4 \cdot 2H_2O$, thioacetamide, deionized water	hydrothermal treatment: 240 °C, reaction time: 24 h, drying: 60 °C	crystal structure of TiO_2 (anatase) and MoS_2, layered structure of MoS_2 particles, molybdenum disulfide particles coated with TiO_2 nanoparticles (3–5 nm)	[33]
$TiCl_4$, $(NH_4)_6Mo_7O_{24} \cdot 4H_2O$, thiourea, ethanol, deionized water	hydrothermal treatment: 220 °C, reaction time: 24 h, drying: 60 °C for 12 h, calcination: 500 °C for 4 h	crystal structure characteristic for TiO_2 (anatase) and MoS_2, 8% loss of mass when heated to 700 °C, particles in the form of a single sheet, 20 nm TiO_2 particles evenly distributed on the surface of MoS_2	[34]
$Ti(OC_4H_9)_4$, MoS_2, ethylene glycol, HCl, deionized water	hydrothermal treatment: 180 °C, reaction time: 12 h	crystal structure of anatase and 2H-MoS_2, BET surface area = 48.2 m^2/g, MoS_2 nanoparticles coated with TiO_2 particles	[96]
$Ti(OC_4H_9)_4$, $Na_2MoO_4 \cdot 2H_2O$, thiourea, HF, deionized water	preparation of TiO_2: hydrothermal treatment: 180 °C, reaction time: 24 h, preparation of TiO_2-MoS_2: hydrothermal treatment: 200 °C, reaction time: 24 h, drying: 105 °C for 24 h, calcination: 400 °C for 4 h	crystal structure of TiO_2 (anatase) and MoS_2 (for the molar ratio Mo:Ti = 7.5%), BET surface area = 86, 87, 98, 92 m^2/g for systems obtained with molar ratios Mo:Ti = 2.5, 5, 7.5% respectively, TiO_2 coated with MoS_2 particles	[97]
TiO_2-P25 (Degussa), $Na_2MoO_4 \cdot 2H_2O$, thiourea, HCl, NaOH, deionized water	preparation of TiO_2: hydrothermal treatment: 180 °C, reaction time: 24 h, preparation of TiO_2-MoS_2: hydrothermal treatment: 180 °C, reaction time: 24 h, drying: 80 °C for 12 h	crystal structure characteristic for TiO_2 and MoS_2, MoS_2 particles in the form of nanofibres with embedded spherical TiO_2 agglomerates	[98]
$TiCl_4$, MoS_2, ethanol, glycerol, sodium hexametaphosphate	obtaining of TiO_2: sonication preparation of TiO_2-MoS_2: hydrothermal treatment: 140 °C, reaction time: 3 h	crystal structure of TiO_2 (anatase), $M_3O_8 \cdot 2H_2O$ (ilsemanite) and 3R-MoS_2 (molybdenite), particle diameter 46 nm, MoS_2 particles coated with spherical TiO_2 particles, 23% mass loss of the TiO_2-MoS_2 hybrid when heated to 900 °C	[99]
Ti, $Na_2MoO_4 \cdot 2H_2O$, thioacetamide, deionized water, acetone, ethanol	preparation of TiO_2: anodic process on steel substrates in DSMO/HF electrolyte, calcination: 500 °C for 4 h, preparation of TiO_2-MoS_2: hydrothermal treatment: 220 °C, reaction time: 24 h, drying: 80 °C for 12 h	crystal structure of anatase and hexagonal MoS_2, particles in the form of nanotubes coated with MoS_2	[100]

Table 3. *Cont.*

Resources	Synthesis Conditions	Physico-Chemical Properties of the Final Product	References
TiF$_4$, Na$_2$MoO$_4$·2H$_2$O, thioacetamide, deionized water	preparation of TiO$_2$: hydrothermal treatment: 140 °C, reaction time: 1.5 h preparation of TiO$_2$-MoS$_2$: hydrothermal treatment: 220 °C, reaction time: 24 h, drying: 60 °C for 12 h	crystal structure of TiO$_2$ (anatase) and rhombohedral MoS$_2$, disappearance of anatase with an increase in the fraction of MoS$_2$, TiO$_2$ spheres covered with MoS$_2$ nanoparticles	[101–103]
Ti(OC$_4$H$_9$)$_4$, MoS$_2$, H$_2$SO$_4$, N,N-dimethylformamide, ethanol, deionized water,	hydrothermal treatment: 200 °C, reaction time: 20 h	crystal structure of TiO$_2$ (anatase) and MoS$_2$, dense covering of molybdenum disulfide particles with titanium dioxide particles of 10 nm diameter	[104]
Ti(OC$_4$H$_9$)$_4$, Na$_2$MoO$_4$·2H$_2$O, thioacetamide, HF	hydrothermal treatment: 200 °C, reaction time: 24 h, drying: 80 °C for 12 h, calcination: 400 °C for 1 h	crystal structure of TiO$_2$ (anatase) and MoS$_2$, number of MoS$_2$ layers in the range 6–9, TiO$_2$ particles applied to MoS$_2$ nanoparticles	[105]

Indoles are among the most versatile and common nitrogen-based heterocyclic scaffolds and are frequently used in the synthesis of various organic compounds [106,107]. Indole-based compounds are very important among heterocyclic structures due to their biological and pharmaceutical activity. However, in order to improve their properties, it is necessary to modify them. Wang et al. [33] described a simple modification of indoles by thiocyanation, using a synthesized TiO_2-MoS_2 hybrid as a photocatalyst. Physico-chemical and structural analysis of this hybrid indicated a defined crystal structure and morphology. X-ray diffraction analysis showed the presence of reflections corresponding to the crystal structure of anatase TiO_2 and MoS_2. TEM images showed a layered structure of molybdenum disulfide with evenly distributed particles of titanium dioxide on the surface. Thiocyanation of indoles was carried out in a photocatalytic process, in which high yields were obtained. The highest photocatalytic activity (93%) was observed for the material synthesized in the molar ratio TiO_2:MoS_2 = 10:1.

Zhu et al. [34] synthesized a hybrid system containing the crystal structures of TiO_2 and MoS_2. Images obtained by scanning and transmission electron microscopy showed TiO_2 particles uniformly distributed on the surface of the MoS_2. Cyclic voltammetry curves showed the TiO_2-MoS_2 hybrid to have good electrochemical properties, with charge and discharge capacities at a current density of 100 mAh/g equal to 643 and 827 mAh/g respectively in the first cycle, and 643 and 674 mAh/g in the second cycle. The capacity of the material also remained very stable over 100 subsequent cycles. It was concluded that the addition of anatase improved the electrochemical properties by facilitating the transport of electrons and ions during charging and discharging.

Ren et al. [96] used the hydrothermal method to obtain a TiO_2-MoS_2 hybrid material with strictly defined photoelectrochemical properties. Measurements showed the hybrids to have superior electrochemical properties to those of pure TiO_2 or MoS_2. The hybrid synthesized in the molar ratio TiO_2:MoS_2 = 20:1 had a photocurrent density of 33 $\mu A/cm^2$ (compared with 8 $\mu A/cm^2$ for TiO_2). Photoelectrochemical results demonstrated a synergy effect between the titanium dioxide and molybdenum disulfide. A wide-ranging analysis of physico-chemical properties was also performed. The synthesized materials produced diffractions peaks corresponding to anatase and 2H-MoS_2. For the hybrid obtained in the molar ratio TiO_2:MoS_2 = 20:1, the BET surface area was measured at 48.2 m^2/g and the pore diameter at 7 nm. Morphological analysis confirmed that the particles of molybdenum disulfide were coated with particles of titanium dioxide.

Molybdenum disulfide photocatalysts are attracting wide attention as they offer a suitable band gap for visible-light harvesting, making the compound a promising earth-abundant photocatalyst for hydrogen production, environmental remediation, and photosynthesis [108,109]. A team of researchers [97] developed a two-stage method for obtaining active TiO_2-MoS_2 photocatalysts by hydrothermal synthesis. X-ray diffractograms revealed the crystal structure of TiO_2 (for all of the materials) and MoS_2 (only for the systems formed in the molar ratio Mo:Ti = 7.5%). The BET surface area was measured at 86, 87, 98 and 92 m^2/g respectively for the systems with molar ratios Mo:Ti = 1%, 2.5%, 5% and 7.5%, and the respective band gaps were determined to be 3.15, 3.14, 3.1 and 2.97 eV. The synthesized TiO_2-MoS_2 hybrids were tested in a process of photodegradation of methylene blue (C.I. Basic Blue 9), and analysis of their photocatalytic activity showed that they produced high yields of degradation of the dye. For the system with Mo:Ti = 7.5% the solution was completely decolored after 60 min, indicating a tripling in the efficiency of photodegradation of the organic pollutant compared with pure TiO_2.

Liu et al. [98] described the synthesis of multifunctional TiO_2-MoS_2 hybrids with photoelectrochemical and photocatalytic properties. The substances used as precursors of TiO_2 and MoS_2 were commercial titanium dioxide (P25 Evonik) and sodium molybdenate. These were mixed to obtain the mass ratios TiO_2:MoS_2 = 80:20, 60:40, 40:60 and 20:80. The systems then underwent hydrothermal processing at 180 °C for 24 h. X-ray diffractograms contained characteristic peaks for TiO_2, Ti_6O_{11} and MoS_2. SEM images (Figure 6) showed that spherical particles of titanium dioxide were deposited on nanofibers of molybdenum disulfide. For the resulting systems, a determination was made of their photoelectrochemical properties and photocatalytic activity in the removal of rhodamine B (C.I. Basic Violet 10). The TiO$_2$-MoS$_2$ hybrid system with the mass ratio TiO$_2$:MoS$_2$ = 60:40 exhibited high photocurrent density (12.5 µA/cm^2). An increase in the content of molybdenum disulfide in the hybrid system led to a deterioration of its photoelectrochemical properties. In the photocatalytic degradation of rhodamine B, the highest photocatalytic activity was exhibited by the system with the mass ratio TiO$_2$:MoS$_2$ = 60:40. Complete decoloration of the dye solution was observed after 90 min.

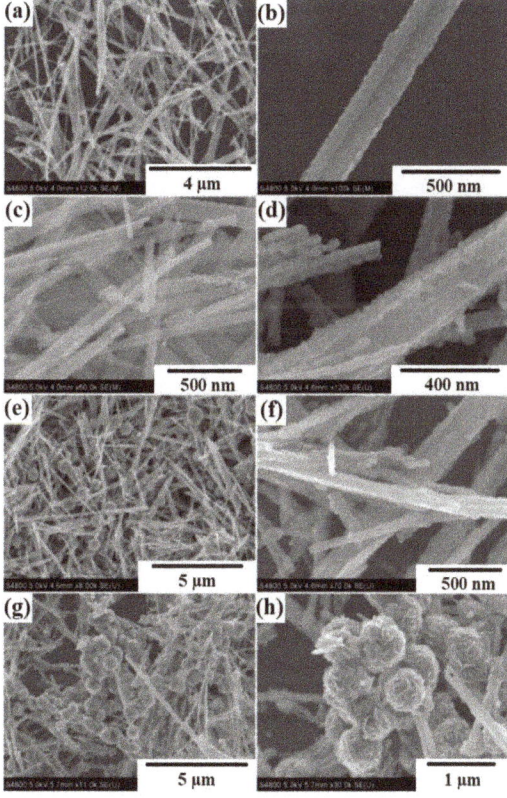

Figure 6. SEM images of TiO$_2$-MoS$_2$ hybrid systems with MoS$_2$ contents of (**a**,**b**) 20%; (**c**,**d**) 40%; (**e**,**f**) 60%; and (**g**,**h**) 80% (created based on [98] with permission from Elsevier Publisher).

Zhang et al. [99] described the use of TiO$_2$-MoS$_2$ binary hybrids in the photocatalytic degradation of methyl orange (C.I. Basic Orange 10). In the first stage, a two-step hydrothermal method was used to synthesize the aforementioned materials. TiO$_2$ nanoparticles were formed by a sonication method. Molybdenum disulfide was then added to the TiO$_2$, and the resulting mixture was subjected to hydrothermal treatment at 140 °C for 3 h. Unlike other previously described hybrids, the material obtained in this study was found to contain the crystal structure of anatase, ilsemanite and molybdenite. Scanning electron microscope images showed that spherical particles of titanium dioxide were deposited on MoS$_2$ nanofibers. In photodegradation tests, the TiO$_2$-MoS$_2$ hybrid system exhibited high photocatalytic efficiency in the decomposition of a model dye solution after 60 min.

Yang et al. [100] synthesized a hybrid system based on TiO$_2$ and MoS$_2$ using a hydrothermal method, with the aim of testing the resulting materials in a process of reduction of Cr(VI). In the first stage of synthesis, titanium dioxide was obtained from titanium in an anodic process. Next, thioacetamide and sodium molybdenate were dissolved in deionized water, the obtained titanium dioxide was added, and hydrothermal treatment was carried out at 220 °C for 24 h. X-ray diffractograms showed the synthesized materials to have the crystal structure of anatase and hexagonal MoS_2. SEM images revealed hollow TiO_2 nanotubes in which MoS_2 particles were deposited. Photoelectrochemical tests demonstrated that the addition of MoS_2 to the TiO_2 structure reduced its resistance, leading to an improvement in the transport of electrons in the catalyst. A high yield of Cr(VI) reduction was obtained using the TiO_2-MoS_2 hybrid system, with a rate of reduction of 103.9 mg/L·min·cm^2 at 0.1 V. Tests of the stability of the reaction process showed that the process efficiency did not deteriorate over five consecutive cycles.

Phenol and its derivatives have been used in many industrial processes, including the production of dyes, polymers and medicines. Due to their high production volumes and widespread application, phenols and their derivatives are released into the environment from industrial and municipal liquid wastes [110,111]. This is particularly undesirable, since phenols pose a serious hazard due to their hematotoxic and hepatotoxic actions [112]. Wang et al. [101] described the use of synthesized TiO_2-MoS_2 hybrid systems in the process of photodegradation of phenol. At the first stage, hollow spheres of TiO_2 were obtained [108,109]. Next, sodium molybdenate was added in the molar ratios TiO_2:Na_2MoO_4 = 30:135 (sample TM3), 60:135 (sample TM6), 90:135 (sample TM9) and 120:135 (sample TM12). Diffractograms obtained for the TiO_2-MoS_2 materials contained characteristic peaks for the crystal structure of anatase TiO_2 and rhombohedral MoS_2. The anatase structure became less visible as the MoS_2 content increased. Scanning electron microscope images (see Figure 7) showed particles of titanium dioxide with close to spherical shapes, hollow inside, with diameters of 600 nm, coated with a layer of molybdenum disulfide. The synthesized materials were tested in a process of photocatalytic degradation of phenol. The hybrids exhibited good photocatalytic properties: the yield of phenol degradation was 78% after 150 min in the presence of the system obtained in the mass ratio TiO_2:Na_2MoO_4 = 90:135. Tests of photocatalytic stability showed that the photoactivity of this material did not decrease significantly over four consecutive catalytic cycles.

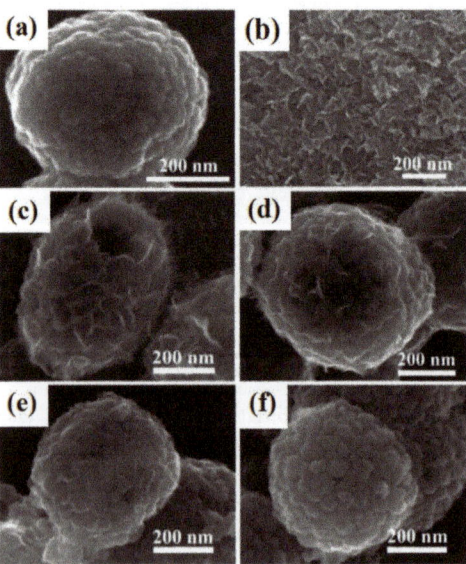

Figure 7. SEM images of (a) TiO_2; (b) MoS_2; (c) TM3; (d) TM6; (e) TM9 and (f) TM12 (created based on [101] with permission from Elsevier Publisher).

Bai et al. [104] were the first to report the combination of exfoliated MoS_2 with TiO_2 nanoparticles and the use of the resulting material in a photocatalytic process. To establish whether the materials could be successfully used as active photocatalysts, a comprehensive physico-chemical analysis was carried out. Recorded diffraction peaks corresponding to the crystal structure of TiO_2 (anatase) and MoS_2. Morphological parameters were investigated by the SEM and TEM techniques. The resulting images showed the particles of molybdenum disulfide to be densely covered with particles of titanium dioxide. Distances between layers were measured at 0.27 nm and 0.35 nm, which confirmed the presence of the structure of MoS_2 and TiO_2 nanocrystals. In tests of photocatalytic activity in the decomposition of rhodamine B (C.I. Basic Violet 10), 100% decomposition of the dye was attained after exposure to light for 30 min in the presence of the TiO_2(anatase)-MoS_2(1T) hybrid system, compared with 40% in the presence of the TiO_2(anatase)-MoS_2(2H) system. In addition, the TiO_2(anatase)-MoS_2(1T) system catalysed the evolution of hydrogen at a rate of 2 mmol/g·h, compared with 0.25 mmol/g·h for TiO_2(anatase)-MoS_2(2H) and 0.5 mmol/g·h for pure TiO_2.

Photocatalytic evolution of H_2 over semiconducting materials is regarded as one of the promising solutions to the growing energy crisis, due to its potential application for clean hydrogen production from water [109,113,114]. Yuan et al. [105] described the use of a single-stage hydrothermal process to synthesize binary hybrid photocatalysts for the evolution of hydrogen. Photocatalytic hydrogen evolution experiments were mainly carried out in a Pyrex glass cell with a top window connected to a gas-closed system. The assessment of photocatalytic properties was carried out using 100 mg of H_2-evolving photocatalyst powder which was suspended in 100 cm^3 of an aqueous solution containing 10% methanol in volume. The resulting materials led to high yields of hydrogen: the highest—2200 µmol/g·h—was recorded for the system with a 0.5% MoS_2 mass fraction (this is 36 times higher than the yield obtained with pure TiO_2). For comparison, TiO_2-Pt, TiO_2-Pd, TiO_2-Rh, TiO_2-Au and TiO_2-Ru were used as catalysts in the same process, giving yields of 1450, 1250 1000, 750 and 550 µmol/g·h respectively. Physico-chemical parameters were analysed by the XRD, TEM and BET techniques. X-ray diffractograms contained characteristic peaks for the crystal structure of anatase TiO_2 and MoS_2. Based on nitrogen sorption, a BET surface area of 105.6 m^2/g was obtained. Transmission electron microscope images showed interplanar distances of 0.19 nm, characteristic for TiO_2, and 0.62 nm, characteristic for hexagonal MoS_2.

A comprehensive review of the literature concerning oxide systems containing titanium dioxide together with zinc oxide or zirconium dioxide, or TiO_2-MoS_2 hybrid systems, synthesized by the hydrothermal method confirms that by the suitable choice of process conditions (precursors, pH of the reaction system, molar ratio of reagents, temperature of thermal treatment) it is possible to achieve effective changes in the properties of the resulting materials. Appropriately selected process parameters make it possible to design the physico-chemical properties of synthesized oxide and hybrid materials, including particle shape and size, degree of crystallinity, phase and surface compositions, and porous structure. Moreover, hydrothermally synthesized materials of TiO_2-ZnO, TiO_2-ZrO_2 and TiO_2-MoS_2 type are finding a wide range of applications in bacterial and fungal degradation and as photocatalysts in the decomposition of certain organic pollutants. Such systems are also used as electrode materials in lithium-ion cells and dye-sensitized solar cells.

2.2. The Sol-Gel Method

The consistent development of technology has led to increased awareness of the importance of chemical synthesis methods, particularly those involving the use of organic precursors. One of the basic types of synthesis in which metallo-organic precursors are used to synthesize binary hybrid systems is the sol-gel method [115–120]. In recent years many definitions of the sol-gel process have been put forward, including that of Dislich and Hinz [121], which covers only multicomponent oxides. According to the most popular and widespread definition, such a process is based on a hydrolysis reaction of metallo-organic precursors of metals (usually alkoxides) in a reaction mixture [31,122–125].

A sol-gel process consists of several stages. In the first, a solution of metal alkoxides is mixed. This is followed by a process of hydrolysis and condensation, after which a sol is obtained [126–132]. In the next stage, gelation takes place—that is, the sol is transformed into a gel. The final stage involves the removal of residues of the liquid phase via a drying process. A block diagram of the sol-gel process appears in Figure 8 [133–137].

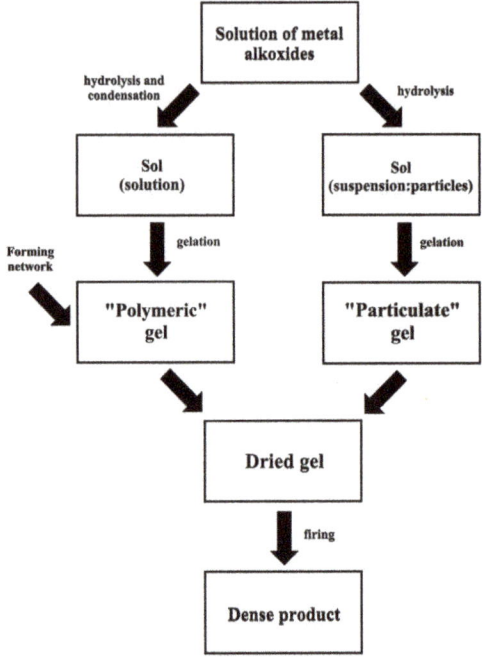

Figure 8. A block diagram of the sol-gel process (created based on [122] with permission from Springer).

Sol-gel processes may be used in many branches of industry; for example, in the production of catalysts or catalyst supports [138], optical and electrical sensors [139], and implant coatings [139,140] Recent reports concern the use of the sol-gel method in the process of immobilization of a wide range of organic materials, including enzymes [141–144].

Compared with other conventional methods of synthesis, the sol-gel method has a number of advantages which enable its use in a wide range of technological processes. Some of these advantages are the ability to control the microstructure of the product, the use of low temperatures and the resulting low energy costs, the high homogeneity of the product, the absence of a need for complex apparatus, and the possibility of synthesizing new materials with designed properties [145,146].

The most significant drawbacks of the sol-gel process include the high costs of the metal precursors, the need to use reagents of high purity, the absence of total control over the process (the evolution of various oxides due to differences in the reactivity of the precursors), and the fusion of the product during calcination [122,138,139].

The sol-gel method may be used to obtain oxide systems containing TiO_2. Titanium dioxide, because of its photocatalytic [147–149] and photoelectrochemical properties [150,151], represents an interesting base for the production of multifunctional oxide systems. There has recently been a marked growth in interest in such materials as TiO_2-ZnO and TiO_2-ZrO_2 oxide systems, in view of the useful properties of zinc oxide (photocatalytic, photoelectrochemical and photoluminescent) and zirconium dioxide (photocatalytic and photoelectrochemical).

In Table 4, an overview is given of the current stage of knowledge concerning the synthesis of TiO_2-ZnO oxide systems using the sol-gel process.

Table 4. Resources, synthesis conditions and physico-chemical properties of TiO$_2$-ZnO hybrid systems obtained by the sol-gel method.

Resources	Synthesis Conditions	Physico-Chemical Properties of the Final Product	References
TiCl$_4$, Zn(CH$_3$COO)$_2$·2H$_2$O, ZnCl$_2$, benzyl alcohol, propan-2-ol, deionized water	ZnO sol: Zn(CH$_3$COO)$_2$·2H$_2$O (sample A) or ZnCl$_2$ (sample B), propan-2-ol TiO$_2$-ZnO system: ZnO sol, TiCl$_4$ mixed in the ratio TiO$_2$:ZnO = 9:1 (sample A); 5:5 (sample B) calcination: 500 °C for 6 h (sample A); 200; 400; 550; 600 °C for 2 h (sample B)	crystal structure of anatase (sample A) and anatase, rutile and zinc titanate (sample B), spherical shaped particles with agglomeration	[1]
TiO$_2$, Zn(CH$_3$COO)$_2$·2H$_2$O, diethylamine, propan-2-ol, deionized water	TiO$_2$ sol: TiO$_2$ compact layer was deposited on an ITO (indium tin oxide) coated glass substrates by RF-sputtering technique with RF-power of 150 W, calcination: 450 °C for 30 min ZnO sol: Zn(CH$_3$COO)$_2$·2H$_2$O, propan-2-ol TiO$_2$-ZnO system: the TiO$_2$ layer was immersed in ZnO sol solution	bands on XPS spectra—Ti2s, Ti3s, Ti2p, Zn2p, Zn3p, Zn3s, O1s, spherical particles, TiO$_2$ particles coated with ZnO (core-shell)	[152]
Ti(OC$_4$H$_9$)$_4$, Zn(CH$_3$COO)$_2$·2H$_2$O, ethylamine, glacial acetic acid, ethanol, deionized water	TiO$_2$ sol: Ti(OC$_4$H$_9$)$_4$, ethanol, acetic acid ZnO sol: Zn(CH$_3$COO)$_2$·2H$_2$O, ethanol, ethylamine, acetic acid TiO$_2$-ZnO system: the sols of the respective oxides were mixed at molar ratio TiO$_2$:ZnO = 7:3, 5:5, 3:7	antistatic properties, homogeneous surface of polyester material coated with a TiO$_2$-ZnO oxide system	[153]
Ti(OC$_4$H$_9$)$_4$, Zn(CH$_3$COO)$_2$·2H$_2$O, glacial acetic acid, diethylamine, ethanol, deionized water	TiO$_2$ sol: Ti(OC$_4$H$_9$)$_4$, ethanol, acetic acid ZnO sol: Zn(CH$_3$COO)$_2$·2H$_2$O, ethanol TiO$_2$-ZnO system: obtained sols were mixed at molar ratio TiO$_2$:ZnO = 1:0; 3:1; 1:3; 0:1, the obtained materials were deposited on carbon steel by immersion in solution, drying: 70 °C for 10 h calcination: 350, 500 °C for 2 h	crystal structure of Fe, TiO$_2$, Fe$_3$O$_4$, ZnFe$_2$O$_4$, morphology - rough surfaces with numerous cracks, EDX spectrum showed bands characteristic for Ti, O, Fe, Zn	[154]
Ti(OC$_4$H$_9$)$_4$, Zn(CH$_3$COO)$_2$·2H$_2$O, hydrochloric acid, anhydrous ethanol, deionized water	TiO$_2$ sol: Ti(OC$_4$H$_9$)$_4$, ethanol, HCl ZnO sol: Zn(CH$_3$COO)$_2$·2H$_2$O, ethanol, deionized water TiO$_2$-ZnO system: materials obtained at molar ratio ZnO:TiO$_2$ = 0.10:0.15; 0.20:0.25; 0.30:0.35, calcination: 450; 480; 500; 550; 600 °C for 1; 1.5; 2; 2.5; 3 h	crystal structure of anatase and wurtzite, particles with spherical shape with a tendency to agglomerate, BET surface area 76.258 m^2/g, pore volume 0.0361 cm^3/g, pore diameter 6.6 nm	[155]
Ti(OC$_4$H$_9$)$_4$, Zn(CH$_3$COO)$_2$·2H$_2$O, nitric acid, ethylene glycol, glycerol, trimethylamine (TEA), anhydrous ethanol, deionized water	TiO$_2$ sol: Ti(OC$_4$H$_9$)$_4$, nitric acid, ethanol ZnO sol: Zn(CH$_3$COO)$_2$·2H$_2$O, ethylene glycol, glycerol, TEA, ethanol TiO$_2$-ZnO system: obtained sols were mixed at molar ratio TiO$_2$:ZnO = 100:0, 75:25, 50:50, 25:75, 0:100, calcination: 500 °C for 1 h	anatase crystal structure (TiO$_2$:ZnO = 100:0, 75:25), amorphous structure (TiO$_2$:ZnO = 50:50, 25:75), wurtzite structure (TiO$_2$:ZnO = 0:100), XPS shows the presence of Ti2p, Zn2p, O1s bands	[156]

Table 4. Cont.

Resources	Synthesis Conditions	Physico-Chemical Properties of the Final Product	References
Ti(OC$_4$H$_9$)$_4$, Zn(CH$_3$COO)$_2$·2H$_2$O, nitric acid, acetic acid, diethylamine, acetylacetone, anhydrous ethanol, deionized water	TiO$_2$ sol: Ti(OC$_4$H$_9$)$_4$, ethanol, acetylacetone, acetic acid, diethylamine ZnO sol: Zn(CH$_3$COO)$_2$·2H$_2$O, ethanol, diethylamine TiO$_2$-ZnO system: obtained sols were mixed at molar ratio TiO$_2$:ZnO = 90:10, 80:20, 70:30, 60:40; calcination: 500 °C for 30 min	anatase crystal structure (TiO$_2$:ZnO = 90:10, 80:20), amorphous structure (TiO$_2$:ZnO = 70:30, 60:40), spherical shaped particles and porous structure, XPS shows bands characteristic for Ti2p, Zn2p, O1s	[157]
Ti[OCH(CH$_3$)$_2$]$_4$, Zn(NO$_3$)$_2$·6H$_2$O, glacial acetic acid, ethanol, deionized water	TiO$_2$ sol: Ti(OC$_4$H$_9$)$_4$, ethanol ZnO sol: Zn(NO$_3$)$_2$·6H$_2$O, ethanol, glacial acetic acid TiO$_2$-ZnO: obtained sols were mixed at molar ratio TiO$_2$:ZnO = 10:3; calcination: 400 °C for 2 h	crystal structure of anatase and zincite, particles with spherical shape and a tendency to agglomerate, EDX shows the presence of bands derived from Ti, Zn, O	[158]
Ti(OC$_4$H$_9$)$_4$, Zn(NO$_3$)$_2$·6H$_2$O, citric acid, deionized water	TiO$_2$-ZnO system: sol-gel auto-ignition method, citric acid (catalyst) Ti(OC$_4$H$_9$)$_4$, Zn(NO$_3$)$_2$·6H$_2$O was heated to a temperature of 300 °C, auto-ignition, calcination: 500 °C for 5 h	crystal structure of anatase, BET surface area = 52, 63.4, 69.9, 64, 36.6 m^2/g respectively for systems prepared with a ZnO mass fraction of 1, 5, 10, 12, 30%, particles of irregular shape with a tendency to agglomerate, EDX shows bands characteristic for Ti, O, Zn	[159]
Ti[OCH(CH$_3$)$_2$]$_4$, Zn(CH$_3$COO)$_2$·2H$_2$O, acetic acid, diethylamine, butanol, propan-2-ol, deionized water	TiO$_2$ sol: Ti(OC$_4$H$_9$)$_4$, butanol, acetic acid ZnO sol: Zn(CH$_3$COO)$_2$·2H$_2$O, propan-2-ol, diethylamine, deionized water TiO$_2$-ZnO system: the glass plate was covered with a layer of TiO$_2$ sol, calcination: 400 °C for 2 h, covering by ZnO layer, calcination: 500, 600 °C for 2 h	crystal structure of anatase and wurtzite, increase of surface roughness with increasing calcination temperature	[160]
Ti(OC$_4$H$_9$)$_4$, Zn(CH$_3$COO)$_2$·2H$_2$O, sodium hydroxide, propan-2ol, deionized water	sol of TiO$_2$: Ti[(OC$_4$H$_9$)$_4$, propan-2-ol sol of ZnO: Zn(CH$_3$COO)$_2$·2H$_2$O, sodium hydroxide (0.5 M), deionized water TiO$_2$-ZnO system: obtained sol was mixed at molar ratio TiO$_2$:ZnO = 0.5:0.25; 1:0.5; 1.5:0.75; 2:1, are designated respectively as TZO1 TZO2, TZO3 and TZO4, calcination: 550 °C for 4 h	crystalline structure of anatase and wurtzite, Raman spectroscopy - bands characteristic for TiO$_2$ and ZnO, spherical shaped particles and rods with tendency to agglomerate, EDX—strands derived from Ti, Zn, O, BET surface area equal to 74.7, 32.2, 12.58, 8.82 m^2/g respectively for samples TZO1 TZO2, TZO3, and TZO4	[161]
Ti(OC$_4$H$_9$)$_4$, Zn(CH$_3$COO)$_2$·2H$_2$O, nitric acid, ethylene glycol, ethanol, deionized water	TiO$_2$ sol: Ti(OC$_4$H$_9$)$_4$, ethanol, nitric acid ZnO sol: Zn(CH$_3$COO)$_2$·2H$_2$O, ethylene glycol TiO$_2$-ZnO system: TiO$_2$ and ZnO sols were mixed at molar ratio TiO$_2$:ZnO = 2:1, calcination: 450 °C for 1 h	particles of spherical shape with no tendency to agglomerate, EDX showed characteristic bands for Ti, Zn, O, energy band gap 3.41 eV	[162]

Table 4. Cont.

Resources	Synthesis Conditions	Physico-Chemical Properties of the Final Product	References
Ti(OC$_4$H$_9$)$_4$, Zn(CH$_3$COO)$_2$·2H$_2$O, nitric acid, diethylamine, acetylacetone, propan-2-ol, deionized water	TiO$_2$ sol: Ti(OC$_4$H$_9$)$_4$, acetylacetone, propan-2-ol, nitric acid ZnO sol: Zn(CH$_3$COO)$_2$·2H$_2$O, propan-2-ol, diethylamine TiO$_2$–ZnO system: obtained sols were mixed in equimolar ratio	irregular shaped particles, EDX showed bands characteristic for Ti, Zn, O	[163]
TiCl$_4$, Zn(CH$_3$COO)$_2$·2H$_2$O, ammonium fluoride, acetonitrile, butanol, acetone, diethylamine, ethanol, deionized water	TiO$_2$: electrode method ZnO sol: Zn(CH$_3$COO)$_2$·2H$_2$O, ethanol diethylamine TiO$_2$–ZnO system: TiO$_2$ (anodic method) was immersed in a solution of TiCl$_4$, calcination: 450 °C for 15 min, TiO$_2$ was covered by ZnO layer, drying: 180 °C for 10 min, calcination: 500 °C for 1 h	particles with shapes similar to nanotubes and spherical, EDX showed characteristic bands for Ti, Zn, O	[164]
Ti[OCH(CH$_3$)$_2$]$_4$, Zn(CH$_3$COO)$_2$·2H$_2$O, acetic acid, propan-2-ol	TiO$_2$–ZnO system: to Ti[OCH(CH$_3$)$_2$]$_4$ was added Zn(CH$_3$COO)$_2$·2H$_2$O dissolved in propan-2-ol (molar ratio Ti:Zn = 5:1), calcination: 400 °C for 4 h	crystal structure of anatase and wurtzite, particles with a spherical shape with agglomeration, BET surface area = 91 m^2/g, pore size 1.49 nm, pore volume 0.343 cm^3/g	[165]

Stoyanova et al. investigated the effect of the type of precursor used, the TiO$_2$:ZnO molar ratio and the calcination temperature on the physico-chemical properties of TiO$_2$-ZnO oxide systems synthesized by the sol-gel method [1]. At the first stage, TiO$_2$-ZnO systems were obtained using zinc acetate (sample A) and zinc chloride (sample B) as the ZnO precursor, and titanium tetrachloride as the precursor of TiO$_2$. The TiO$_2$:ZnO molar ratio was 90:10 (sample A) or 50:50 (sample B). The materials prepared from zinc acetate were calcined at 500 °C for 6 h, and those made using zinc chloride were calcined at 200, 400, 550 and 600 °C for 2 h. The diffractograms contained peaks corresponding to anatase in sample A, and to anatase, rutile and zinc titanate in sample B. It was demonstrated that the crystal structure of the final product was affected by the precursor type, the molar ratio of the reagents and the temperature of thermal processing. SEM images showed that the synthesized materials consisted of particles with a spherical shape and a marked tendency to agglomerate. Authors applied the obtained materials in a process of photodegradation of malachite green (C.I. Basic Green 4). They also determined the antibacterial properties of the materials against the Gram-positive *E. coli* (ATCC 25922). The results of the dye photodegradation experiment showed that both the material obtained from zinc acetate and that made using zinc chloride exhibited high photocatalytic activity: total degradation occurred after 40 min in the presence of sample B and after 170 min in the presence of sample A. In the tests with *E. coli*, the TiO$_2$-ZnO system obtained using zinc chloride demonstrated 100% antibacterial activity after 120 min, while the system made using zinc acetate achieved only a 30% yield in the same amount of time. Under UV light, it was found that the bacteria were removed completely after 15 min in the presence of sample A and after 30 min in the presence of sample B.

In Section 2.1 it was described how a TiO$_2$-ZnO system synthesized by the hydrothermal method could be used in dye-sensitized solar cells [56]. Shanmugam et al. [152] also described the use of a binary oxide system in a DSSC, but applied the sol-gel method of synthesis. Electrochemical characterization of the cell showed it to have a current density of 13.1 mA/cm^3 and a photoconversion efficiency of 32% (in darkness). The maximum equivalent quantum efficiency (EQE) of the DSSC built from a TiO$_2$-ZnO oxide system was measured at 85%, compared with 67% for a cell based on TiO$_2$ alone.

In recent years there have been reports concerning the potential use of non-toxic and inexpensive titania nanoparticles for imparting multifunctional properties to various textile materials [166–168]. Chu et al. [153] described the effect of the addition of a titania-based binary material on the antistatic properties of polyester fabrics. It was shown by physico-chemical analysis that the resulting materials had good stability in water, and that the stability improved as the molar fraction of ZnO was increased. SEM images of polyester fibres covered with TiO$_2$-ZnO showed that the fabric was covered uniformly by a layer of the oxide material. The best antistatic properties were exhibited by materials obtained with TiO$_2$:ZnO molar ratios of 5:5 and 7:3. Mechanical characterization of the materials showed that they offered greater permeability to gases than pure TiO$_2$ or ZnO. A whiteness quality test showed that the white colouring of the obtained TiO$_2$-ZnO systems was 10–30% lower than that of the reference samples.

Recent literature reports indicate that both titania [169,170] and zinc oxide [171] can be used as anti-corrosion coatings. Yu et al. [154] presented a simple synthesis of TiO$_2$-ZnO oxide systems using the sol-gel method and described their properties as anti-corrosion coatings. In the first stage, TiO$_2$ sol was obtained from titanium tetrabutoxide, ethanol, glacial acetic acid and water, and ZnO sol from zinc acetate, diethylamine, ethanol and water. These sols were mixed together in the molar ratios TiO$_2$:ZnO = 0:1, 1:3, 1:1, 3:1, 1:0. The materials were applied to carbon steel using an immersion technique. The steel with the oxide layer was then dried at 70 °C for 10 h. Finally the materials were calcined at 350 and 500 °C for 2 h. The procedure was repeated two or four times to obtain coatings consisting of two, four or eight layers. The carbon steel plates coated with oxide materials were subjected to comprehensive physico-chemical analysis. The materials were shown to have the crystal structure of Fe, TiO$_2$, Fe$_3$O$_4$, Fe$_2$O$_3$ and ZnFe$_2$O$_4$. X-ray diffractograms contained peaks originating from the iron present in the carbon steel. The material obtained at a TiO$_2$:ZnO molar ratio of 1:0 was characterized by an anatase crystal structure. For the product calcined at 500 °C the forms Fe$_3$O$_4$ and Fe$_2$O$_3$ were observed, which indicates oxidation of the steel. SEM images showed that the materials calcined at 350 °C had numerous cracks, creating a non-uniform layer on the surface of the steel. An increase in the ZnO content in the oxide system was found to improve the uniformity of the coating. Potentiodynamic polarization curves were obtained to determine the anticorrosive properties of the coatings: the results showed that significant protection against corrosion was provided in contact with sea water. The best anticorrosive properties were obtained for the systems with TiO_2:ZnO ratios of 1:1 and 1:3, thermally treated at 500 °C.

To obtain a specific crystal structure of materials synthesized by the sol-gel method, it is necessary to carry out a calcination process. Many literature reports refer to the effect of calcination on photocatalytic properties [172–174]. Wang et al. [155] conducted a simple synthesis of a TiO_2-ZnO oxide system using the sol-gel method and a calcination process. A TiO_2 sol was obtained from titanium tetrabutoxide, anhydrous ethanol and hydrochloric acid, and a ZnO sol was obtained from zinc acetate, anhydrous ethanol and deionized water. Oxide systems were obtained by mixing the TiO_2 and ZnO sols at the molar ratios ZnO:TiO_2 = 0.1:0.15, 0.2:0.25, 0.3:0.35. In the final stage the oxide materials were calcined at 450, 480, 500, 550 and 600 °C for 1, 1.5, 2, 2.5 and 3 h. Since the physico-chemical properties of titania-based materials determine their photocatalytic ability, analysis was performed using such techniques as X-ray diffraction and scanning electron microscopy. The results of both XRD and SEM indicate the formation of the crystalline structure of anatase and wurtzite, and the spherical shape of the particles of the materials obtained. Their photocatalytic activity was evaluated with respect to the photodegradation of methyl orange (C.I. Basic Orange 10). All of the materials exhibited good degradation efficiency. The highest yield of dye degradation (93% after 5 h) was achieved by the material obtained at a ZnO:TiO_2 molar ratio of 0.2:0.25.

Due to the increase in the number of cars and the emission of harmful gases in city centers, it is apparent that there is a need to remove pollutants such as nitrogen oxides (NO_x). Many researchers, such as Dalton et al. [175], Karapati et al. [176] and Todorova et al. [177], have reported the use of titanium dioxide as an effective NO_x photocatalyst. Giannakopoulou et al. [156] described the synthesis of an effective binary oxide photocatalyst of nitrogen oxides using the sol-gel method. At the first stage of synthesis, TiO_2 sol was obtained by mixing titanium tetrabutoxide with anhydrous ethanol, nitric acid and distilled water. The ZnO sol was produced using zinc acetate, ethylene glycol, glycerine, TEA and anhydrous ethanol. The sols were mixed together at the molar ratios TiO_2:ZnO = 100:0, 75:25, 50:50, 25:75 and 0:100. Thin foils of the resulting binary oxide photocatalysts were produced using an immersion technique. The final step was calcination at 500 °C for 1 h. Analysis of X-ray diffractograms showed that the TiO_2:ZnO molar ratio affected the parameters of the crystal structure. An increase in the ZnO content leads to the vanishing of the diffraction reflections characteristic of TiO_2. The oxide materials with TiO_2:ZnO molar ratios of 100:0 and 75:25 contained a single diffraction peak corresponding to anatase. When the ratio was 50:50 or 25:75 an amorphous structure was observed, while the system with 100% ZnO had the crystal structure of wurtzite. A test of photocatalytic activity showed that the highest activity in the oxidation of NO_x was achieved by the materials obtained with TiO_2:ZnO molar ratios of 100:0 and 75:25. An increase in the content of ZnO led to a significant drop in photocatalytic activity, possibly due to the weak crystallinity of the materials obtained.

The surface wettability of a material is one of the properties that determine its potential applications. Chen et al. [157] reported the synthesis of a superhydrophilic TiO_2-ZnO oxide system in the form of a thin foil. It was produced using a TiO_2 sol made from titanium tetrabutoxide, anhydrous ethanol, diethylamine (DEA) and nitric acid, and a ZnO sol obtained using zinc acetate, anhydrous ethanol and DEA. The precursor solutions were mixed together in the molar ratios TiO_2:ZnO = 90:10, 80:20, 70:30, 60:40, and then underwent calcination at 500 °C for 30 min. The results relating to the crystal structure were similar to those reported by Giannakopoulou et al. [156]. An increase in the molar fraction of ZnO led to the vanishing of the anatase structure, and in the case of the materials with TiO_2:ZnO ratios of 70:30 and 60:40, this resulted in the presence of an amorphous structure. Atomic force microscope (AFM) images (Figure 9) showed that the obtained composite foils had a microporous structure. Contact angle measurements showed all of the materials to be superhydrophilic, with contact angles in the range of 1.8–3.2° and diffusion of a drop of water on the surface of the material within 800 ms. A test involving the photodegradation of methylene blue (C.I. Basic Blue 9) showed that the addition of a small quantity (10%) of ZnO improved the material's photocatalytic activity: the degradation yield was 80% after exposure to light for 280 min.

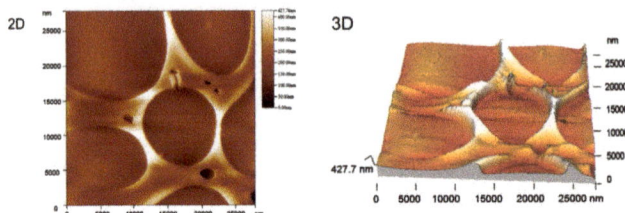

Figure 9. AFM images of the oxide system (90)TiO$_2$-(10)ZnO (created based on [157] with permission from Elsevier Publisher).

Pournuroz et al. [158] analyzed the degradation of the dyes C.I. Reactive Red 195 (RR195) and C.I. Reactive Blue 19 (RB19) in the presence of a TiO$_2$-ZnO oxide system. In the synthesis process sols of TiO$_2$ and ZnO were obtained, which were mixed together in the molar ratio TiO$_2$:ZnO = 10:3. The resulting material exhibited high photocatalytic activity. Complete degradation of the RR195 and RB19 dyes was observed after exposure to light for 40 and 60 min respectively, in the presence of the synthesized material.

The sol-gel auto-combustion synthesis method (also called low-temperature self-combustion, auto-ignition or self-propagation, as well as gel-thermal decomposition) combines the chemical sol-gel and combustion processes [178,179]. This method has shown great potential in the preparation of novel nanomaterials. Al-Mayman et al. [159] described the use of a sol-gel auto-ignition technique to synthesize a TiO$_2$-ZnO oxide system with defined photocatalytic properties. At the first stage, TiO$_2$ and ZnO precursors were dissolved in deionized water, and citric acid was added. The reaction mixture was then heated to 100 °C to evaporate off water. The resulting gel was ignited by heating to a temperature of approximately 300 °C. In the final stage, the TiO$_2$-ZnO system was calcined in air at 500 °C for 5 h. Materials were obtained with the general formula Ti$_{1-x}$Zn$_x$O$_2$, for x = 0%, 1%, 5%, 10%, 12% and 30%. Their diffractograms contained peaks corresponding to the crystal structure of anatase. Microstructural and morphological analysis indicated particles with a spherical shape and a marked tendency to agglomerate. The synthesized TO$_2$-ZnO oxide materials had surface areas of 52, 63.4, 69.9, 64 and 36.6 m^2/g when the mass fraction of ZnO was 1%, 5%, 10%, 12% and 30% respectively. The wide band gaps of titanium dioxide and zinc oxide make them promising materials for applications in the process of photocatalytic hydrogen production. All of the obtained materials were shown to be capable of evolving hydrogen under UV light. The highest yield was obtained from the material with the formula Ti$_{90}$Zn$_{10}$O$_2$ (1.048 mmol/h of hydrogen after 9 h of irradiation).

Zulkiflee and Hussin [160] investigated the effect of the temperature of thermal processing on the physico-chemical properties of a TiO$_2$-ZnO oxide system obtained by the sol-gel method. At the first stage a TiO$_2$ sol was obtained from titanium tetrabutoxide, acetic acid and butanol, and a ZnO sol from zinc acetate, diethylamine and propan-2-ol. To obtain thin layers of the oxide material, glass plates were immersed in the titanium dioxide sol and calcined at 400 °C. The plates were then immersed in ZnO sol and calcined at 500 and 600 °C for 2 h. X-ray diffraction results showed that the materials obtained had the crystal structure of anatase and wurtzite. Atomic force microscope (AFM) observations demonstrated that an increase in the calcination temperature caused an increase in the roughness of the material surface. The band gaps of the synthesized materials were measured at 3.6 and 3.7 eV for the samples TiO$_2$(400 °C)-ZnO(500 °C) and TiO$_2$(400 °C)-ZnO(600 °C).

Prasannalakshmi and Shanmugam [161] used a TiO$_2$-ZnO oxide system in the process of photocatalytic degradation of brilliant green (C.I. Basic Green 1) and methylene blue (C.I. Basic Blue 9).

In the first stage of the synthesis, a TiO$_2$ sol was obtained by hydrolysis of titanium tetraisopropoxide, and a ZnO sol was prepared from zinc acetate and 0.5 M sodium hydroxide. The sols were mixed together in the molar ratios TiO$_2$:ZnO = 0.5:0.25, 1:0.5, 1.5:0.75, 2:1, to produce samples labelled TZO1 TZO2, TZO3 and TZO4. The final stage of the process was calcination at 550 °C for 4 h. In analysis of the crystal structure, all of the oxide systems gave diffraction reflections corresponding to anatase and wurtzite. The XRD results were confirmed using HRTEM analysis. HRTEM images showed distances between planes characteristic of (101) TiO$_2$ and (101) ZnO. The surface areas, determined by low-temperature nitrogen adsorption and calculated based on BET theory, were 74.7, 32.2, 12.58 and 8.82 m^2/g for the samples TZO1 TZO2, TZO3 and TZO4 respectively. All of the synthesized materials exhibited high photocatalytic activity, and each of the binary samples (TZO1, TZO2, TZO3, TZO4) had superior photocatalytic properties to those of pure TiO$_2$ or ZnO.

Materials such as titania [180,181] and zinc oxide [182] are successfully used for the construction of p-silicon heterojunction photodiodes. Al-Hazami and Yakuphanoglu [162] used a binary combination of these materials to build a p-type diode. A layer of the TiO$_2$-ZnO oxide system was applied to a prepared silicone base, after which the material was dried and then calcined in air at 450 °C for 1 h. SEM results showed particles with a spherical shape with no tendency to agglomerate. The band gap was measured at 3.41 eV. Photoelectrochemical tests showed that the p-type diode covered with a thin foil of the TiO$_2$-ZnO system had photovoltaic and photoconductive properties. Tests under different luminous intensities confirmed a rapid reaction to changes in the studied parameters and good process repeatability.

Taking account of the known antibacterial properties of titanium dioxide and zinc oxide, Armin et al. [163] carried out a wide-ranging analysis of possible applications of a TiO$_2$-ZnO binary system. The obtained material was used in electrochemical, photocatalytic and antibacterial tests. Dynamic potentiometric curves showed that TiO$_2$-ZnO coatings offered high resistance at low currents. The curves are contained within a very wide range of potentials, from -1.4 to 1.1 V. Based on the measured parameters, a corrosion index was computed for coated and uncoated materials. The material coated with TiO$_2$-ZnO had a corrosion protection index of 98.6%, compared with only 82% for the uncoated material. Antibacterial tests showed that the TiO$_2$-ZnO material suppressed the growth of populations of the organisms *C. albicans* ($1.5 \cdot \times 10^5$) and *E. coli* ($1.5 \cdot \times 10^7$). In an experiment involving the photodegradation of methylene blue (C.I. Basic Blue 9), the TiO$_2$-ZnO oxide system produced a high yield of dye decomposition (70% after exposure to UV light for 6 h).

Chamanzadeh et al. [164] described the synthesis of a TiO$_2$-ZnO oxide system with potential applications in DSSCs. Similar properties of a TiO$_2$-ZnO binary oxide material were reported by Shanmugam et al. [152]. A DSSC was built using the commercial dye N719 (di-tetrabutylammonium cis-bis(isothiocyanato) bis(2,2'-bipyridyl-4,4'-dicarboxylato) ruthenium(II)). Based on obtained electrochemical curves it was shown that the ZnO layer improves adsorption of the dye and reduces the electron recombination rate. It was also found that the materials offered high photovoltaic efficiency (PCE = 8.3%) and charge accumulation at a level of 99% (for the material TiO$_2$-(1)ZnO).

The combination of several catalytic techniques can provide greater possibilities of removing dangerous contaminants. The observed improvement in catalytic activity may be linked to a synergy effect, as has been widely described in literature [183–185]. The combination of techniques such as photocatalysis and sonocatalysis can significantly improve the efficiency of degradation of organic impurities using TiO$_2$ [186,187]. Fatimah and Novirasari [165] described a simple synthesis of a binary photoactive material based on titanium dioxide. A combination of different catalytic techniques was studied with the aim of improving the catalytic properties. The material obtained was found to exhibit high photoactivity. A synergy effect was observed between photocatalytic degradation (yield 47%) and sonocatalytic degradation (yield 37%): the yield of phenol degradation in a sonophotocatalytic process was as high as 99%.

The following part of this review will discuss binary oxide systems based on TiO$_2$ and ZrO$_2$, again synthesized using the sol-gel method. A summary is given in Table 5.

Table 5. Resources, synthesis conditions and physico-chemical properties of TiO_2-ZrO_2 oxide systems obtained by the sol-gel method.

Resources	Synthesis Conditions	Physico-Chemical Properties of the Final Product	References
$Ti(OC_4H_9)_4$, $ZrOCl_2 \cdot 8H_2O$, nitric acid, polyethylene glycol (PEG), $(PEO)_{20}(PPO)_{70}(PEO)_{20}$ (Pluronic P123, M_W = 5800, Aldrich), ethanol, deionized water	TiO_2-ZrO_2 oxide system: single-step synthesis, mixed with $Ti(C_4H_9)_4$: PEG:P123 with $ZrOCl_2 \cdot 8H_2O$ (molar ratio Ti:Zr = 1:0.1), aging: 24 h, calcination: 800 °C for 5 h	crystal structure of anatase, tetragonal zirconia rutile, BET surface area = 148.9; 138.5; 136.9 m^2/g for TiO_2-ZrO_2(P123 + PEG); TiO_2-ZrO_2(PEG); TiO_2-ZrO_2(P123)	[29]
$Ti[OCH(CH_3)_2]_4$, $Zr[OCH(CH_3)_2]_4$, nitric acid, deionized water	TiO_2-ZrO_2 oxide systems: M1 and M2 M1: $Ti_{0.9}Zr_{0.1}O_2$, hydrolysis of precursors in an aqueous environment M2: $Ti_{0.9}Zr_{0.1}O_2$, polymerization of precursors in propan-2-ol drying: 100 °C, calcination: 350 °C for 2 h	crystal structure of anatase and brookite, Raman spectroscopy showed anatase-specific bands, BET surface area = 313 m^2/g (M1); 269 m^2/g (M2)	[188]
$Ti[OCH(CH_3)_2]_4$, $Zr[OCH(CH_3)_2]_4$, ammonia water, propan-2-ol, distilled water	TiO_2 sol: $Ti[OCH(CH_3)_2]_4$, propan-2-ol, ammonia water ZrO_2 sol: $Zr[OCH(CH_3)_2]_4$, propan-2-ol, ammonia water TiO_2-ZrO_2 oxide systems: sol was mixed with ZrO_2 in a molar fraction of 3%, 6%, 13%, 37%, drying: 100 °C for 24 h, calcination: 550, 700 °C for 5 h	crystal structure of anatase and rutile, spherical particle shape with agglomeration, BET surface area = 26, 37, 40, 172 m^2/g for materials calcined at 550 °C and 26, 29, 30, 36 m^2/g for materials calcined at 700 °C, for systems with ZrO_2 molar fractions 3%, 6%, 13%, 37% respectively	[189]
$Ti[OCH(CH_3)_2]_4$, $Zr(CH_3COO)_2 \cdot 2H_2O$, hydrochloric acid, hydroxypropylcellulose, deionized water	ZrO_2 sol: $Zr(CH_3COO)_2 \cdot 2H_2O$, hydroxypropylcellulose TiO_2 sol: $Ti[OCH(CH_3)_2]_4$, HCl, water TiO_2-ZrO_2: the sol was mixed maintaining the ratio Ti:Zr = 0.5:0.5, drying: 150 °C for 1 h, calcination: 300, 500, 700, 900 °C for 1 h	crystal structure of anatase and zirconium dioxide, crystallite size 5.8 and 8.5 nm for calcination temperatures of 500 °C and 900 °C, particles with a spherical shape with a tendency to agglomerate	[190]
$Ti[OCH(CH_3)_2]_4$, $Zr[OCH(CH_3)_2]_4$, poly(ethylene oxide), $(PEO)_{20}(PPO)_{70}(PEO)_{20}$ (Pluronic P123, BASF), acetylacetone, hydrochloric acid, deionized water	TiO_2 sol: $Ti[OCH(CH_3)_2]_4$, HCl, water ZrO_2 sol: $Zr[OCH(CH_3)_2]_4$, HCl, water TiO_2-ZrO_2: the sol was mixed, aging: 12 h, calcination: 400, 500 °C for 2 h	crystal structure of anatase, incorporation of zirconium dioxide into titania network forming $Ti_{1-x}Zr_xO_2$ with anatase crystal structure, Raman spectroscopy showed anatase and ZrO_2 characteristic bands, spherical shaped particles densely packed with diameter 10–15 nm	[191]

Table 5. Cont.

Resources	Synthesis Conditions	Physico-Chemical Properties of the Final Product	References
Ti[OCH(CH$_3$)$_2$]$_4$, Zr[OCH(CH$_3$)$_2$]$_4$, lauramine hydrochloride, acetyl acetone, diethylamine, deionized water	TiO$_2$-ZrO$_2$ oxide system: single-step synthesis, mixed Ti[OCH(CH$_3$)$_2$]$_4$, Zr[OCH(CH$_3$)$_2$]$_4$ acetylacetone and laurylamine, aging: 7 days, calcination: 500–900 °C	crystal structure of anatase, rutile and tetragonal zirconia, particles with spherical shape with agglomeration	[27]
Ti[OCH(CH$_3$)$_2$]$_4$, Zr[OCH(CH$_3$)$_2$]$_4$, nitric acid, ethanol, deionized water	TiO$_2$-ZrO$_2$: one-stage synthesis, hydrolysis of Ti[OCH(CH$_3$)$_2$]$_4$, Zr[OCH(CH$_3$)$_2$]$_4$ in ethanol:water	Fourier-transform infrared spectroscopy (FTIR) spectroscopy showed bands characteristic for TiO$_2$ and ZrO$_2$, distances between planes 3.533 Å	[192]
Ti[OCH(CH$_3$)$_2$]$_4$, Zr[OCH(CH$_3$)$_2$]$_4$, diethylamine (DEA), isoeugenol (ISOH), propan-1-ol, deionized water	TiO$_2$-ZrO$_2$ oxide system: one-stage synthesis, Ti[OCH(CH$_3$)$_2$]$_4$, Zr[OCH(CH$_3$)$_2$]$_4$ mixed with a chelating complex (diethylamine or isoeugenol), calcination: 500, 650 °C in air and 350, 400 °C in a nitrogen atmosphere	amorphous structure (materials calcined at 500 °C), crystal structure of anatase and rhombic TiO$_2$-ZrO$_2$ (materials calcined at 650 °C), BET surface area = 7.8 m^2/g (TiO$_2$-ZrO$_2$-ISOH) and 8.3 m^2/g (TiO$_2$-ZrO$_2$-DEA) for samples calcined in a nitrogen atmosphere, 240 m^2/g (TiO$_2$-ZrO$_2$-ISOH) and 186 m^2/g (TiO$_2$-ZrO$_2$-DEA) for samples calcined in air	[193]
TiF$_4$, Zr(NO$_3$)$_4$, anodic aluminium oxide (AAO), ethanol, sulphuric acid, deionized water	TiO$_2$-ZrO$_2$ oxide system: single-step synthesis, AAO was immersed in Zr(NO$_3$)$_4$, drying: 80 °C for 1 h, immersed in TiF$_4$ for 9 min, calcination: 600 °C for 3 h	crystal structure of anatase and monoclinic zirconia, nanotube-shaped particles densely packed, BET surface area = 47.4 m^2/g	[26]
Ti(OC$_4$H$_9$)$_4$, Zr(OC$_4$H$_9$)$_4$, butan-1-ol, ammonium hydroxide, deionized water	TiO$_2$-ZrO$_2$ oxide system: single-step synthesis, hydrolysis of Ti(OC$_4$H$_9$)$_4$, Zr(OC$_4$H$_9$)$_4$ in butan-1-ol:water, ammonium hydroxide was added to neutralize the pH, drying: 120 °C for 12 h, calcination: 500 °C for 12 h	crystal structure of anatase, monoclinic and tetragonal zirconia, Raman spectroscopy showed anatase-specific bands, BET surface area = 68, 62 and 60 m^2/g for systems obtained at molar ratio TiO$_2$:ZrO$_2$ = 3:1, 1:1, 1:3	[194]
Ti[OCH(CH$_3$)$_2$]$_4$, Zr[OCH(CH$_3$)$_2$]$_4$, ethanol, deionized water	TiO$_2$-ZrO$_2$: single-step synthesis, hydrolysis of Ti(OC$_4$H$_9$)$_4$, Zr(OC$_4$H$_9$)$_4$ in ethanol:water, drying: 100 °C for 6 h, calcination: 550 °C for 2 h	crystal structure of anatase, monoclinic and tetragonal zirconia, spherical shaped particles with a pronounced tendency to agglomerate, BET surface area 32.5, 42.5, 36 m^2/g for 67%TiO$_2$–33%ZrO$_2$, 50%TiO$_2$–50%ZrO$_2$, 33%TiO$_2$–67%ZrO$_2$	[195]

The use of polymer matrices such as the commercial Pluronic P123 or F127 (poly(ethylene glycol)-block-poly(propylene glycol)-block-poly(ethylene glycol)—PEG-PPG-PEG) is becoming more and more popular in the synthesis of titania-based materials [169–198]. Fan et al. [29] developed a method of synthesizing a TiO_2-ZrO_2 oxide system using Pluronic123 (P123) and Macrogol20000 (PEG) matrices. In the first stage, Pluronic P123 and Macrogol20000 were added to titanium tetrabutoxide. Then a zirconium precursor was added to the mixture, to achieve the molar ratio Ti:Zr = 1:0.1. The product underwent aging for 24 h at room temperature, and was then dried and calcined at 800 °C for 5 h. Diffractograms showed characteristic peaks for anatase, rutile and tetragonal zirconium dioxide. BET surface areas of 136.9, 138.5 and 148.9 m^2/g were measured for the systems TiO_2-ZrO_2(P123 + PEG), TiO_2-ZrO_2(PEG) and TiO_2-ZrO_2(P123) respectively. The synthesized oxide materials were tested in the photodegradation of rhodamine B (C.I. Basic Violet 10) in UV light and in darkness. The materials demonstrated high photocatalytic efficiency: 91.93%, 91% and 90% for TiO_2-ZrO_2(P123 + PEG), TiO_2-ZrO_2(PEG) and TiO_2-ZrO_2(P123) respectively.

Environmental pollution by harmful organic compounds is a severe problem for both humans and animals. Acetone is one of the most serious air pollutants in indoor environments, and its catalytic decomposition to less harmful compounds is a problem of interest to many research centres worldwide. Hernandez-Alonso et al. [188] obtained a TiO_2-ZrO_2 oxide system in the form of thin films, which they tested for photocatalytic activity in the degradation of acetone. Two methods of synthesis were investigated: sample M1, in which $Ti_{0.9}Zr_{0.1}O_2$ was obtained by simultaneous hydrolysis of titanium tetraisopropoxide and zirconium tetrapropoxide; and sample M2, in which $Ti_{0.9}Zr_{0.1}O_2$ was produced by polymerization of metal alkoxides in n-propanol. Thin layers of the binary oxide materials may be used successfully to coat Raschig rings, for example, and applied in various catalytic processes. The results of photo-oxidation of acetone confirmed the good photocatalytic properties of the synthesized materials. The method of synthesis (M1 or M2) was not found to affect the photocatalytic activity.

Kraleva et al. [189] described a simple method of obtaining a TiO_2-ZrO_2 oxide system using metal alkoxides (titanium tetraisopropoxide and zirconium tetrapropoxide). In the first stage the TiO_2 and ZrO_2 precursors underwent hydrolysis in a solution of propan-2-ol and ammonia, and the resulting sols were subjected to aging for 24 h. The TiO_2-ZrO_2 system was formed by mixing the sols, with the molar contribution of zirconium dioxide equal to 3, 6, 13 and 37 (the samples were denoted TiZrn, where n was the molar contribution of ZrO_2). The resulting systems were dried at 100 °C for 24 h, and calcined at 550 and 700 °C for 5 h. The final binary oxide materials had a similar crystal structure to those described in earlier work. X-ray diffractograms contained characteristic peaks for anatase and rutile. For the system TiZr37 calcined at 700 °C, the presence of srilankite ($Ti_{0.63}Zr_{0.37}O_x$) was also bserved. The calcination temperature was found to have a significant effect on the crystaline structure, with anatase-rutile transformation at 700 °C. TEM images (Figure 10) showed particles with a spherical shape, having a marked tendency to agglomerate.

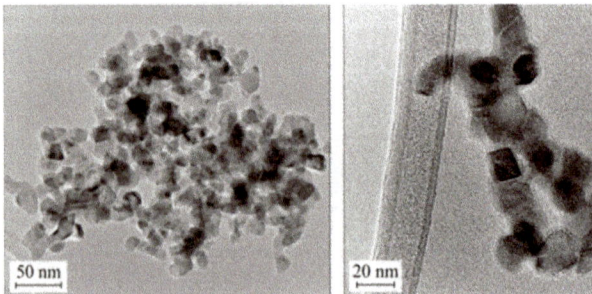

Figure 10. HRTEM images of the oxide system TiZr6 calcined at 700 °C (created based on [189] with permission from Springer).

The BET surface areas of the TiO_2-ZrO_2 systems calcined at different temperatures were 26, 37, 40 and 172 m^2/g for 550 °C and 26, 29, 30 and 36 m^2/g for 700 °C, respectively for the systems denoted TiZr3, TiZr6, TiZr13 and TiZr37. The calcination temperature was found to affect the porous structure parameters: with an increase in temperature, the BET surface area decreased (this is related to the effect of sintering).

Atmospheric pollution is one of the most urgent present-day problems. A priority task in the synthesis of new nanomaterials is the possibility of controlling exhaust emissions. Titania-based gas sensors for the quantitative detection of various toxic and harmful gases have been widely developed in view of their high response, outstanding selectivity, excellent repeatability, and good stability. Mohammadi and Fray [190] described the use of a system combining titanium dioxide and zirconium dioxide in detecting nitrogen(IV) oxide and carbon(II) oxide. Results obtained for the sensoric properties of the materials showed that the highest response to the applied concentrations of NO_2 (2 ppm) and CO (100 ppm) was given by the system that had been calcined at 500 °C (the response was 9.1 for CO and 5.1 for NO_2 at a temperature of 150 °C). Detailed analysis of the morphological and crystal structures was carried out using SEM and XRD. In the crystalline structure analysis, diffraction peaks for anatase and zirconium dioxide were obtained. Based on the X-ray analysis, the average crystallite size was computed to be 5.8 nm (500 °C) and 8.5 nm (900 °C). The SEM images showed particles with a spherical shape, having a marked tendency to agglomerate. The average particle size was 20 nm (500 °C) and 36 nm (900 °C).

Naumenko et al. [191] described the synthesis of a TiO_2-ZrO_2 oxide system using a sol-gel method, with titanium tetraisopropoxide and zirconium tetrapropoxide as precursors of the oxides. The first stage of the synthesis was the hydrolysis of $Ti[OCH(CH_3)_2]_4$ and $Zr[OCH(CH_3)_2]_4$ in a mixture of deionized water and hydrochloric acid. The prepared sols were then mixed, with the $Ti[OCH(CH_3)_2]_4$:$Zr[OCH(CH_3)_2]_4$ molar ratio ranging from 1:0 to 0.7:0.3. The resulting materials underwent aging for 12 h, and were then calcined at 400 and 500 °C in air. The products were found to contain only the crystal structure of anatase. Detailed analysis of the results obtained for crystal structure parameters showed a change in the parameters when the molar fraction of ZrO_2 was 0.3, with values of a = 3.82 Å, c = 9.726 Å (ASTM 84-1286: a = 3.7822 Å, c = 9.5023 Å). This may indicate that the ZrO_2 was incorporated into the TiO_2 structure to form $Ti_{1-x}Zr_xO_2$. AFM images showed the TiO_2 particles to have diameters of 10–15 nm. An increase in the molar fraction of ZrO_2 led to the appearance of particles with diameter 20 nm.

It is well known that in order to obtain a binary oxide material based on titania with strictly defined properties, it is necessary to choose appropriate synthesis conditions. Parameters such as the molar ratio or the temperature of thermal treatment can have a decisive influence on the resulting material. Kokporka et al. [27] investigated the effect of the molar ratio of the oxides and the temperature of calcination on the physico-chemical properties of a TiO_2-ZrO_2 system. They also determined the photocatalytic activity of the materials in the evolution of hydrogen. Diffractograms showed the presence of characteristic peaks for the crystal structures of anatase and rutile. The crystal structure parameters were found to be affected by the temperature of thermal treatment and by the molar ratio. Scanning electron microscope images showed particles of spherical shape with a marked tendency to agglomerate. The BET surface area, determined by low-temperature nitrogen sorption, was 130 m^2/g for the systems 0.95TiO_2-0.05ZrO_2 and 0.2TiO_2-0.8ZrO_2, and 190 m^2/g for 0.4TiO_2-0.6ZrO_2 and 0.6TiO_2-0.4ZrO_2. In the tests of photocatalytic activity, high yields of hydrogen were obtained in the presence of TiO_2-ZrO_2. The highest yield (0.61 mL/h·g) was obtained using the oxide system 0.95TiO_2-0.05ZrO_2 calcined at 800 °C.

Due to their high chemical and thermal stability, both Titania [199] and Zirconia [200] are used as matrices for the application of nanoparticles. Karthika et al. [192] described an application of a TiO_2-ZrO_2 binary material as a matrix for Sm^{3+} and CdS nanoparticles. TEM images showed a crystallite size of 7.8 nm and an interplanar distance of 3.533 Å, which corresponds to the CdS (100) plane. FTIR spectra contained bands for Ti–O–Ti at the wavenumber 440 cm^{-1}, Zr–O–Ti at 663 cm^{-1}

and Zr–O–Zr at 841 cm^{-1}. Optical absorption and emission spectra confirmed the presence of CdS nanoparticles along with Sm^{3+} ions in the TiO_2-ZrO_2 matrices.

Due to the ever faster development of industry, it is necessary to obtain effective membranes for gas separation processes, such as the separation of CO_2 and purification of organic gases under normal temperature and pressure conditions [201]. Therefore, interest in inexpensive inorganic materials such as Titania [202] and zirconia [201,203] is increasing. Fukumato et al. [193] described the use of the chelating complexes DEA and isoeugenol (ISOH) in the synthesis of a TiO_2-ZrO_2 oxide system. The goal of the study was to obtain amorphous microporous membranes. The single-stage synthesis involved the hydrolysis of titanium tetraisopropoxide and zirconium tetrapropoxide in a water solution of propan-1-ol. Two forms of material were obtained: with the addition of diethylamine (TiO_2-ZrO_2-DEA) and with the addition of isoeugenol (TiO_2-ZrO_2-ISOH). The synthesis product was used to prepare a membrane on a base made from Al_2O_3 and SiO_2-ZrO_2. The final stage of the process was calcination of the TiO_2-ZrO_2 system in membrane form at 500 and 600 °C in air and at 350 and 400 °C in a nitrogen atmosphere. The TiO_2-ZrO_2 membranes were tested for permeability to the gases He, N_2, CO_2 and N_2. The oxide systems calcined in air exhibited higher permeability than those calcined in nitrogen. This is probably due to residues of the chelating compounds present in the nitrogen-calcined systems. The permeabilities of the TiO_2-ZrO_2-ISOH membranes were higher than those of the TiO_2-ZrO_2-DEA membranes, and were equal to 110, 48 and 50 [10^{-7} mol/m^2·s·Pa] for He, CO_2 and N_2 respectively. Figure 11 shows SEM images of the membrane.

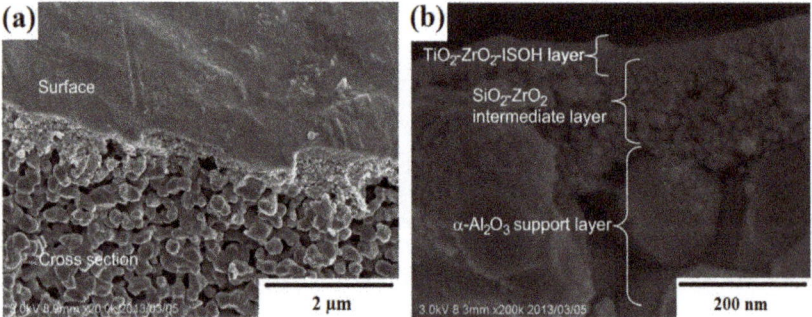

Figure 11. SEM images of surface and cross-section views of a TiO_2–ZrO_2–ISOH membrane calcined at 350 °C under N_2: (**a**) surface and cross-section views at 20,000×; (**b**) cross-section view at 200,000× (created based on [193] with permission from Elsevier Publisher).

Anodized aluminium oxide (AAO), due to properties such as its nano-sized, self-organized, hexagonal porous structure [204], is an interesting material as a substrate for titania-based materials. Such materials can be used, for example, as sensors for hydrogen detection [204–206]. Qu et al. [26] investigated the use of anodic aluminium oxide as a matrix to obtain a photoactive TiO_2-ZrO_2 oxide system. In the first stage, AAO was immersed in zirconium nitrate for 12 h at 80 °C, after which it was dried for 1 h at 80 °C. The ZrO_2-grafted AAO was then immersed in TiF_4 for 9 min at 60 °C. In the final stage, the product was calcined in air for 3 h at 600 °C. Diffractograms showed the presence of characteristic diffraction peaks for anatase and for monoclinic zirconium dioxide. SEM images showed particles in the form of densely packed nanotubes with a tendency to agglomerate. Analysis of porous structure parameters showed the BET surface area of the oxide system to be 47.4 m^2/g. The band gap of the system was measured at 3.22 eV. The synthesized material was tested as a catalyst in the photodegradation of methyl orange (C.I. Basic Orange 10). A high yield of photodegradation of the dye was obtained, amounting to 96% after 180 min, compared with 83% for TiO_2 and 45% for ZrO_2.

Another team of researchers [194] described the use of a TiO_2-ZrO_2 oxide system to obtained hydroxymethylfufural (HMF) from biomass. HMF is a precursor of materials such as polyesters and polymers based on furfural, as well as diesel fuels [207,208]. The binary oxide system was obtained in a single-stage sol-gel process, with the simultaneous hydrolysis of titanium tetrabutoxide and zirconium tetrabutoxide. The material was then dried at 120 °C for 12 h and calcined at 500 °C for 12 h. Diffractograms showed the presence of peaks corresponding to the structures of anatase and of monoclinic and tetragonal zirconium dioxide. When the molar fraction of ZrO_2 was increased, the anatase structure became less visible. Measurements of the porous structure parameters gave BET surface areas of 68, 62 and 60 m^2/g and pore volumes of 0.27, 0.18 and 0.13 cm^3/g respectively for the systems TiO_2-ZrO_2(3/1), TiO_2-ZrO_2(1/1) and TiO_2-ZrO_2(1/3). Tests of glucose conversion gave yields of 72–86%, the main product being HMF; its maximum yield was obtained using the system TiO_2-ZrO_2(1/1). The use of tetrahydrofuran (THF) as a co-solvent and NaCl as the aqueous phase caused a tripling of the HMF conversion yield. The HMF yield varied depending on the co-catalysts: Amberlyst 70 (85.6%) > Nafion NR50 (73.4%) > $Cs_{3.5}SiW$ (35.2%) > $Cs_{2.5}PW$ (31.8%).

One of the pollutants of greatest importance in drinking water treatment is humic acid [209]. Many attempts have been made to use low-pressure driven membrane systems for the removal of the largest fractions of humic acid. The recently widely described photocatalytic membrane reactor is an interesting alternative to conventional techniques [210–212]. Khan et al. [195] described the synthesis of a TiO_2-ZrO_2 binary oxide system as an active photocatalyst for removing humic acid in membrane reactors. The TiO_2-ZrO_2 system achieved 85% efficiency in the removal of total organic carbon. The effect of double-positive ions (Ca^{2+}) on the efficiency of the process was also investigated: addition of these ions caused an increase in the efficiency of removal of total organic carbon, possibly due to the reduced electrostatic repulsion between humic acid and the TiO_2-ZrO_2 system. The last part of the study concerned the effect of UV radiation, which has oxidizing properties, on a PVDF membrane placed in a reactor. FTIR spectra demonstrated that the structure of the PVDF membrane did not undergo an oxidation process under irradiation with UV light.

Literature reports on the synthesis of TiO_2-ZnO and TiO_2-ZrO_2 oxide systems by sol-gel routes confirm that this technique provides the possibility of controlling the process through appropriate selection of process conditions, and thus enables control of the physico-chemical properties of the final product (well-defined particle size, crystal structure and surface activity). A suitable choice of process conditions (applied reagents such as organic and inorganic precursors, solvent type, type of hydrolysing and/or nucleating agents, molar ratio of oxides, pH of the reaction system, rate and direction of reagent dosing, time and temperature of the process, conditions of final heat treatment, etc.) can lead to oxide materials based on titanium dioxide together with zinc oxide or zirconium dioxide with precisely designed structure and texture. This has a significant impact on the possible applications of the produced systems in such fields as photocatalysis, electrochemistry and adsorption.

2.3. The Electrospinning Method

Another developing method of synthesis of oxide systems containing titanium dioxide is the electrospinning method, which represents a continuation of research into electrosprays. The first work on an electrospray was reported by Rayleigh, who observed that a droplet subjected to high voltage disintegrates due to the potential gradient. This line of research was taken up by Zeylen and Dole [213,214], who described droplets dissociated in an electric field to form an aerosol. The physical principles behind the two processes are similar, the only difference being the form of the product obtained: nanofibers in the case of electrospinning, and nanodroplets in the case of electrosprays. The apparatus used to obtain nanofibers by electrospinning consists of a high voltage source in the range 1–30 kV, a solution feeder (a syringe and needle), a pump and a collector [215–218]. Schematic diagrams of vertical and horizontal electrospinning processes appear in Figure 12.

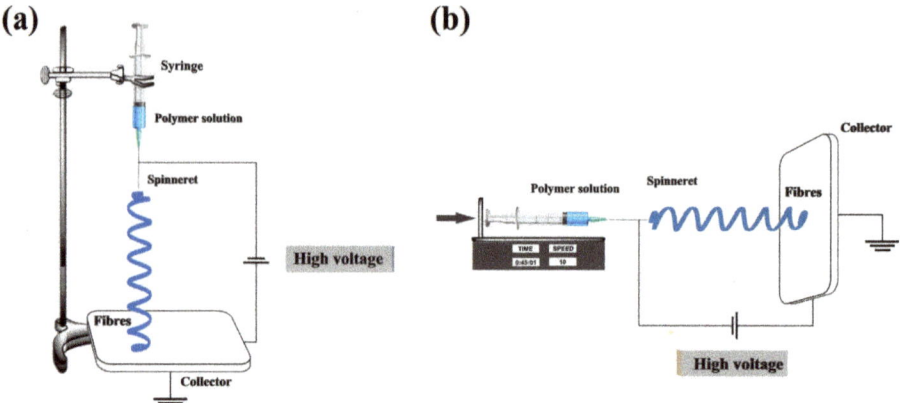

Figure 12. Schematic diagrams of (**a**) vertical and (**b**) horizontal electrospinning processes (created based on [219] with permission from Elsevier Publisher).

In electrospinning, a solution is placed in a syringe, and then a high voltage is applied to the needle or nozzle, producing electrostatic forces which cause movement of the charges in the solution towards the collector. A dosing pump, as well as exceeding of the critical value of the field intensity, initiates the outward flow of the solution towards the collector. Under the influence of the applied voltage, the liquid is stretched into a thin stream. As the stream falls, the solvent evaporates, and the resulting nanofibers are accumulated on the collector [220–222].

As in the case of the other methods that we have described, such as the hydrothermal and sol-gel methods, the parameters of the electrospinning process may affect the physico-chemical properties of the final product [223–226]. Table 6 illustrates the process parameters impacting the result of electrospinning, relating to the solution, the apparatus and the external environment.

Table 6. Process parameters affecting the electrospinning method.

	Process Parameters	
Solution	Apparatus	External Environment
type of solvent concentration -	applied voltage distance of nozzle from collector diameter and length of nozzle	temperature humidity -

The chief advantages of the electrospinning method include the relatively simple apparatus and the fast production time. Another advantage of this method is the ability to produce nanofibers, for both laboratory and industrial use. The greatest drawbacks, however, are the instability of the stream, which affects the diameter of the resulting nanofibers, the fact that individual fibers may stick together, and the need for the polymer used in electrospinning to be soluble in the reaction medium [227–231].

Because of its advantages, electrospinning is becoming an interesting alternative method for obtaining oxide systems containing titanium dioxide. There has recently been a marked increase in interest in the use of electrospinning to produce oxide systems such as TiO_2-ZnO and TiO_2-ZrO_2, or TiO_2-MoS_2 hybrid systems.

In the next part of this review, a survey will be made of work in which TiO_2-ZnO oxide systems were obtained by the electrospinning method (Table 7).

Table 7. Resources, synthesis conditions and physico-chemical properties of TiO$_2$–ZnO hybrid systems obtained by the electrospinning method.

Resources	Synthesis Conditions	Physico-Chemical Properties of the Final Product	References
Ti[OCH(CH$_3$)$_2$]$_4$, Zn(CH$_3$COO)$_2$·2H$_2$O, polyvinylpyrrolidone, acetic acid, ethanol, deionized water	solution of TiO$_2$ and ZnO: Zn(CH$_3$COO)$_2$·2H$_2$O, acetic acid and Ti[OCH(CH$_3$)$_2$]$_4$ dissolved in a mixture of ethanol and polyvinylpyrrolidone TiO$_2$–ZnO electrospinning: voltage: +7 kV, spinning speed 0.3 cm^3/h calcination: 500 °C for 3 h	nanofibre shaped particles, X-ray photoelectron spectroscopy (XPS) shows bands derived from Ti2p, O1s, Zn2p	[232]
Ti[OCH(CH$_3$)$_2$]$_4$, Zn(C$_2$H$_5$)$_2$, dimethylformamide (DMF), acetic acid, poly(vinyl acetate), deionized water	preparation of TiO$_2$ solution: Ti[OCH(CH$_3$)$_2$]$_4$, acetic acid, dimethylformamide, polyvinyl acetate TiO$_2$ electrospinning: voltage: +10 kV, spinning speed 0.2 cm^3/h, calcination: 600 °C for 8 h TiO$_2$–ZnO oxide system: ZnO was applied by atomic layer deposition (ALD)	crystal structure of anatase and wurtzite, nanofibre particles of core-shell structure, average nanofibre diameter 250 nm, increase of ZnO coating thickness 0.66 nm per ALD cycle	[233]
Ti[OCH(CH$_3$)$_2$]$_4$, ZnCl$_2$, polyvinylpyrrolidone, (PEO)$_{20}$(PPO)$_{70}$(PEO)$_{20}$ (Pluronic P123), ethanol, deionized water	preparation of TiO$_2$ and ZnO precursors: polyvinylpyrrolidone dissolved in ethanol, Ti[OCH(CH$_3$)$_2$]$_4$ solution added to aqueous ZnCl$_2$ solution, calcination: 500 °C for 4 h TiO$_2$–ZnO electrospinning: voltage: +20 kV	crystal structure of anatase, wurtzite and zinc titanate (ZnTiO$_3$), particles with diameter 20 nm, BET specific surface area = 203 m^2/g, pore volume 0.24 cm^3/g	[234]
Ti[OCH(CH$_3$)$_2$]$_4$, Zn(CH$_3$COO)$_2$·2H$_2$O, cellulose acetate, dimethylformamide (DMF), acetic acid, acetone, deionized water	preparation of TiO$_2$ and ZnO precursors: Ti[OCH(CH$_3$)$_2$]$_4$, Zn(CH$_3$COO)$_2$·2H$_2$O cellulose acetate was dissolved in a mixture of DMF:acetone = 1:2 (v/v) TiO$_2$–ZnO electrospinning: voltage: +10 kV, spinning speed 10 µL/min, calcination: 500 °C for 5 h	crystal structure of anatase, wurtzite and zinc titanate (ZnTiO$_3$), diameter of nanofibres 85–200 nm, cylindrical nanofibres made of granular nanoparticles, XPS showed characteristic bands Ti2p, O1s and Zn2p	[235]
Ti[OCH(CH$_3$)$_2$]$_4$, ZnO, poly(vinyl acetate) (PVA), dimethylformamide (DMF), acetic acid, deionized water	preparation of TiO$_2$ and ZnO precursors: polyvinyl acetate was dissolved in DMF, added Ti[OCH(CH$_3$)$_2$]$_4$ and ZnO TiO$_2$–ZnO electrospinning: voltage: +12 kV, drying: 80 °C for 12 h, calcination: 600 °C for 1 h	crystal structure of wurtzite, nanofibres with a smooth surface, EDX showed presence of elements Ti, Zn, O	[236]
Ti[OCH(CH$_3$)$_2$]$_4$, Zn(CH$_3$COO)$_2$·2H$_2$O, polyvinylpyrrolidone, dimethylformamide (DMF), ethanol, deionized water	preparation of precursors TiO$_2$ and ZnO: Ti[OCH(CH$_3$)$_2$]$_4$ and Zn(CH$_3$COO)$_2$·2H$_2$O dissolved in DMF, polyvinylpyrrolidone was added TiO$_2$–ZnO electrospinning: voltage: +10 kV, calcination: 400 °C for 2 h	crystal structure of anatase and wurtzite, nanofibres composed of individual nanoparticles, XPS showed presence of specific bands for Ti2p, Zn2p, O1s	[237]
TiO$_2$ (P25), Zn(CH$_3$COO)$_2$·2H$_2$O, hexamethyltetramine, ethanol, demineralized water	electrospinning of TiO$_2$ fibres: P25 solution and polymer in ethanol, applied voltage: +15 eV, TiO$_2$ fibres mixed with solution 1 heating: 120 °C for 1 h TiO$_2$ nanofibres were immersed in solution 2 (aqueous zinc acetate solution and hexamethyl tetramine) and maintained at 85 °C for 24 h, calcination: 500 °C for 1 h	crystal structure of anatase, rutile and wurtzite, TiO$_2$ fibres coated with ZnO nanorods, EDX spectrum contained specific bands characteristic for Ti, O, Zn, Raman spectrum contained bands characteristic for TiO$_2$, ZnO and ZnTiO$_3$	[238]

Ammonia is a useful chemical material, widely used in many industries. However, it is a common material in industrial effluents and very dangerous to the environment and humans [239,240]. It attacks the respiratory system, skin, and eyes, and at a concentration higher than 300 ppm may lead to death [241,242]. Therefore, it is necessary to develop new materials to detect this type of impurity. Nanofibers possess favorable properties such as small diameters and large specific surface areas. For gas sensors, surface area is one of the most important factors influencing sensing performance [243]. Wang et al. [232] described the use of the electrospinning method for the saturation of a binary TiO_2-ZnO material with sensory properties. The first stage was the preparation of a solution of precursors of the oxides and acetic acid with the addition of poly(vinylpyrrolidone). The solution was then transferred to a syringe pump and a voltage of +7 kV was applied. The spinning rate was 0.3 cm^3/h. The oxide material was calcined at 500 °C for 3 h, and then immersed in a 0.1 mol/L solution of $FeCl_3$ for 30 min, dried, and subjected to the action of polypyrrole (PPy) for 5 h, to produce the oxide system TiO_2-ZnO-PPy. Analysis of XPS spectra showed the presence of Ti2p, O1s and Zn2p bands. SEM images of the PPy-modified materials showed the oxide system to have the form of black nanofibers with diameter 180 nm and lengths of several millimeters. The unmodified fibers had a uniform smooth surface. TEM images showed the oxide material to have a core-shell structure, where the thickness of the PPy shell was 7 nm. The TiO_2-ZnO-PPy system was tested as an ammonia sensor. It was found to have a fast response rate and a detection limit of 60 ppb. The results obtained were probably influenced by the reduction of diffusion resistance caused by the porous PPy shell.

Park et al. [233] described a combination of the methods of electrospinning and atomic layer deposition (ALD) in the synthesis of a TiO_2-ZnO binary oxide system. In the first stage, TiO_2 nanofibers were obtained by electrospinning from a solution of titanium tetraisopropoxide and poly(vinyl acetate), with an applied voltage of +10 kV and a constant spinning rate of 0.2 cm^3/h. The resulting nanofibers were calcined at 600 °C for 8 h. The TiO_2 fibers were then coated with ZnO using the ALD method. The number of ALD cycles was in the range 50–400. Diffractograms showed the TiO_2-ZnO system to contain the crystal structure of anatase and wurtzite. The HRTEM results were consistent with those of XRD. SEM images showed that the increment in the ZnO layer per ALD cycle was 0.66 nm. Based on EDS (Energy-dispersive X-ray spectroscopy) mapping a core-shell structure was demonstrated, with a core of TiO_2 and a shell of ZnO. As in the work of Wang et al. [232], the obtained TiO_2-ZnO nanofibers were tested as a chemical sensor. Park et al. [233] assessed the sensory properties of their product using oxygen. It was found to have good properties for the detection of oxygen, and the sensor's response varied with changes in the oxygen concentration. The tested chemical sensor demonstrated high stability and repeatability.

Surveys of the subject literature have already shown that binary oxide materials such as TiO_2-ZnO obtained by the hydrothermal method [62] and by the sol-gel method [156,161,163] have good photocatalytic properties in the degradation of methylene blue (C.I. Basic Blue 9). Materials of the same type synthesized by other methods can be expected to exhibit similar properties. A team of researchers [234] described the synthesis of a TiO_2-ZnO system by electrospinning and its use to catalyze the photodegradation of C.I. Basic Blue 9. X-ray diffraction results showed the presence of reflection corresponding to the structure of anatase, wurtzite and zinc titanate. From TEM images, the mean particle diameter was determined to be 20 nm. Using low-temperature nitrogen sorption, the BET surface area was measured at 203 m^2/g and the pore volume at 0.24 cm^3/g. The TiO_2-ZnO oxide system synthesized by electrospinning demonstrated high photocatalytic activity, at a similar level to the same material produced using other methods.

Rhodamine B is widely used as a colorant in textiles and foodstuffs, and is also a well-known water tracer fluorescent [244]. Many scientific studies have demonstrated the harmful effects of rhodamine B on humans and the environment. Its carcinogenicity, reproductive and developmental toxicity, neurotoxicity and chronic toxicity to humans and animals have been experimentally proven [245–247]. Due to the harmful effects of this material, it is necessary to develop new nanomaterials enabling its effective degradation [248,249]. Liu et al. [235] used cellulose acetate as a matrix for the synthesis of TiO$_2$-ZnO oxide systems with TiO$_2$:ZnO mass ratios of 87.6:12.4, 84.2:15.8 and 77.9:22.1. XRD results showed the presence of diffraction peaks corresponding to anatase, wurtzite and zinc titanate. Based on TEM images, the interplanar distance was measured at 3.56 Å, corresponding to dimension d of the (101) plane of the anatase crystal structure. XPS spectra showed the presence of Ti2p, Zn2p and O1s bands. The obtained materials were used in the process of photodegradation of rhodamine B (C.I. Basic Violet 10) and phenol. All of the oxide systems demonstrated higher photocatalytic activity than pure TiO$_2$. The highest photodegradation yield was obtained in the presence of the material synthesized in the molar ratio TiO$_2$:ZnO = 84.2:15.8. The researchers suggested that excessive ZnO may restrict access to recombination sites, thus reducing the photocatalytic performance.

Kanjwal et al. [236] compared TiO$_2$-ZnO binary oxide materials synthesized by the hydrothermal and electrospinning methods. In the electrospinning process, vinyl acetate was dissolved in DMF, and then titanium tetraisopropoxide, zinc oxide and acetic acid were added. The solution was transferred to a syringe, and a voltage of +12 kV was applied. The material was dried at 80 °C for 24 h and calcined at 600 °C for 1 h. Hydrothermal synthesis was carried out at 150 °C for 1 h, using nanofibers of TiO$_2$ and zinc nitrate. Diffractograms revealed the characteristic crystal structures of wurtzite (after electrospinning) and anatase (after hydrothermal synthesis). SEM and TEM images showed that the electrospun oxide system consisted of nanofibers with a smooth surface, while in the case of the material obtained hydrothermally, ZnO nanoparticles were observed on the TiO$_2$ surface. Physico-chemical analysis of the binary oxide materials showed marked differences in their morphological and crystal structure. Their photocatalytic properties were assessed in the degradation of rhodamine B (C.I. Basic Violet 10). Both the electrospun and hydrothermally synthesized materials demonstrated superior photocatalytic activity to pure TiO$_2$ or ZnO. The hydrothermally treated materials produced higher degradation yields (100% after 105 min of exposure to light) than the electrospun materials; this may be linked to differences in the crystal structures of the oxide systems.

Today, various chemical pollutants such as dyes are attracting increased attention owing to their significant effects on public health and the environment. The efficient degradation of this type of pollution from wastewater is becoming an urgent and challenging problem. Many centers are conducting research on the use of electrospinning for the synthesis of titania-based materials and their use in wastewater photodegradation [250–252]. Li et al. [237] developed a synthesis process based on electrospinning for a photocatalytic TiO$_2$-ZnO oxide system. The resulting system demonstrated very good photocatalytic properties in the degradation of rhodamine B (C.I. Basic Violet 10), producing a higher degradation yield than pure TiO$_2$ or ZnO. It also exhibited high photocatalytic stability over five consecutive cycles. More detailed physico-chemical analysis was performed using the XRD, SEM and TEM techniques. X-ray diffraction results demonstrated the presence of characteristic diffraction peaks corresponding to anatase and wurtzite. The SEM and TEM images showed smooth nanofibers consisting of single nanoparticles.

Araújo et al. [238] synthesized a TiO$_2$-ZnO oxide material using the methods of electrospinning and atomic layer deposition. In the first stage, a solution of TiO$_2$ (P25) and poly(methacrylic acid-co-methyl methacrylate) 1:1 block copolymer in ethanol was subjected to a constant voltage of +15 kV for 5 min. Next, zinc acetate was dissolved in water and mixed for 15 min at 85 °C, after which triethylamine was added and the mixture incubated for 3 h (solution 1). At the same time, zinc acetate was dissolved in deionized water with the addition of hexamethyltetramine (solution 2). The obtained TiO$_2$ nanofibers were immersed in solution 1, dried at 120 °C for 1 h, then added to solution 2 and kept at 85 °C for 24 h. The synthesized oxide material was calcined for one hour at 500 °C. Diffractograms revealed the presence of characteristic reflections for the crystalline structure of anatase, rutile and wurtzite. SEM images (Figure 13a–d) showed that the TiO$_2$ nanofibers were coated with ZnO nanorods. The EDX spectrum (Figure 13e–f) showed characteristic bands for Ti, O and Zn. The materials were tested as a sensor for the detection of moisture. Materials with one and three layers of ZnO produced a high response to changes in humidity. The three-layer material exhibited significantly greater sensitivity in the humidity range 40–100%. It also demonstrated significantly better parameters in moisture detection than the sensors described in a previous study [253]. The researchers suggest that the increase in ZnO content improves the porous structure, and thus increases the sensitivity [253].

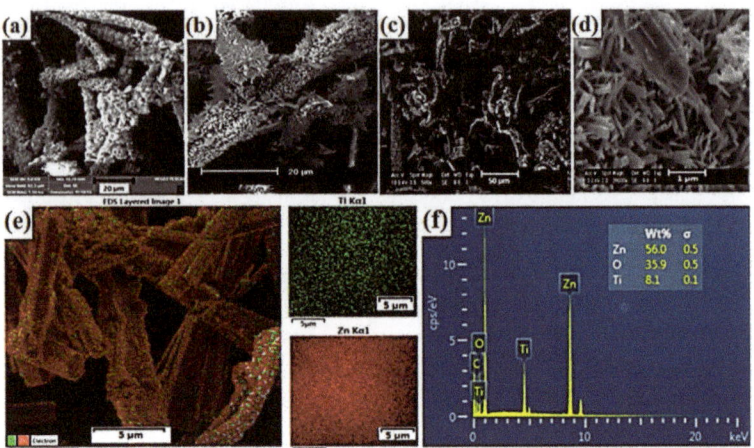

Figure 13. Images of TiO$_2$-ZnO oxide materials: (**a**,**b**) SEM for samples before calcination; (**c**,**d**) SEM for calcined materials; (**e**) EDS mapping and (**f**) EDX spectrum (created based on [238] with permission from Springer).

Table 8 contains a review of the literature concerning the synthesis of TiO$_2$-ZrO$_2$ oxide systems using the electrospinning method.

Humidity is a very common component in our environment that can affect not only human life, but also many industrial processes. Therefore, the measurement and control of humidity are of great importance. The design of a successful humidity sensor is a complex process, as the material used to build the sensor must offer properties including linear response, high sensitivity, fast response time, chemical and physical stability, wide operating range and low cost. Su et al. [254] developed a simple synthesis method involving electrospinning of a TiO$_2$-ZrO$_2$ oxide system with defined moisture detection properties. The impedance was shown to differ by four orders of magnitude, from 10^5 to 10^1 kΩ, in the humidity range 11–97%, which indicates correct operation of the sensor. Good repeatability was demonstrated for the obtained results, and the influence of temperature on the sensor was shown to be small. To determine the physico-chemical characteristics of the sensor material, analysis was performed using XRD, SEM and TEM. The XRD results indicated the crystal structure of anatase and tetragonal zirconium dioxide. Based on the SEM images it was concluded that the binary oxide material contained fibers originating from both ZrO$_2$ and TiO$_2$. The diameters of the nanofibers were found to be in the range 240–400 nm in the case of ZrO$_2$, and 60–200 nm in the case of TiO$_2$. Transmission electron microscopy results (Figure 14) showed the nanofibers to be built of numerous small particles.

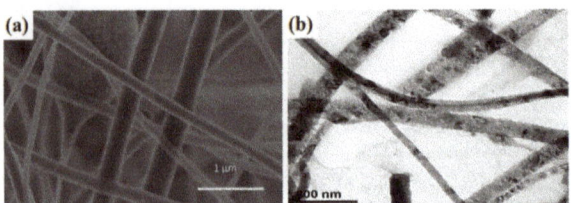

Figure 14. (**a**) SEM image and (**b**) TEM image of the TiO$_2$-ZrO$_2$ oxide system (created based on [254] with permission from Elsevier Publisher).

Table 8. Resources, synthesis conditions and physico-chemical properties of TiO_2-ZrO_2 oxide systems obtained by the electrospinning method.

Resources	Synthesis Conditions	Physico-Chemical Properties of the Final Product	References
$Ti(O_4H_9)_4$, $ZrOCl_2$, polyvinylpyrrolidone, acetic acid, ethanol, deionized water	solution of ZrO_2:$ZrOCl_2$ was dissolved in ethanol:water, poly(vinylpyrrolidone) was added, stirred for 12 h at room temperature TiO_2:$Ti(O_4H_9)_4$ solution was hydrolysed in ethanol:acetic acid, polyvinylpyrrolidone solution was added, stirred for 6 h at room temperature TiO_2-ZrO_2 electrospinning: voltage for the ZrO_2 stream: +5 kV, voltage for TiO_2: −7 kV, calcination: 600 °C for 3 h	crystal structure of anatase and tetragonal zirconia, presence of nanofibers with different diameters, (130 nm and 70 nm for ZrO_2 and TiO_2 respectively)	[254]
$Ti[OCH(CH_3)_2]_4$, $Zr[OCH(CH_3)_2]_4$, polyvinylpyrrolidone, phosphoric acid, acetic acid, ethanol, deionized water	TiO_2 and ZrO_2 precursor solution: polyvinylpyrrolidone dissolved in ethanol, mixed for 1 h, acetic acid, $Ti[CH(CH_3)_2]_4$, $Zr[OCH(CH_3)_2]_4$ (atomic ratio Ti:Zr = 1:1) TiO_2-ZrO_2 electrospinning: voltage 14 kV, spinning speed 30 µL/min, calcination: 500, 700, 1000 °C for 6 h	amorphous structure for material calcined at 500 °C, crystal structure of srilankite ($Ti_{0.5}Zr_{0.5}O_2$) for calcination temperature 700 and 1000 °C, BET surface area = 75 m^2/g for material calcined at 500 °C, average fibre diameter 497 nm (before calcination) and 344 nm (after calcination at 500 °C)	[255]

Recent years have seen a marked increase in interest in proton exchange membrane fuel cells (PEMFCs). The development of the ionomer membrane is crucial for the reliability of such cells and their high-volume commercialization. This type of membrane must attain a high proton conductivity and a low gas permeability, offering at the same time high mechanical, chemical and thermal stability, a low degree of deformation due to water absorption, and low costs of production so that it can be used globally [256–258]. Lee et al. [255] described the use of a TiO_2-ZrO_2 binary oxide system to prepare aquivion (KNF Neuberger GmbH) membrane for use in a single fuel cell. In the course of comprehensive physico-chemical analysis, diffractograms were obtained which indicated an amorphous structure for the material calcined at 500 °C, and the crystal structure of srilankite ($Ti_{0.5}Zr_{0.5}O_2$) for the oxide systems calcined at 700 and 1000 °C. In view of these results, further analysis was carried out only on the material with an amorphous structure, since that material had the highest specific surface area among the synthesized oxide systems. Based on SEM images, the average fiber diameter was measured at 497 nm before calcination and 344 nm after calcination. The synthesized TiO_2-ZrO_2 oxide system calcined at 500 °C was modified with 85% phosphoric acid to increase its proton conductivity. The XPS spectrum before modification showed two bands at 530.35 and 532.8 eV, corresponding to O–M bonds (M = Ti, Zr) and O–H bonds. On the spectrum of the modified material a new band appeared at 531.36 eV, originating from P–O bonds. The use of the oxide material to build the membrane eliminated the effect of cracking, and the membrane modified with phosphate groups showed improved proton conductivity compared with pure TiO_2-ZrO_2 and aquivion. The composite membrane demonstrated good durability, as confirmed by the results of accelerated lifetime (ALT) measurements.

The above review of the literature on the formation of TiO_2-ZnO and TiO_2-ZrO_2 oxide systems by the electrospinning method confirms that this is a new and promising method which leads to the effective synthesis of materials in the form of nanofibers with defined chemical composition, porosity, geometry and fiber dimensions. Moreover, the method does not place excessive limitations on researchers. The sole problem is the selection of optimum process conditions, including the type and concentration of solvent, temperature, solution viscosity, average molecular weight of polymer, applied voltage, and flow rate. It has been shown that electrospun TiO_2-ZnO and TiO_2-ZrO_2 fibers may be used, among other things, as sensors for the detection of various types of gas, as photocatalysts in the degradation of certain types of organic pollutants, and as membranes for the production of fuel cells.

3. Conclusions

The literature review that has been presented here reflects the search for innovative combinations of TiO_2 with other materials, which may cause morphological changes in the crystalline phases of titanium dioxide and changes in its electron structure, leading to—among other things—improvement in its photocatalytic activity and spectral sensitivity. Studies being carried out at multiple research centers worldwide are largely focused on the choice of appropriate methods for the synthesis or modification of titanium dioxide, and on the selection of suitable components to be combined with TiO_2 to produce advanced multifunctional hybrid materials.

It is widely known that the physico-chemical as well as the morphological-structural properties of materials based on titanium dioxide, such as particle shape and size, degree of crystallinity, surface composition, pore size distribution, specific surface area, etc., are strongly dependent on the suitable specification of synthesis conditions, such as the nature and composition of the precursors, solvent and complexing/templating agent, and the conditions of hydrolysis and calcination. Suitably chosen process conditions for the synthesis of oxide and hybrid materials based on TiO_2 make it possible to design their physico-chemical properties.

Issues relating to the creation of a new generation of modified forms of TiO_2 are of huge importance, particularly in view of the growing need for active photocatalysts functioning in visible light, as well as high-performance electrode materials.

Author Contributions: A.K.—described a review on obtaining hybrid systems using the hydrothermal method and electrospinning; K.S.-C.—prepared a review on the synthesis of oxide systems using the sol-gel method; T.J.—performed critical revision and supervised all research aspects.

Funding: Presented research was financed within Poznan University of Technology research grant No. 03/32/DSPB/0806/2018.

Conflicts of Interest: The authors declare no conflicts of interest.

References

1. Stoyanova, A.; Hitkova, H.; Bachvarova-Nedelcheva, A.; Iordanova, R.; Ivanova, N.; Sredkova, M. Synthesis and antibacterial activity of TiO_2/ZnO nanocomposites prepared via nonhydrolytic route. *J. Chem. Technol. Metall.* **2013**, *48*, 154–161.
2. Bach, U.; Corr, D.; Lupo, D.; Pichot, F.; Ryan, M. Nanomaterials-based electrochromics for paper-quality displays. *Adv. Mater.* **2002**, *14*, 845–848. [CrossRef]
3. Gouma, P.I.; Mills, M.J.; Sandhage, K.H. Fabrication of free-standing titania-based gas sensors by the oxidation of metallic titanium foils. *J. Am. Ceram. Soc.* **2000**, *83*, 1007–1009. [CrossRef]
4. Oey, C.C.; Djurišić, A.B.; Wang, H.; Man, K.K.Y.; Chan, W.K.; Xie, M.H.; Leung, Y.H.; Pandey, A.; Nunzi, J.-M.; Chui, P.C. Polymer-TiO_2 solar cells: TiO_2 interconnected network for improved cell performance. *Nanotechnology* **2006**, *17*, 706–713. [CrossRef]
5. Polleux, J.; Gurlo, A.; Barsan, N.; Weimar, U.; Antonietti, M.; Niederberger, M. Template-free synthesis and assembly of single-crystalline tungsten oxide nanowires and their gas-sensing properties. *Angew. Chem. Int. Ed.* **2006**, *45*, 261–265. [CrossRef] [PubMed]
6. McCullagh, C.; Robertson, J.M.C.; Bahnemann, D.W.; Robertson, P.K.J. The application of TiO_2 photocatalysis for disinfection of water contaminated with pathogenic micro-organisms: A review. *Res. Chem. Intermed.* **2007**, *33*, 359–375. [CrossRef]
7. Sobczyński, A.; Dobosz, A. Water purification by photocatalysis on semiconductors. *Pol. J. Environ. Stud.* **2001**, *10*, 195–205.
8. Xu, X.; Wang, J.; Tian, J.; Wang, X.; Dai, J.; Liu, X. Hydrothermal and post-heat treatments of TiO_2/ZnO composite powder and its photodegradation behavior on methyl orange. *Ceram. Int.* **2011**, *37*, 2201–2206. [CrossRef]
9. Liao, D.L.; Badour, C.A.; Liao, B.Q. Preparation of nanosized TiO_2/ZnO composite catalyst and its photocatalytic activity for degradation of methyl orange. *J. Photochem. Photobiol. A* **2008**, *194*, 11–19. [CrossRef]
10. Siwinska-Stefanska, K.; Kubiak, A.; Kurc, B.; Moszynski, D.; Goscianska, J.; Jesionowski, T. An active anode material based on titania and zinc oxide hybrids fabricated via hydrothermal route: Comperehensive physicochemical and electrochemical evaluations. *J. Electrochem. Soc.* **2018**, *165*, A3056–A3066. [CrossRef]
11. Siwinska-Stefanska, K.; Paukszta, D.; Piasecki, A.; Jesionowski, T. Synthesis and physicochemical characteristics of titanium dioxide doped with selected metals. *Physicochem. Probl. Miner. Process.* **2014**, *50*, 265–276. [CrossRef]
12. Raghupathi, K.R.; Koodali, R.T.; Manna, A.C. Size-dependent bacterial growth inhibition and mechanism of antibacterial activity of zinc oxide nanoparticles. *Langmuir* **2011**, *27*, 4020–4028. [CrossRef] [PubMed]
13. Xia, Y.; Wang, J.; Chen, R.; Zhou, D.; Xiang, L. A review on the fabrication of hierarchical ZnO nanostructures for photocatalysis application. *Crystals* **2016**, *6*, 148. [CrossRef]
14. Wang, J.; Gao, L. Hydrothermal synthesis and photoluminescence properties of ZnO nanowires. *Solid State Commun.* **2004**, *132*, 269–271. [CrossRef]
15. Zhang, J.; Wang, S.; Xu, M.; Wang, Y.; Zhu, B.; Zhang, S.; Huang, W.; Wu, S. Hierarchically porous ZnO architectures for gas sensor application. *Cryst. Growth Des.* **2009**, *9*, 3532–3537. [CrossRef]
16. Wang, L.; Kang, Y.; Liu, X.; Zhang, S.; Huang, W.; Wang, S. ZnO nanorod gas sensor for ethanol detection. *Sens. Actuators B Chem.* **2012**, *162*, 237–243. [CrossRef]
17. Kolodziejczak-Radzimska, A.; Jesionowski, T. Zinc oxide-from synthesis to application: A review. *Materials* **2014**, *7*, 2833–2881. [CrossRef] [PubMed]
18. Takenaka, S.; Shimizu, T.; Otsuka, K. Complete removal of carbon monoxide in hydrogen-rich gas stream through methanation over supported metal catalysts. *Int. J. Hydrogen Energy* **2004**, *29*, 1065–1073. [CrossRef]

19. Yang, Y.; Yang, H.; Yang, M.; Ilu, Y.; Shen, G.; Yu, R. Amperometric glucose biosensor based on a surface treated nanoporous ZrO_2/Chitosan composite film as immobilization matrix. *Anal. Chim. Acta* **2004**, *525*, 213–220. [CrossRef]
20. Liu, Q.; Long, S.; Lv, H.; Wang, W.; Niu, J.; Huo, Z.; Chen, J.; Liu, M. Controllable growth of nanoscale conductive filaments in solid-electrolyte-based ReRAM by using a metal nanocrystal covered bottom electrode. *ACS Nano* **2010**, *4*, 6162–6188. [CrossRef] [PubMed]
21. Apostolescu, N.; Geiger, B.; Hizbullah, K.; Jan, M.T.; Kureti, S.; Reichert, D.; Schott, F.; Weisweiler, W. Selective catalytic reduction of nitrogen oxides by ammonia on iron oxide catalysts. *Appl. Catal. B* **2006**, *62*, 104–114. [CrossRef]
22. Wysokowski, M.; Szalaty, T.J.; Jesionowski, T.; Motylenko, M.; Rafaja, D.; Koltsov, I.; Stöcker, H.; Bazhenov, V.V.; Ehrlich, H.; Stelling, A.L.; et al. Extreme biomimetic approach for synthesis of nanocrystalline chitin-$(Ti,Zr)O_2$ multiphase composites. *Mater. Chem. Phys.* **2017**, *188*, 115–124. [CrossRef]
23. Emeline, A.; Kataeva, G.V.; Litke, A.S.; Rudakova, A.V.; Ryabchuk, V.K.; Serpone, N. Spectroscopic and photoluminescence studies of a wide band gap insulating material: Powdered and colloidal ZrO_2 sols. *Laungmir* **1998**, *7463*, 5011–5022. [CrossRef]
24. Siwińska-Stefańska, K.; Kurc, B. A composite TiO_2-SiO_2-ZrO_2 oxide system as a high-performance anode material for lithium-ion batteries. *J. Electrochem. Soc.* **2017**, *164*, A728–A734. [CrossRef]
25. Jangra, S.L.; Stalin, K.; Dilbaghi, N.; Kumar, S.; Tawale, J.; Singh, S.P.; Pasricha, R. Antimicrobial activity of zirconia (ZrO_2) nanoparticles and zirconium complexes. *J. Nanosci. Nanotechnol.* **2012**, *12*, 7105–7112. [CrossRef] [PubMed]
26. Qu, X.; Xie, D.; Cao, L.; Du, F. Synthesis and characterization of TiO_2/ZrO_2 coaxial core-shell composite nanotubes for photocatalytic applications. *Ceram. Int.* **2014**, *40*, 12647–12653. [CrossRef]
27. Kokporka, L.; Onsuratoom, S.; Puangpetch, T.; Chavadej, S. Sol-gel-synthesized mesoporous-assembled TiO_2-ZrO_2 mixed oxide nanocrystals and their photocatalytic sensitized H_2 production activity under visible light irradiation. *Mater. Sci. Semicond. Process.* **2013**, *16*, 667–678. [CrossRef]
28. Zhou, W.; Liu, K.; Fu, H.; Pan, K.; Zhang, L.; Wang, L.; Sun, C. Multi-modal mesoporous TiO_2-ZrO_2 composites with high photocatalytic activity and hydrophilicity. *Nanotechnology* **2008**, *19*, 035610. [CrossRef] [PubMed]
29. Fan, M.; Hu, S.; Ren, B.; Wang, J.; Jing, X. Synthesis of nanocomposite TiO_2/ZrO_2 prepared by different templates and photocatalytic properties for the photodegradation of Rhodamine B. *Powder Technol.* **2013**, *235*, 27–32. [CrossRef]
30. Liu, H.; Su, Y.; Hu, H.; Cao, W.; Chen, Z. An ionic liquid route to prepare mesoporous ZrO_2-TiO_2 nanocomposites and study on their photocatalytic activities. *Adv. Powder Technol.* **2013**, *24*, 683–688. [CrossRef]
31. Siwińska-Stefańska, K.; Kubiak, A.; Piasecki, A.; Goscianska, J.; Nowaczyk, G.; Jurga, S.; Jesionowski, T. TiO_2-ZnO binary oxide systems: Comprehensive characterization and tests of photocatalytic activity. *Materials* **2018**, *11*, 841. [CrossRef]
32. Tian, J.; Chen, L.; Dai, J.; Wang, X.; Yin, Y.; Wu, P. Preparation and characterization of TiO_2, ZnO, and TiO_2/ZnO nanofilms via sol-gel process. *Ceram. Int.* **2009**, *35*, 2261–2270. [CrossRef]
33. Wang, L.; Wang, C.; Liu, W.; Chen, Q.; He, M. Visible-light-induced aerobic thiocyanation of indoles using reusable TiO_2/MoS_2 nanocomposite photocatalyst. *Tetrahedron Lett.* **2016**, *57*, 1771–1774. [CrossRef]
34. Zhu, X.; Yang, C.; Xiao, F.; Wang, J.; Su, X. Synthesis of nano-TiO_2-decorated MoS_2 nanosheets for lithium ion batteries. *New J. Chem.* **2015**, *39*, 683–688. [CrossRef]
35. Roy, R. Accelerating the kinetics of low-temperature inorganic syntheses. *J. Solid State Chem.* **1994**, *111*, 11–17. [CrossRef]
36. Rabenau, A. The role of hydrothermal synthesis in preparative chemistry. *Angew. Chem. Int.* **1985**, *24*, 1026–1040. [CrossRef]
37. Wang, X.; Li, Y. Selected-control hydrothermal synthesis of α- and β-MnO_2 single crystal nanowires. *J. Am. Chem. Soc.* **2002**, *124*, 2880–2881. [CrossRef] [PubMed]
38. Wang, N.; Cai, Y.; Zhang, R.Q. Growth of nanowires. *Mater. Sci. Eng. R Rep.* **2008**, *60*, 1–51. [CrossRef]
39. Yu, J.; Wang, G.; Cheng, B.; Zhou, M. Effects of hydrothermal temperature and time on the photocatalytic activity and microstructures of bimodal mesoporous TiO_2 powders. *Appl. Catal. B Environ.* **2007**, *69*, 171–180. [CrossRef]

40. Jung, J.; Perrut, M. Particle design using supercritical fluids: Literature and patent survey. *J. Supercrit. Fluids* **2001**, *20*, 179–219. [CrossRef]
41. Li, W.J.; Shi, E.W.; Zhong, W.Z.; Yin, Z.W. Growth mechanism and growth habit of oxide crystals. *J. Cryst. Growth* **1999**, *203*, 186–196. [CrossRef]
42. Bavykin, D.V.; Parmon, V.N.; Lapkin, A.A.; Walsh, F.C. The effect of hydrothermal conditions on the mesoporous structure of TiO_2 nanotubes. *J. Mater. Chem.* **2004**, *14*, 3370–3377. [CrossRef]
43. Byrappa, K.; Yoshimura, M. Hydrothermal technology—Principles and applications. In *Handbook of Hydrothermal Technology*, 2nd ed.; William Andrew Publishing LLC.: Norwich, NY, USA, 2013; pp. 1–52.
44. Byrappa, K.; Yoshimura, M. Physical chemistry of hydrothermal growth of crystals. In *Handbook of Hydrothermal Technology*, 2nd ed.; William Andrew Publishing LLC.: Norwich, NY, USA, 2013; pp. 139–175.
45. Shen, J.; Yan, B.; Shi, M.; Ma, H.; Li, N.; Ye, M. One step hydrothermal synthesis of TiO_2-reduced graphene oxide sheets. *J. Mater. Chem.* **2011**, *21*, 3415–3421. [CrossRef]
46. Riman, R.E.; Suchanek, W.L.; Lencka, M.M. Hydrothermal crystallization of ceramics. *Ann. Chim. Sci. Mater.* **2002**, *27*, 15–36. [CrossRef]
47. Zhang, Y.X.; Li, G.H.; Jin, Y.X.; Zhang, Y.; Zhang, J.; Zhang, L.D. Hydrothermal synthesis and photoluminescence of TiO_2 nanowires. *Chem. Phys. Lett.* **2002**, *365*, 300–304. [CrossRef]
48. Yuan, Z.Y.; Su, B.L. Titanium oxide nanotubes, nanofibers and nanowires. *Colloids Surf. A Physicochem. Eng. Asp.* **2004**, *241*, 173–183. [CrossRef]
49. Shandilya, M.; Rai, R.; Singh, J. Review: Hydrothermal technology for smart materials. *Adv. Appl. Ceram.* **2016**, *115*, 354–376. [CrossRef]
50. Zhang, L.; Jeem, M.; Okamoto, K.; Watanabe, S. Photochemistry and the role of light during the submerged photosynthesis of zinc oxide nanorods. *Sci. Rep.* **2018**, *8*, 177. [CrossRef] [PubMed]
51. Ding, S.; Luan, D.; Boey, F.Y.C.; Chen, J.S.; Lou, X.W. SnO_2 nanosheets grown on graphene sheets with enhanced lithium storage properties. *Chem. Commun.* **2011**, *47*, 7155–7157. [CrossRef] [PubMed]
52. Komarneni, S.; Roy, R.; Li, Q.H. Microwave-hydrothermal synthesis of ceramic powders. *Mater. Res. Bull.* **1992**, *27*, 1393–1405. [CrossRef]
53. Tompsett, G.A.; Conner, W.C.; Yngvesson, K.S. Microwave synthesis of nanoporous materials. *Chem. Phys. Chem.* **2006**, *7*, 296–319. [CrossRef] [PubMed]
54. Nakata, K.; Fujishima, A. TiO_2 photocatalysis: Design and applications. *J. Photochem. Photobiol.* **2012**, *13*, 169–189. [CrossRef]
55. Pelaez, M.; Nolan, N.T.; Pillai, S.C.; Seery, M.K.; Falaras, P.; Kontos, A.G.; Dunlop, P.S.M.; Hamilton, J.W.J.; Byrne, J.A.; O'Shea, K.; et al. A review on the visible light active titanium dioxide photocatalysts for environmental applications. *Appl. Catal. B Environ.* **2012**, *125*, 331–349. [CrossRef]
56. Cheng, P.; Wang, Y.; Xu, L.; Sun, P.; Su, Z.; Jin, F.; Liu, F.; Sun, Y.; Lu, G. 3D TiO_2/ZnO composite nanospheres as an excellent electron transport anode for efficient dye-sensitized solar cells. *RSC Adv.* **2016**, *6*, 51320–51326. [CrossRef]
57. Cheng, P.; Du, S.; Cai, Y.; Liu, F.; Sun, P.; Zheng, J.; Lu, G. Tripartite layered photoanode from hierarchical anatase TiO_2 urchin-like spheres and P25: A candidate for enhanced efficiency dye sensitized solar cells. *J. Phys. Chem. C* **2013**, *117*, 24150–24156. [CrossRef]
58. Li, Y. Synthesis and characterization of TiO_2 doped ZnO microtubes. *Chin. J. Chem. Phys.* **2010**, *23*, 358–362. [CrossRef]
59. Vlazan, P.; Ursu, D.H.; Irina-Moisescu, C.; Miron, I.; Sfirloaga, P.; Rusu, E. Structural and electrical properties of TiO_2/ZnO core–shell nanoparticles synthesized by hydrothermal method. *Mater. Charact.* **2015**, *101*, 153–158. [CrossRef]
60. Zhang, M.; An, T.; Liu, X.; Hu, X.; Sheng, G.; Fu, J. Preparation of a high-activity ZnO/TiO_2 photocatalyst via homogeneous hydrolysis method with low temperature crystallization. *Mater. Lett.* **2010**, *64*, 1883–1886. [CrossRef]
61. Wang, Y.; Zhu, S.; Chen, X.; Tang, Y.; Jiang, Y.; Peng, Z.; Wang, H. One-step template-free fabrication of mesoporous ZnO/TiO_2 hollow microspheres with enhanced photocatalytic activity. *Appl. Surf. Sci.* **2014**, *307*, 263–271. [CrossRef]

62. Rusu, E.; Ursaki, V.; Gutul, T.; Vlazan, P.; Siminel, A. Characterization of TiO$_2$ nanoparticles and ZnO/TiO$_2$ composite obtained by hydrothermal method. In *3rd International Conference on Nanotechnologies and Biomedical Engineering*; Sontea, V., Tiginyanu, I., Eds.; Springer: Singapore, 2016; Volume 55, pp. 93–96. [CrossRef]
63. Chen, D.; Zhang, H.; Hu, S.; Li, J. Preparation and enhanced photoelectrochemical performance of coupled bicomponent. *J. Phys. Chem. C* **2008**, *112*, 117–122. [CrossRef]
64. Wang, C.; Hwang, W.; Chang, K.; Ko, H.; Hsi, C.; Huang, H.; Wang, M. Formation and morphology of Zn$_2$Ti$_3$O$_8$ powders using hydrothermal process without dispersant agent or mineralizer. *Int. J. Mol. Sci.* **2011**, *12*, 935–945. [CrossRef] [PubMed]
65. Ashok, C.; Venkateswara Rao, K. ZnO/TiO$_2$ nanocomposite rods synthesized by microwave-assisted method for humidity sensor application. *Superlattices Microstruct.* **2014**, *76*, 46–54. [CrossRef]
66. Divya, K.S.; Marilyn, M.X.; Vandana, P.V.; Rethu, V.N.; Suresh, M. A quaternary TiO$_2$/ZnO/RGO/Ag nanocomposite with enhanced visible light photocatalytic performance. *New J. Chem.* **2017**, *41*, 6445–6454. [CrossRef]
67. Cho, I.S.; Chen, Z.; Forman, A.J.; Kim, D.R.; Rao, P.M.; Jaramillo, T.F.; Zheng, X. Branched TiO$_2$ nanorods for photoelectrochemical hydrogen production. *Nano Lett.* **2011**, *11*, 4978–4984. [CrossRef] [PubMed]
68. Macák, J.M.; Tsuchiya, H.; Schmuki, P. High-aspect-ratio TiO$_2$ nanotubes by anodization of titanium. *Angew. Chem.* **2005**, *44*, 2100–2102. [CrossRef] [PubMed]
69. Hu, Y.S.; Kienle, L.; Guo, Y.G.; Maier, J. High lithium electroactivity of nanometer-sized rutile TiO$_2$. *Adv. Mater.* **2006**, *18*, 1421–1426. [CrossRef]
70. Yang, P.; Xiao, X.; Li, Y.; Ding, Y.; Qiang, P.; Tan, X.; Mai, W.; Lin, Z.; Wu, W.; Li, T.; et al. Hydrogenated ZnO core-shell nanocables for flexible supercapacitors and self-powered systems. *ACS Nano* **2013**, *7*, 2617–2626. [CrossRef] [PubMed]
71. Keis, K.; Magnusson, E.; Lindstr, H.; Lindquist, S.; Hagfeldt, A.; Lindström, H.; Lindquist, S.; Hagfeldt, A. A 5% efficient photoelectrochemical solar cell based on nanostructured ZnO electrodes. *Sol. Energy Mater. Sol. Cells* **2002**, *73*, 51–58. [CrossRef]
72. Zhang, L.; Jiang, Y.; Ding, Y.; Povey, M.; York, D. Investigation into the antibacterial behaviour of suspensions of ZnO nanoparticles (ZnO nanofluids). *J. Nanopart. Res.* **2007**, *9*, 479–489. [CrossRef]
73. Wang, C.; Ao, Y.; Wang, P.; Hou, J.; Qian, J. Preparation, characterization and photocatalytic activity of the neodymium-doped TiO$_2$ hollow spheres. *Appl. Surf. Sci.* **2010**, *257*, 227–231. [CrossRef]
74. Chen, J.; Nie, X.; Shi, H.; Li, G.; An, T. Synthesis of TiO$_2$ hollow sphere multimer photocatalyst by etching titanium plate and its application to the photocatalytic decomposition of gaseous styrene. *Chem. Eng. J.* **2013**, *228*, 834–842. [CrossRef]
75. Sui, Y.; Yang, H.; Fu, W.; Xu, J.; Chang, L.; Zhu, H.; Yu, Q.; Li, M.; Zou, G. Preparation and characterization of hollow glass microspheres/ZnO composites. *J. Alloys Compd.* **2009**, *469*, 2–6. [CrossRef]
76. Das, S.; Chatterjee, S.; Pramanik, S.; Devi, P.S.; Kumar, G.S. A new insight into the interaction of ZnO with calf thymus DNA through surface defects. *J. Photochem. Photobiol. B Biol.* **2018**, *178*, 339–347. [CrossRef] [PubMed]
77. Ghosh Chaudhuri, R.; Paria, S. Core/shell nanoparticles: Classes, properties, synthesis mechanisms, characterization, and applications. *Chem. Rev.* **2012**, *112*, 2373–2433. [CrossRef] [PubMed]
78. Rafatullah, M.; Sulaiman, O.; Hashim, R.; Ahmad, A. Adsorption of methylene blue on low-cost adsorbents: A review. *J. Hazard. Mater.* **2010**, *177*, 70–80. [CrossRef] [PubMed]
79. Garg, V.K.; Amita, M.; Kumar, R.; Gupta, R. Basic dye (methylene blue) removal from simulated wastewater by adsorption using Indian rosewood sawdust: A timber industry waste. *Dyes Pigments* **2004**, *63*, 243–250. [CrossRef]
80. Murugan, A.V.; Samuel, V.; Ravi, V. Synthesis of nanocrystalline anatase TiO$_2$ by microwave hydrothermal method. *Mater. Lett.* **2006**, *60*, 479–480. [CrossRef]
81. Cabello, G.; Davoglio, R.A.; Pereira, E.C. Microwave-assisted synthesis of anatase-TiO$_2$ nanoparticles with catalytic activity in oxygen reduction. *J. Electroanal. Chem.* **2017**, *794*, 36–42. [CrossRef]
82. Leahy, J.J. ZrO$_2$-modified TiO$_2$ nanorod composite: Hydrothermal synthesis, characterization and application in esterification of organic acid. *Mater. Chem. Phys.* **2014**, *145*, 82–89. [CrossRef]
83. Tomar, L.J.; Chakrabarty, B.S. Synthesis, structural and optical properties of TiO$_2$-ZrO$_2$ nanocomposite by hydrothermal method. *Adv. Mater. Lett.* **2013**, *4*, 64–67. [CrossRef]

84. Kubo, K.; Hosokawa, S.; Furukawa, S.; Inoue, M. Synthesis of ZrO_2-TiO_2 solid solutions by various synthetic methods in the region of high zirconium contents. *J. Mater. Sci.* **2008**, 2198–2205. [CrossRef]
85. Hirano, M.; Nakahara, C.; Ota, K.; Tanaike, O.; Inagaki, M. Photoactivity and phase stability of ZrO_2-doped anatase-type TiO_2 directly formed as nanometer-sized particles by hydrolysis under hydrothermal conditions. *J. Solid State Chem.* **2003**, *170*, 39–47. [CrossRef]
86. Tomar, L.J.; Bhatt, P.J.; Desai, R.K.; Chakrabarty, B.S.; Panchal, C.J. Improved conversion efficiency of dye sensitized solar cell using Zn doped TiO_2-ZrO_2 nanocomposite. *AIP Conf. Proc.* **2016**, *1731*, 50132. [CrossRef]
87. Yao, B.; Han, X.; Ying, L.; Peng, C.; Zhang, C. Hydrothermal synthesis and photocatalytic activity of TiO_2-ZrO_2 hybrid composite microspheres. *Mater. Sci. Forum* **2016**, *852*, 257–263. [CrossRef]
88. Caillot, T.; Salama, Z.; Chanut, N.; Cadete Santos Aires, F.J.; Bennici, S.; Auroux, A. Hydrothermal synthesis and characterization of zirconia based catalysts. *J. Solid State Chem.* **2013**, *203*, 79–85. [CrossRef]
89. Ueda, M.; Sasaki, Y.; Ikeda, M.; Ogawa, M.; Fujitani, W.; Nakano, T. Chemical-hydrothermal synthesis of bioinert ZrO_2-TiO_2 films on pure Ti substrates and proliferation of osteoblast-like cells. *Mater. Trans.* **2009**, *50*, 2147–2153. [CrossRef]
90. Kim, D.; Kim, J.; Kim, K.; Cho, S.; Seo, M.; Kim, M.; Lee, J.; Kim, J.; Kim, K.; Cho, S.; et al. Preparations of titanium composite electrodes from commercial inorganic pigment and its application to light scattering layers on dye-sensitized solar cells. *Mol. Cryst. Liq. Cryst.* **2011**, *539*, 156–165. [CrossRef]
91. Kumar, S.G.; Devi, L.G. Review on modified TiO_2 photocatalysis under UV/visible light: Selected results and related mechanisms on interfacial charge carrier transfer dynamics. *J. Phys. Chem. A* **2011**, *115*, 13211–13241. [CrossRef] [PubMed]
92. Kristianto, Y.; Taufik, A.; Saleh, R. Preparation and catalytic performance of ZrO_2- nanographene platelets composites. *J. Phys. Conf. Ser.* **2016**, *776*, 012040. [CrossRef]
93. Moazami, A.; Montazer, M. A novel multifunctional cotton fabric using ZrO_2 NPs/urea/CTAB/MA/SHP· Introducing flame retardant, photoactive and antibacterial properties. *J. Text. Inst.* **2016**, *107*, 1253–1263. [CrossRef]
94. Sudrajat, H.; Babel, S.; Sakai, H.; Takizawa, S. Rapid enhanced photocatalytic degradation of dyes using novel N-doped ZrO_2. *J. Environ. Manag.* **2016**, *165*, 224–234. [CrossRef] [PubMed]
95. Shakeel Ahmad, M.; Pandey, A.K.; Abd Rahim, N. Advancements in the development of TiO_2 photoanodes and its fabrication methods for dye sensitized solar cell (DSSC) applications. A review. *Renew. Sustain. Energy Rev.* **2017**, *77*, 89–108. [CrossRef]
96. Ren, X.; Qi, X.; Shen, Y.; Xiao, S.; Xu, G.; Zhang, Z.; Huang, Z.; Zhong, J. 2D co-catalytic MoS_2 nanosheets embedded with 1D TiO_2 nanoparticles for enhancing photocatalytic activity. *J. Phys. D Appl. Phys.* **2016**, *49*, 315304. [CrossRef]
97. Zhang, J.; Huang, L.; Lu, Z.; Jin, Z.; Wang, X.; Xu, G.; Zhang, E.; Wang, H.; Kong, Z.; Xi, J.; et al. Crystal face regulating MoS_2/TiO_2 (001) heterostructure for high photocatalytic activity. *J. Alloys Compd.* **2016**, *688*, 840–848. [CrossRef]
98. Liu, H.; Lv, T.; Zhu, C.; Su, X.; Zhu, Z. Efficient synthesis of MoS_2 nanoparticles modified TiO_2 nanobelts with enhanced visible-light-driven photocatalytic activity. *J. Mol. Catal. A Chem.* **2015**, *396*, 136–142. [CrossRef]
99. Zhang, W.; Xiao, X.; Zheng, L.; Wan, C. Fabrication of TiO_2/MoS_2 composite photocatalyst and its photocatalytic mechanism for degradation of methyl orange under visible light. *Can. J. Chem. Eng.* **2015**, *93*, 1594–1602. [CrossRef]
100. Yang, L.; Zheng, X.; Liu, M.; Luo, S.; Luo, Y.; Li, G. Fast photoelectro-reduction of Cr(VI) over MoS_2/TiO_2 nanotubes on Ti wire. *J. Hazard. Mater.* **2017**, *329*, 230–240. [CrossRef] [PubMed]
101. Wang, J.; Wei, B.; Xu, L.; Gao, H.; Sun, W.; Che, J. Multilayered MoS_2 coated TiO_2 hollow spheres for efficient photodegradation of phenol under visible light irradiation. *Mater. Lett.* **2016**, *179*, 42–46. [CrossRef]
102. Yang, H.G.; Zeng, H.C. Preparation of hollow anatase TiO_2 nanospheres via Ostwald ripening. *J. Phys. Chem. B* **2004**, *108*, 3492–3495. [CrossRef]
103. Yan, C.; Rosei, F. Hollow micro/nanostructured materials prepared by ion exchange synthesis and their potential applications. *New. J. Chem.* **2014**, *5*, 1883–1904. [CrossRef]
104. Bai, S.; Wang, L.; Chen, X.; Du, J.; Xiong, Y. Chemically exfoliated metallic MoS_2 nanosheets: A promising supporting co-catalyst for enhancing the photocatalytic performance of TiO_2 nanocrystals. *Nano Res.* **2015**, *8*, 175–183. [CrossRef]

105. Yuan, Y.; Ye, Z.; Lu, H.; Hu, B.; Li, Y.; Chen, D.; Zhong, J.; Yu, Z.; Zou, Z. Constructing anatase TiO_2 nanosheets with exposed (001) facets/layered MoS_2 two-dimensional nanojunctions for enhanced solar hydrogen generation. *ACS Catal.* **2016**, *6*, 532–541. [CrossRef]
106. Yasumitsu, T.; Sundo, K.; Moon Woo, C.; Masazumi, I. Tiocyanation of indoles. *J. Heterocycl. Chem.* **1978**, *3*, 425–427. [CrossRef]
107. Mohammadi Ziarani, G.; Moradi, R.; Ahmadi, T.; Lashgari, N. Recent advances in the application of indoles in multicomponent reactions. *RSC Adv.* **2018**, *8*, 12069–12103. [CrossRef]
108. Shi, X.; Fujitsuka, M.; Majima, T. Electron transfer dynamics of quaternary sulfur semiconductor/MoS_2 layer-on-layer for efficient visible-light H_2 evolution. *Appl. Catal. B* **2018**, *235*, 9–16. [CrossRef]
109. Li, Z.; Meng, X.; Zhang, Z. Recent development on MoS_2-based photocatalysis: A review. *J. Photochem. Photobiol. C Photochem. Rev.* **2018**, *35*, 39–55. [CrossRef]
110. Kang, X.; Song, X.Z.; Han, Y.; Cao, J.; Tan, Z. Defect-engineered TiO_2 hollow spiny nanocubes for phenol degradation under visible light irradiation. *Sci. Rep.* **2018**, *8*, 1–10. [CrossRef] [PubMed]
111. Kermani, M.; Kakavandi, B.; Farzadkia, M.; Esrafili, A.; Jokandan, S.F.; Shahsavani, A. Catalytic ozonation of high concentrations of catechol over $TiO_2@Fe_3O_4$ magnetic core-shell nanocatalyst: Optimization, toxicity and degradation pathway studies. *J. Clean. Prod.* **2018**, *192*, 597–607. [CrossRef]
112. Norman, M.; Żółtowska-Aksamitowska, S.; Zgoła-Grześkowiak, A.; Ehrlich, H.; Jesionowski, T. Iron(III) phthalocyanine supported on a spongin scaffold as an advanced photocatalyst in a highly efficient removal process of halophenols and bisphenol A. *J. Hazard. Mater.* **2018**, *347*, 78–88. [CrossRef] [PubMed]
113. Song, S.; Wang, J.; Peng, T.; Fu, W.; Zan, L. MoS_2-MoO_{3-x} hybrid cocatalyst for effectively enhanced H_2 production photoactivity of $AgIn_5S_8$ nano-octahedrons. *Appl. Catal. B* **2018**, *228*, 39–46. [CrossRef]
114. Rozenfeld, S.; Teller, H.; Schechter, M.; Farber, R.; Krichevski, O.; Schechter, A.; Cahan, R. Exfoliated molybdenum di-sulfide (MoS_2) electrode for hydrogen production in microbial electrolysis cell. *Bioelectrochemistry* **2018**, *123*, 201–210. [CrossRef] [PubMed]
115. Chen, A.; Holt-hindle, P. Platinum-based nanostructured materials: Synthesis, properties, and applications. *Chem. Rev.* **2010**, *110*, 3767–3804. [CrossRef] [PubMed]
116. Zhang, X.Y.; Li, H.P.; Cui, X.L.; Lin, Y. Graphene/TiO_2 nanocomposites: Synthesis, characterization and application in hydrogen evolution from water photocatalytic splitting. *J. Mater. Chem.* **2010**, *20*, 2801–2806. [CrossRef]
117. Klapiszewski, Ł.; Królak, M.; Jesionowski, T. Silica synthesis by the sol-gel method and its use in the preparation of multifunctional biocomposites. *Cent. Eur. J. Chem.* **2014**, *12*, 173–184. [CrossRef]
118. Ciesielczyk, F.; Przybysz, M.; Zdarta, J.; Piasecki, A.; Paukszta, D.; Jesionowski, T. The sol-gel approach as a method of synthesis of $xMgO \cdot ySiO_2$ powder with defined physicochemical properties including crystalline structure. *J. Sol-Gel Sci. Technol.* **2014**, *71*, 501–513. [CrossRef]
119. Baccile, N.; Fischer, A.; Julián-López, B.; Grosso, D.; Sanchez, C. Core-shell effects of functionalized oxide nanoparticles inside long-range meso-ordered spray-dried silica spheres. *J. Sol-Gel Sci. Technol.* **2008**, *47*, 119–123. [CrossRef]
120. Faustini, M.; Grosso, D.; Boissière, C.; Backov, R.; Sanchez, C. "Integrative sol-gel chemistry": A nanofoundry for materials science. *J. Sol-Gel Sci. Technol.* **2014**, *70*, 216–226. [CrossRef]
121. Dislich, H.; Hinz, P. History and principles of the sol-gel process, and some new multicomponent oxide coatings. *J. Non-Cryst. Solids* **1982**, *48*, 11–16. [CrossRef]
122. Carter, B.; Norton, G. Sols, Gels, and Organic Chemistry. In *Ceramic Materials Science and Engineering*, 2nd ed.; Springer: New York, NY, USA, 2007; pp. 400–411, ISBN 9780387462707.
123. Ciesielczyk, F.; Szczekocka, W.; Siwińska-Stefańska, K.; Piasecki, A.; Paukszta, D.; Jesionowski, T. Evaluation of the photocatalytic ability of a sol-gel-derived MgO-ZrO_2 oxide material. *Open Chem.* **2017**, *15*, 7–18. [CrossRef]
124. Letailleur, A.A.; Ribot, F.; Boissière, C.; Teisseire, J.; Barthel, E.; Desmazières, B.; Chemin, N.; Sanchez, C. Sol-gel derived hybrid thin films: The chemistry behind processing. *Chem. Mater.* **2011**, *23*, 5082–5089. [CrossRef]
125. Faustini, M.; Nicole, L.; Ruiz-Hitzky, E.; Sanchez, C. History of organic-inorganic hybrid materials: Prehistory, art, science, and advanced applications. *Adv. Funct. Mater.* **2018**, *28*, 1704158. [CrossRef]
126. Gleiter, H. Nanocrystalline materials. *Prog. Mater. Sci.* **1989**, *33*, 223–315. [CrossRef]

127. Livage, J.; Henry, M.; Sanchez, C. Sol-gel chemistry of transition metal oxides. *Prog. Solid State Chem.* **1988**, *18*, 259–341. [CrossRef]
128. Schottner, G. Hybrid sol-gel-derived polymers: Applications of multifunctional materials. *Chem. Mater.* **2001**, *13*, 3422–3435. [CrossRef]
129. Akpan, U.G.; Hameed, B.H. Parameters affecting the photocatalytic degradation of dyes using TiO_2-based photocatalysts: A review. *J. Hazard. Mater.* **2009**, *170*, 520–529. [CrossRef] [PubMed]
130. Miao, Z.; Xu, D.; Ouyang, J.; Guo, G.; Zhao, X.; Tang, Y. Electrochemically induced sol-gel preparation of single-crystalline TiO_2 nanowires. *Nano Lett.* **2002**, *2*, 717–720. [CrossRef]
131. Siwinska-Stefanska, K.; Zdarta, J.; Paukszta, D.; Jesionowski, T. The influence of addition of a catalyst and cheating agent on the properties of titanium dioxide synthesized via the sol-gel method. *J. Sol-Gel Sci. Technol.* **2015**, *75*, 264–278. [CrossRef]
132. Ciesielczyk, F.; Bartczak, P.; Zdarta, J.; Jesionowski, T. Active $MgO-SiO_2$ hybrid material for organic dye removal: A mechanism and interaction study of the adsorption of C.I. Acid Blue 29 and C.I. Basic Blue 9. *J. Environ. Manag.* **2017**, *204*, 123–135. [CrossRef] [PubMed]
133. Fu, G.; Vary, P.S.; Lin, C.-T. Anatase TiO_2 Nanocomposites for antimicrobial coatings. *J. Phys. Chem. B* **2005**, *109*, 8889–8898. [CrossRef] [PubMed]
134. Nakashima, T.; Kimizuka, N. Interfacial synthesis of hollow TiO_2 microspheres in ionic liquids. *J. Am. Chem. Soc.* **2003**, *125*, 6386–6387. [CrossRef] [PubMed]
135. Liu, X.; Chu, P.K.; Ding, C. Surface modification of titanium, titanium alloys, and related materials for biomedical applications. *Mater. Sci. Eng. R Rep.* **2004**, *47*, 49–121. [CrossRef]
136. Lettmann, C.; Hildenbrand, K.; Kisch, H.; Macyk, W.; Maier, W.F. Visible light photodegradation of 4-chlorophenol with a coke-containing titanium dioxide photocatalyst. *Appl. Catal. B* **2001**, *32*, 215–227. [CrossRef]
137. Li, G.; Li, L.; Boerio-Goates, J.; Woodfield, B.F. High purity anatase TiO_2 nanocrystals: Near room-temperature synthesis, grain growth kinetics, and surface hydration chemistry. *J. Am. Chem. Soc.* **2005**, *127*, 8659–8666. [CrossRef] [PubMed]
138. Pierre, A.C. *Introduction to Sol-Gel Processing*, 1st ed.; Springer: New York, NY, USA, 1998; pp. 205–247, ISBN 9780792381211.
139. Klein, L.; Aparicio, M.; Jitianu, A. *Handbook of Sol-Gel Science and Technology*, 2nd ed.; Springer: Cham, Switzerland, 2018; ISBN 9783319320991.
140. Srikanth, B.; Goutham, R.; Badri Narayan, R.; Ramprasath, A.; Gopinath, K.P.; Sankaranarayanan, A.R. Recent advancements in supporting materials for immobilised photocatalytic applications in waste water treatment. *J. Environ. Manag.* **2017**, *200*, 60–78. [CrossRef] [PubMed]
141. Long, J.; Zhang, B.; Li, X.; Zhan, X.; Xu, X.; Xie, Z.; Jin, Z. Effective production of resistant starch using pullulanase immobilized onto magnetic chitosan/Fe_3O_4 nanoparticles. *Food Chem.* **2018**, *239*, 276–286. [CrossRef] [PubMed]
142. Cui, H.F.; Wu, W.W.; Li, M.M.; Song, X.; Lv, Y.; Zhang, T.T. A highly stable acetylcholinesterase biosensor based on chitosan-TiO_2-graphene nanocomposites for detection of organophosphate pesticides. *Biosens. Bioelectron.* **2018**, *99*, 223–229. [CrossRef] [PubMed]
143. Kołodziejczak-Radzimska, A.; Zdarta, J.; Jesionowski, T. Physicochemical and catalytic properties of acylase I from *aspergillus melleus* immobilized on amino- and carbonyl-grafted Stöber silica. *Biotechnol. Prog.* **2018**, *34*, 767–777. [CrossRef] [PubMed]
144. Kolodziejczak-Radzimska, A. Functionalized Stöber silica as a support in immobilization process of lipase from *Candida rugosa*. *Physicochem. Probl. Miner. Process.* **2017**, *53*, 878–892. [CrossRef]
145. Brinker, C.J.; Scherer, G.W. Particulate Sol and Gels. In *Sol-Gel Science, the Physics and Chemistry of Sol-Gel Processing*, 1st ed.; Academic Press, INC.: New York, NY, USA, 1990; pp. 235–297, ISBN 9780080571034.
146. Nogami, M. Semiconductor-doped sol-gel optics. In *Sol-Gel Optics: Processing and Applications*, 1st ed.; Klein, L.C., Ed.; Springer: New York, NY, USA, 1994; pp. 329–344.
147. Challagulla, S.; Tarafder, K.; Ganesan, R.; Roy, S. Structure sensitive photocatalytic reduction of nitroarenes over TiO_2. *Sci. Rep.* **2017**, *7*, 8783. [CrossRef] [PubMed]
148. Ye, Y.; Feng, Y.; Bruning, H.; Yntema, D.; Rijnaarts, H.H.M. Photocatalytic degradation of metoprolol by TiO_2 nanotube arrays and UV-LED: Effects of catalyst properties, operational parameters, commonly present water constituents, and photo-induced reactive species. *Appl. Catal. B Environ.* **2018**, *220*, 171–181. [CrossRef]

149. MiarAlipour, S.; Friedmann, D.; Scott, J.; Amal, R. TiO$_2$/porous adsorbents: Recent advances and novel applications. *J. Hazard. Mater.* **2018**, *341*, 404–423. [CrossRef] [PubMed]
150. Sopha, H.; Krbal, M.; Ng, S.; Prikryl, J.; Zazpe, R.; Yam, F.K.; Macak, J.M. Highly efficient photoelectrochemical and photocatalytic anodic TiO$_2$ nanotube layers with additional TiO$_2$ coating. *Appl. Mater. Today* **2017**, *9*, 104–110. [CrossRef]
151. Kmentova, H.; Kment, S.; Wang, L.; Pausova, S.; Vaclavu, T.; Kuzel, R.; Han, H.; Hubicka, Z.; Zlamal, M.; Olejnicek, J.; et al. Photoelectrochemical and structural properties of TiO$_2$ nanotubes and nanorods grown on FTO substrate: Comparative study between electrochemical anodization and hydrothermal method used for the nanostructures fabrication. *Catal. Today* **2017**, *287*, 130–136. [CrossRef]
152. Shanmugam, M.; Bills, B.; Baroughi, M.F.; Galipeau, D. Electron transport in dye sensitized solar cells with TiO$_2$/ZnO core- shell photoelectrode. In Proceedings of the 35th IEEE Photovoltaic Specialists Conference, Honolulu, HI, USA, 20–25 June 2016. [CrossRef]
153. Chu, Y.; Wang, Q.; Cui, S. TiO$_2$ and ZnO water sol preparation by sol-gel method and application on polyester fabric of antistatic finishing. *Adv. Mater. Res.* **2011**, *331*, 270–274. [CrossRef]
154. Yu, D.; Wang, J.; Tian, J.; Xu, X.; Dai, J.; Wang, X. Preparation and characterization of TiO$_2$/ZnO composite coating on carbon steel surface and its anticorrosive behavior in seawater. *Compos. Part B* **2013**, *46*, 135–144. [CrossRef]
155. Wang, L.; Fu, X.; Han, Y.; Chang, E.; Wu, H.; Wang, H.; Li, K.; Qi, X.; Wang, L.; Fu, X.; et al. Preparation, characterization, and photocatalytic activity of TiO$_2$/ZnO nanocomposites. *J. Nanomater.* **2013**, *2013*, 321459. [CrossRef]
156. Giannakopoulou, T.; Todorova, N.; Giannouri, M.; Yu, J.; Trapalis, C. Optical and photocatalytic properties of composite TiO$_2$/ZnO thin films. *Catal. Today* **2014**, *230*, 174–180. [CrossRef]
157. Chen, Y.; Zhang, C.; Huang, W.; Yang, C.; Huang, T.; Situ, Y.; Huang, H. Synthesis of porous ZnO/TiO$_2$ thin films with superhydrophilicity and photocatalytic activity via a template-free sol-gel method. *Surf. Coat. Technol.* **2014**, *258*, 531–538. [CrossRef]
158. Pournuroz, Z.; Khosravi, M.; Giahi, M.; Sohrabi, M. Compare removal of Reactive Red 195 and Blue 19 by nano TiO$_2$-ZnO. *Orient. J. Chem.* **2015**, *31*, 2–6. [CrossRef]
159. Al-Mayman, S.I.; Al-Johani, M.S.; Mokhtar, M.M.; Al-Zeghayer, Y.S.; Ramay, S.M.; Al-Awadi, A.S.; Soliman, M.A. TiO$_2$-ZnO photocatalysts synthesized by sol-gel auto-ignition technique for hydrogen production. *Int. J. Hydrogen Energy* **2016**, *42*, 5016–5025. [CrossRef]
160. Zulkiflee, N.S.; Hussin, R. Effect of temperature on TiO$_2$/ZnO nanostructure thin films. *Mater. Sci. Forum* **2016**, *840*, 262–266. [CrossRef]
161. Prasannalakshmi, P.; Shanmugam, N. Fabrication of TiO$_2$/ZnO nanocomposites for solar energy driven photocatalysis. *Mater. Sci. Semicond. Process.* **2017**, *61*, 114–124. [CrossRef]
162. Al-Hazami, F.E.; Yakuphanoglu, F. Photoconducting and photovoltaic properties of ZnO:TiO$_2$ composite/p-silicon heterojunction photodiode. *Silicon* **2017**, *10*, 781–787. [CrossRef]
163. Armin, S.; Estekhraji, Z.; Amiri, S. Synthesis and characterization of anti-fungus, anti-corrosion and self-cleaning hybrid nanocomposite coatings based on sol-gel process. *J. Inorg. Organomet. Polym. Mater.* **2017**, *27*, 883–891. [CrossRef]
164. Chamanzadeh, Z.; Noormohammadi, M.; Zahedifar, M. Enhanced photovoltaic performance of dye sensitized solar cell using TiO$_2$ and ZnO nanoparticles on top of free standing TiO$_2$ nanotube arrays. *Mater. Sci. Semicond. Process.* **2017**, *61*, 107–113. [CrossRef]
165. Fatimah, I. Preparation of TiO$_2$-ZnO and its activity test in sonophotocatalytic degradation of phenol. *IOP Conf. Ser. Mater. Sci. Eng.* **2016**, *107*, 012003. [CrossRef]
166. Bozzi, A.; Yuranova, T.; Guasaquillo, I.; Laub, D.; Kiwi, J. Self-cleaning of modified cotton textiles by TiO$_2$ at low temperatures under daylight irradiation. *J. Photochem. Photobiol. A Chem.* **2005**, *174*, 156–164. [CrossRef]
167. Bozzi, A.; Yuranova, T.; Kiwi, J. Self-cleaning of wool-polyamide and polyester textiles by TiO$_2$-rutile modification under daylight irradiation at ambient temperature. *J. Photochem. Photobiol. A Chem.* **2005**, *172*, 27–34. [CrossRef]
168. Mihailović, D.; Šaponjić, Z.; Radoičić, M.; Radetić, T.; Jovančić, P.; Nedeljković, J.; Radetić, M. Functionalization of polyester fabrics with alginates and TiO$_2$ nanoparticles. *Carbohydr. Polym.* **2010**, *79*, 526–532. [CrossRef]

169. Rashvand, M.; Ranjbar, Z.; Rastegar, S. Preserving anti-corrosion properties of epoxy based coatings imultaneously exposed to humidity and UV-radiation using nano zinc oxide. *J. Electrochem. Soc.* **2012**, *159*, 129–132. [CrossRef]
170. Li, Q.; Yang, X.; Zhang, L.; Wang, J.; Chen, B. Corrosion resistance and mechanical properties of pulse electrodeposited Ni-TiO$_2$ composite coating for sintered NdFeB magnet. *J. Alloys Compd.* **2009**, *482*, 339–344. [CrossRef]
171. Yu, Z.; Di, H.; Ma, Y.; He, Y.; Liang, L.; Lv, L.; Ran, X.; Pan, Y.; Luo, Z. Preparation of graphene oxide modified by titanium dioxide to enhance the anti-corrosion performance of epoxy coatings. *Surf. Coat. Technol.* **2015**, *276*, 471–478. [CrossRef]
172. You, X.; Chen, F.; Zhang, J. Effects of calcination on the physical and photocatalytic properties of TiO$_2$ powders prepared by sol-gel template method. *J. Sol-Gel Sci. Technol.* **2005**, *34*, 181–187. [CrossRef]
173. Chen, Y.; Dionysiou, D.D. Effect of calcination temperature on the photocatalytic activity and adhesion of TiO$_2$ films prepared by the P-25 powder-modified sol-gel method. *J. Mol. Catal. A Chem.* **2006**, *244*, 73–82. [CrossRef]
174. Wu, N.L.; Lee, M.S.; Pon, Z.J.; Hsu, J.Z. Effect of calcination atmosphere on TiO$_2$ photocatalysis in hydrogen production from methanol/water solution. *J. Photochem. Photobiol. A Chem.* **2004**, *163*, 277–280. [CrossRef]
175. Dalton, J.S.; Janes, P.A.; Jones, N.G.; Nicholson, J.A.; Hallam, K.R.; Allen, G.C. Photocatalytic oxidation of NO$_x$ gases using TiO$_2$: A surface spectroscopic approach. *Environ. Pollut.* **2002**, *120*, 415–422. [CrossRef]
176. Karapati, S.; Giannakopoulou, T.; Todorova, N.; Boukos, N.; Antiohos, S.; Papageorgiou, D.; Chaniotakis, E.; Dimotikali, D.; Trapalis, C. TiO$_2$ functionalization for efficient NO$_x$ removal in photoactive cement. *Appl. Surf. Sci.* **2014**, *319*, 29–36. [CrossRef]
177. Todorova, N.; Giannakopoulou, T.; Karapati, S.; Petridis, D.; Vaimakis, T.; Trapalis, C. Composite TiO$_2$/clays materials for photocatalytic NO$_x$ oxidation. *Appl. Surf. Sci.* **2014**, *319*, 113–120. [CrossRef]
178. Thompson, S.; Shirtcliffe, N.J.; O'Keefe, E.S.; Appleton, S.; Perry, C.C. Synthesis of SrCo$_x$Ti$_x$Fe$_{(12-2x)}$O$_{19}$ through sol-gel auto-ignition and its characterisation. *J. Magn. Magn. Mater.* **2005**, *292*, 100–107. [CrossRef]
179. Sutka, A.; Mezinskis, G. Sol-gel auto-combustion synthesis of spinel-type ferrite nanomaterials. *Front. Mater. Sci.* **2012**, *6*, 128–141. [CrossRef]
180. Hendi, A.A.; Yakuphanoglu, F. Graphene doped TiO$_2$/p-silicon heterojunction photodiode. *J. Alloys Compd.* **2016**, *665*, 418–427. [CrossRef]
181. Yu, H.; Li, X.; Quan, X.; Chen, S.; Zhang, Y. Effective utilization of visible light (including $\lambda > 600$ nm) in phenol degradation with p-silicon nanowire/TiO$_2$ core/shell heterojunction array cathode. *Environ. Sci. Technol.* **2009**, *43*, 7849–7855. [CrossRef] [PubMed]
182. Kang, H.; Park, J.; Choi, T.; Jung, H.; Lee, K.H.; Im, S.; Kim, H. N-ZnO:N/p-Si nanowire photodiode prepared by atomic layer deposition. *Appl. Phys. Lett.* **2012**, *100*, 2010–2014. [CrossRef]
183. Silva, C.G.; Faria, J.L. Photocatalytic oxidation of benzene derivatives in aqueous suspensions: Synergic effect induced by the introduction of carbon nanotubes in a TiO$_2$ matrix. *Appl. Catal. B Environ.* **2010**, *101*, 81–89. [CrossRef]
184. Liu, G.; Zhao, Y.; Sun, C.; Li, F.; Lu, G.Q.; Cheng, H.M. Synergistic effects of B/N doping on the visible-light photocatalytic activity of mesoporous TiO$_2$. *Angew. Chem. Int. Ed.* **2008**, *47*, 4516–4520. [CrossRef] [PubMed]
185. Hoang, S.; Berglund, S.P.; Hahn, N.T.; Bard, A.J.; Mullins, C.B. Enhancing visible light photo-oxidation of water with TiO$_2$ nanowire arrays via cotreatment with H$_2$ and NH$_3$: Synergistic effects between Ti^{3+} and N. *J. Am. Chem. Soc.* **2012**, *134*, 3659–3662. [CrossRef] [PubMed]
186. Zhang, K.; Zhang, F.J.; Chen, M.L.; Oh, W.C. Comparison of catalytic activities for photocatalytic and sonocatalytic degradation of methylene blue in present of anatase TiO$_2$-CNT catalysts. *Ultrason. Sonochem.* **2011**, *18*, 765–772. [CrossRef] [PubMed]
187. Wang, J.; Guo, B.; Zhang, X.; Zhang, Z.; Han, J.; Wu, J. Sonocatalytic degradation of methyl orange in the presence of TiO$_2$ catalysts and catalytic activity comparison of rutile and anatase. *Ultrason. Sonochem.* **2005**, *12*, 331–337. [CrossRef] [PubMed]
188. Hernández-Alonso, M.D.; Tejedor-Tejedor, I.; Coronado, J.M.; Soria, J.; Anderson, M.A. Sol-gel preparation of TiO$_2$-ZrO$_2$ thin films supported on glass rings: Influence of phase composition on photocatalytic activity. *Thin Solid Films* **2006**, *502*, 125–131. [CrossRef]
189. Kraleva, E.; Saladino, M.L.; Matassa, R.; Caponetti, E.; Enzo, S.; Spojakina, A. Phase formation in mixed TiO$_2$-ZrO$_2$ oxides prepared by sol-gel method. *J. Struct. Chem.* **2011**, *52*, 330–339. [CrossRef]

190. Mohammadi, M.R.; Fray, D.J. Synthesis and characterisation of nanosized TiO_2-ZrO_2 binary system prepared by an aqueous sol-gel process: Physical and sensing properties. *Sens. Actuators B Chem.* **2011**, *155*, 568–576. [CrossRef]
191. Naumenko, A.; Gnatiuk, I.; Smirnova, N.; Eremenko, A. Characterization of sol-gel derived TiO_2/ZrO_2 films and powders by Raman spectroscopy. *Thin Solid Films* **2012**, *520*, 4541–4546. [CrossRef]
192. Karthika, S.; Prathibha, V.; Ann, M.K.A.; Viji, V.; Biju, P.R.; Unnikrishnan, N.V. Structural and spectroscopic studies of Sm^{3+}/CdS nanocrystallites in sol-gel TiO_2-ZrO_2 matrix. *J. Electron. Mater.* **2014**, *43*, 447–451. [CrossRef]
193. Fukumoto, T.; Yoshioka, T.; Nagasawa, H.; Kanezashi, M.; Tsuru, T. Development and gas permeation properties of microporous amorphous TiO_2-ZrO_2-organic composite membranes using chelating ligands. *J. Membr. Sci.* **2014**, *461*, 96–105. [CrossRef]
194. Atanda, L.; Silahua, A.; Mukundan, S.; Shrotri, A.; Torres-Torres, G.; Beltramini, J. Catalytic behaviour of TiO_2-ZrO_2 binary oxide synthesized by sol-gel process for glucose conversion to 5-hydroxymethylfurfural. *RSC Adv.* **2015**, *5*, 80346–80352. [CrossRef]
195. Khan, S.; Kim, J.; Sotto, A.; Van der Bruggen, B. Humic acid fouling in a submerged photocatalytic membrane reactor with binary TiO_2-ZrO_2 particles. *J. Ind. Eng. Chem.* **2015**, *21*, 779–786. [CrossRef]
196. Zukalová, M.; Zukal, A.; Kavan, L.; Nazeeruddin, M.K.; Liska, P.; Grätzel, M. Organized mesoporous TiO_2 films exhibiting greatly enhanced performance in dye-sensitized solar cells. *Nano Lett.* **2005**, *5*, 1789–1792. [CrossRef] [PubMed]
197. Marien, C.B.D.; Marchal, C.; Koch, A.; Robert, D.; Drogui, P. Sol-gel synthesis of TiO_2 nanoparticles: Effect of Pluronic P123 on particle's morphology and photocatalytic degradation of paraquat. *Environ. Sci. Pollut. Res.* **2017**, *24*, 12582–12588. [CrossRef] [PubMed]
198. Choi, S.Y.; Mamak, M.; Coombs, N.; Chopra, N.; Ozin, G.A. Thermally stable two-dimensional hexagonal mesoporous nanocrystalline anatase, meso-nc-TiO_2: Bulk and crack-free thin film morphologies. *Adv. Funct. Mater.* **2004**, *14*, 335–344. [CrossRef]
199. Vatanpour, V.; Madaeni, S.S.; Khataee, A.R.; Salehi, E.; Zinadini, S.; Monfared, H.A. TiO_2 embedded mixed matrix PES nanocomposite membranes: Influence of different sizes and types of nanoparticles on antifouling and performance. *Desalination* **2012**, *292*, 19–29. [CrossRef]
200. Ramachandra, M.; Abhishek, A.; Siddeshwar, P.; Bharathi, V. Hardness and wear resistance of ZrO_2 nano particle reinforced Al nanocomposites produced by powder metallurgy. *Procedia Mater. Sci.* **2015**, *10*, 212–219. [CrossRef]
201. Shojai, F.; Mäntylä, T.A. Structural stability of yttria doped zirconia membranes in acid and basic aqueous solutions. *J. Eur. Ceram. Soc.* **2001**, *21*, 37–44. [CrossRef]
202. Puhlfürß, P.; Voigt, A.; Weber, R.; Morbé, M. Microporous TiO_2 membranes with a cut off <500 Da. *J. Membr. Sci.* **2000**, *174*, 123–133. [CrossRef]
203. Fan, J.; Ohya, H.; Suga, T.; Ohashi, H.; Yamashita, K.; Tsuchiya, S.; Aihara, M.; Takeuchi, T.; Negishi, Y. High flux zirconia composite membrane for hydrogen separation at elevated temperature. *J. Membr. Sci.* **2000**, *170*, 113–125. [CrossRef]
204. Lu, C.; Chen, Z. High-temperature resistive hydrogen sensor based on thin nanoporous rutile TiO_2 film on anodic aluminum oxide. *Sens. Actuators B Chem.* **2009**, *140*, 109–115. [CrossRef]
205. Mather, G.C.; Marques, F.M.B.; Frade, J.R. Detection mechanism of TiO_2 -based ceramic H_2 sensors. *J. Eur. Ceram. Soc.* **1999**, *19*, 887–891. [CrossRef]
206. Tang, H.; Prasad, K.; Sanjines, R.; Levy, F. TiO_2 anatase thin-films as gas sensors. *Sens. Actuators B-Chem.* **1995**, *26*, 71–75. [CrossRef]
207. Kuster, B.F.M. 5-Hydroxymethylfurfural (HMF). A review focussing on its manufacture. *Starch-Stärke* **1990**, *42*, 314–321. [CrossRef]
208. Rosatella, A.A.; Simeonov, S.P.; Frade, R.F.M.; Afonso, C.A.M. 5-Hydroxymethylfurfural (HMF) as a building block platform: Biological properties, synthesis and synthetic applications. *Green Chem.* **2011**, *13*, 754–793. [CrossRef]
209. Ho, D.P.; Vigneswaran, S.; Ngo, H.H. Photocatalysis-membrane hybrid system for organic removal from biologically treated sewage effluent. *Sep. Purif. Technol.* **2009**, *68*, 145–152. [CrossRef]

210. Yang, N.; Wen, X.; Waite, T.D.; Wang, X.; Huang, X. Natural organic matter fouling of microfiltration membranes: Prediction of constant flux behavior from constant pressure materials properties determination. *J. Membr. Sci.* **2011**, *366*, 192–202. [CrossRef]
211. Katrib, A.; Benadda, A.; Sobczak, J.W.; Maire, G. XPS and catalytic properties of the bifunctional supported $MoO_2(H_x)_{ac}$ on TiO_2 for the hydroisomerization reactions of hexanes and 1-hexene. *Appl. Catal. A Gen.* **2003**, *242*, 31–40. [CrossRef]
212. Fu, J.; Ji, M.; Wang, Z.; Jin, L.; An, D. A new submerged membrane photocatalysis reactor (SMPR) for fulvic acid removal using a nano-structured photocatalyst. *J. Hazard. Mater.* **2006**, *131*, 238–242. [CrossRef] [PubMed]
213. Zeleny, J. Instability of electrified liquid surfaces. *Phys. Rev.* **1917**, *10*, 1–6. [CrossRef]
214. Dole, M.; Hines, R.L.; Mack, L.L.; Mobley, R.C.; Ferguson, L.D.; Alice, M.B. Gas phase macroions. *Macromolecules* **1968**, *1*, 96–97. [CrossRef]
215. Ramaseshan, R.; Sundarrajan, S.; Jose, R.; Ramakrishna, S. Nanostructured ceramics by electrospinning. *J. Appl. Phys.* **2007**, *102*, 111101. [CrossRef]
216. Dai, Y.; Cobley, C.M.; Zeng, J.; Sun, Y.; Xia, Y. Synthesis of anatase TiO_2 nanocrystals with exposed (001) facets. *Nano Lett.* **2009**, *9*, 1–5. [CrossRef] [PubMed]
217. Formo, E.; Lee, E.; Campbell, D.; Xia, Y. Functionalization of electrospun TiO_2 nanofibers with Pt nanoparticles and nanowires for catalytic applications. *Nano Lett.* **2008**, *8*, 668–672. [CrossRef] [PubMed]
218. Inagaki, M.; Yang, Y.; Kang, F. Carbon nanofibers prepared via electrospinning. *Adv. Mater.* **2012**, *24*, 2547–2566. [CrossRef] [PubMed]
219. Bhardwaj, N.; Kundu, S.C. Electrospinning: A fascinating fiber fabrication technique. *Biotechnol. Adv.* **2010**, *28*, 325–347. [CrossRef] [PubMed]
220. Liu, Z.; Sun, D.D.; Guo, P.; Leckie, J.O. An efficient bicomponent TiO_2/SnO_2 nanofiber photocatalyst fabricated by electrospinning with a side-by-side dual spinneret method. *Nano Lett.* **2007**, *7*, 1081–1085. [CrossRef] [PubMed]
221. Lu, X.; Wang, C.; Wei, Y. One-dimensional composite nanomaterials: Synthesis by electrospinning and their applications. *Small* **2009**, *5*, 2349–2370. [CrossRef] [PubMed]
222. Thavasi, V.; Singh, G.; Ramakrishna, S. Electrospun nanofibers in energy and environmental applications. *Energy Environ. Sci.* **2008**, *1*, 205–221. [CrossRef]
223. Anselme, K.; Davidson, P.; Popa, A.M.; Giazzon, M.; Liley, M.; Ploux, L. The interaction of cells and bacteria with surfaces structured at the nanometre scale. *Acta Biomater.* **2010**, *6*, 3824–3846. [CrossRef] [PubMed]
224. Ding, B.; Wang, M.; Wang, X.; Yu, J.; Sun, G. Electrospun nanomaterials for ultrasensitive sensors. *Mater. Today* **2010**, *13*, 16–27. [CrossRef]
225. Agarwal, S.; Greiner, A.; Wendorff, J.H. Functional materials by electrospinning of polymers. *Prog. Polym. Sci.* **2013**, *38*, 963–991. [CrossRef]
226. Long, Y.Z.; Li, M.M.; Gu, C.; Wan, M.; Duvail, J.L.; Liu, Z.; Fan, Z. Recent advances in synthesis, physical properties and applications of conducting polymer nanotubes and nanofibers. *Prog. Polym. Sci.* **2011**, *36*, 1415–1442. [CrossRef]
227. Huang, Z.M.; Zhang, Y.Z.; Kotaki, M.; Ramakrishna, S. A review on polymer nanofibers by electrospinning and their applications in nanocomposites. *Compos. Sci. Technol.* **2003**, *63*, 2223–2253. [CrossRef]
228. Spasova, M.; Mincheva, R.; Paneva, D.; Manolova, N.; Rashkov, I. Perspectives on: Criteria for complex evaluation of the morphology and alignment of electrospun polymer nanofibers. *J. Bioact. Compat. Polym.* **2006**, *21*, 465–479. [CrossRef]
229. Reneker, D.H.; Chun, I. Nanometre diameter fibres of polymer, produced by electrospinning. *Nanotechnology* **1996**, *7*, 216–223. [CrossRef]
230. Fong, H.; Liu, W.; Wang, C.S.; Vaia, R.A. Generation of electrospun fibers of nylon 6 and nylon 6-montmorillonite nanocomposite. *Polymer* **2001**, *43*, 775–780. [CrossRef]
231. Mitarai, T.; Shander, A.; Tight, M.; Fabrics, N.; Martin, F.N.; English, J.T. An introduction to electrospinning and nanofibres. *J. Cardiothorac. Vasc. Anesth.* **2013**, *27*, 1–2. [CrossRef]
232. Wang, Y.; Jia, W.; Strout, T.; Schempf, A.; Zhang, H.; Li, B.; Cui, J.; Lei, Y. Ammonia gas sensor using polypyrrole-coated TiO_2/ZnO nanofibers. *Electroanalysis* **2009**, *21*, 1432–1438. [CrossRef]
233. Park, J.Y.; Choi, S.W.; Lee, J.W.; Lee, C.; Kim, S.S. Synthesis and gas sensing properties of TiO_2-ZnO core-shell nanofibers. *J. Am. Ceram. Soc.* **2009**, *92*, 2551–2554. [CrossRef]

234. Wang, H.Y.; Yang, Y.; Li, X.; Li, L.J.; Wang, C. Preparation and characterization of porous TiO$_2$/ZnO composite nanofibers via electrospinning. *Chin. Chem. Lett.* **2010**, *21*, 1119–1123. [CrossRef]
235. Liu, R.; Ye, H.; Xiong, X.; Liu, H. Fabrication of TiO$_2$/ZnO composite nanofibers by electrospinning and their photocatalytic property. *Mater. Chem. Phys.* **2010**, *121*, 432–439. [CrossRef]
236. Kanjwal, M.A.; Barakat, N.A.M.; Sheikh, F.A.; Park, S.J.; Kim, H.Y. Photocatalytic activity of ZnO-TiO$_2$ hierarchical nanostructure prepared by combined electrospinning and hydrothermal techniques. *Macromol. Res.* **2010**, *18*, 233–240. [CrossRef]
237. Li, J.; Yan, L.; Wang, Y.; Kang, Y.; Wang, C.; Yang, S. Fabrication of TiO$_2$/ZnO composite nanofibers with enhanced photocatalytic activity. *J. Mater. Sci. Mater. Electron.* **2016**, *27*, 7834–7838. [CrossRef]
238. Araújo, E.S.; Libardi, J.; Faia, P.M.; de Oliveira, H.P. Humidity-sensing properties of hierarchical TiO$_2$:ZnO composite grown on electrospun fibers. *J. Mater. Sci. Mater. Electron.* **2017**, *28*, 1–9. [CrossRef]
239. Karunagaran, B.; Uthirakumar, P.; Chung, S.J.; Velumani, S.; Suh, E.K. TiO$_2$ thin film gas sensor for monitoring ammonia. *Mater. Charact.* **2007**, *58*, 680–684. [CrossRef]
240. Shavisi, Y.; Sharifnia, S.; Hosseini, S.N.; Khadivi, M.A. Application of TiO$_2$/perlite photocatalysis for degradation of ammonia in wastewater. *J. Ind. Eng. Chem.* **2014**, *20*, 278–283. [CrossRef]
241. Saha, D.; Deng, S. Characteristics of ammonia adsorption on activated alumina. *J. Chem. Eng. Data* **2010**, *55*, 5587–5593. [CrossRef]
242. Yuzawa, H.; Mori, T.; Itoh, H.; Yoshida, H. Reaction mechanism of ammonia decomposition to nitrogen and hydrogen over metal loaded titanium oxide photocatalyst. *J. Phys. Chem. C* **2012**, *116*, 4126–4136. [CrossRef]
243. Pang, Z.; Yang, Z.; Chen, Y.; Zhang, J.; Wang, Q.; Huang, F.; Wei, Q. A room temperature ammonia gas sensor based on cellulose/TiO$_2$/PANI composite nanofibers. *Colloids Surf. A Physicochem. Eng. Asp.* **2016**, *494*, 248–255. [CrossRef]
244. Nagaraja, R.; Kottam, N.; Girija, C.R.; Nagabhushana, B.M. Photocatalytic degradation of Rhodamine B dye under UV/solar light using ZnO nanopowder synthesized by solution combustion route. *Powder Technol.* **2012**, *215–216*, 91–97. [CrossRef]
245. Kornbrust, D.; Barfknecht, T. Testing of 24 food, drug, cosmetic, and fabric dyes in the in vitro and the in vivo/in vitro rat hepatocyte primary culture/DNA repair assays. *Environ. Mutagen* **1985**, *7*, 101–120. [CrossRef] [PubMed]
246. Snawder, J.E.; Lipscomb, J.C. Interindividual variance of cytochrome P450 forms in human hepatic microsomes: Correlation of individual forms with xenobiotic metabolism and implications in risk assessment. *Regul. Toxicol. Pharm.* **2000**, *32*, 200–209. [CrossRef] [PubMed]
247. Mirsalis, J.C.; Tyson, C.K.; Steinmetz, K.L.; Loh, E.K.; Hamilton, C.M.; Bakke, J.P.; Spalding, J.W. Measurement of unscheduled DNA synthesis and S-phase synthesis in rodent hepatocytes following in vivo treatment: Testing of 24 compounds. *Environ. Mol. Mutagen* **1989**, *14*, 155–164. [CrossRef] [PubMed]
248. Ding, L.; Zou, B.; Gao, W.; Liu, Q.; Wang, Z.; Guo, Y.; Wang, X.; Liu, Y. Adsorption of Rhodamine-B from aqueous solution using treated rice husk-based activated carbon. *Colloids Surf. A Physicochem. Eng. Asp.* **2014**, *446*, 1–7. [CrossRef]
249. Jain, R.; Mathur, M.; Sikarwar, S.; Mittal, A. Removal of the hazardous dye rhodamine B through photocatalytic and adsorption treatments. *J. Environ. Manag.* **2007**, *85*, 956–964. [CrossRef] [PubMed]
250. Liang, Z.; Li, Q.; Li, F.; Zhao, S. Microstructure of SiO$_2$/TiO$_2$ hybrid electrospun nanofibers and their application in dye degradation. *Res. Chem. Intermed.* **2016**, *42*, 7017–7029. [CrossRef]
251. Hou, C.; Jiao, T.; Xing, R.; Chen, Y.; Zhou, J.; Zhang, L. Preparation of TiO$_2$ nanoparticles modified electrospun nanocomposite membranes toward efficient dye degradation for wastewater treatment. *J. Taiwan Inst. Chem. Eng.* **2017**, *78*, 118–126. [CrossRef]
252. Doh, S.J.; Kim, C.; Lee, S.G.; Lee, S.J.; Kim, H. Development of photocatalytic TiO$_2$ nanofibers by electrospinning and its application to degradation of dye pollutants. *J. Hazard. Mater.* **2008**, *154*, 118–127. [CrossRef] [PubMed]
253. Faia, P.M.; Furtado, C.S. Effect of composition on electrical response to humidity of TiO$_2$:ZnO sensors investigated by impedance spectroscopy. *Sens. Actuators B Chem.* **2013**, *181*, 720–729. [CrossRef]
254. Su, M.; Wang, J.; Du, H.; Yao, P.; Zheng, Y.; Li, X. Characterization and humidity sensitivity of electrospun ZrO$_2$:TiO$_2$ hetero-nanofibers with double jets. *Sens. Actuators B Chem.* **2012**, *161*, 1038–1045. [CrossRef]

255. Lee, C.; Park, J.H.; Jeon, Y.; Park, J.I.; Einaga, H.; Truong, Y.B.; Kyratzis, I.L.; Mochida, I.; Choi, J.; Shul, Y.G. Phosphate modified TiO_2/ZrO_2 nanofibrous web composite membrane for enhanced performance and durability of high temperature PEM fuel cells. *Energy Fuels* **2017**, *31*, 4–11. [CrossRef]
256. Boaretti, C.; Pasquini, L.; Sood, R.; Giancola, S.; Donnadio, A.; Roso, M.; Modesti, M.; Cavaliere, S. Mechanically stable nanofibrous sPEEK/Aquivion® composite membranes for fuel cell applications. *J. Membr. Sci.* **2018**, *545*, 66–74. [CrossRef]
257. Xiao, P.; Li, J.; Tang, H.; Wang, Z.; Pan, M. Physically stable and high performance Aquivion/ePTFE composite membrane for high temperature fuel cell application. *J. Membr. Sci.* **2013**, *442*, 65–71. [CrossRef]
258. Marrony, M.; Beretta, D.; Ginocchio, S.; Nedellec, Y.; Subianto, S.; Jones, D.J. Lifetime prediction approach applied to the aquivion™ short side chain perfluorosulfonic acid ionomer membrane for intermediate temperature proton exchange membrane fuel cell application. *Fuel Cells* **2013**, *13*, 1146–1154. [CrossRef]

© 2018 by the authors. Licensee MDPI, Basel, Switzerland. This article is an open access article distributed under the terms and conditions of the Creative Commons Attribution (CC BY) license (http://creativecommons.org/licenses/by/4.0/).

Article

Development of Alumina–Mesoporous Organosilica Hybrid Materials for Carbon Dioxide Adsorption at 25 °C

Chamila Gunathilake [1,2], Rohan S. Dassanayake [3], Chandrakantha S. Kalpage [1] and Mietek Jaroniec [2,*]

1. Department of Chemical and Processing Engineering, Faculty of Engineering, University of Peradeniya, Peradeniya 20400, Sri Lanka; chamilag@pdn.ac.lk (C.G.); csk@eng.pdn.ac.lk (C.S.K.)
2. Department of Chemistry and Biochemistry, Kent State University, Kent, OH 44242, USA
3. Department of Chemistry, Ithaca College, Ithaca, NY 14850, USA; rdassanayake@ithaca.edu
* Correspondence: jaroniec@kent.edu; Tel.: +1-330-672-3790

Received: 27 September 2018; Accepted: 13 November 2018; Published: 16 November 2018

Abstract: Two series of alumina (Al_2O_3)–mesoporous organosilica (Al–MO) hybrid materials were synthesized using the co-condensation method in the presence of Pluronic 123 triblock copolymer. The first series of Al–MO samples was prepared using aluminum nitrate nanahydrate (Al–NN) and aluminum isopropoxide (Al–IP) as alumina precursors, and organosilanes with three different bridging groups, namely tris[3-(trimethoxysilyl)propyl]isocyanurate, 1,4-bis(triethoxysilyl)benzene, and bis(triethoxysilyl)ethane. The second series was obtained using the aforementioned precursors in the presence of an amine-containing 3-aminopropyltriethoxysilane to introduce, also, hanging groups. The Al–IP-derived mesostructures in the first series showed the well-developed porosity and high specific surface area, as compared to the corresponding mesostructures prepared in the second series with 3-aminopropyltriethoxysilane. The materials obtained from Al–NN alumina precursor possessed enlarged mesopores in the range of 3–17 nm, whereas the materials synthesized from Al–IP alumina precursor displayed relatively low pore widths in the range of 5–7 nm. The Al–IP-derived materials showed high CO_2 uptakes, due to the enhanced surface area and microporosity in comparison to those observed for the samples of the second series with AP hanging groups. The Al–NN- and Al–IP-derived samples exhibited the CO_2 uptakes in the range of 0.73–1.72 and 1.66–2.64 mmol/g at 1 atm pressure whereas, at the same pressure, the Al–NN and Al–IP-derived samples with 3-aminopropyl hanging groups showed the CO_2 uptakes in the range of 0.72–1.51 and 1.70–2.33 mmol/g, respectively. These data illustrate that Al–MO hybrid materials are potential adsorbents for large-scale CO_2 capture at 25 °C.

Keywords: alumina; CO_2 capture; porous hybrid adsorbents; mesoporous organosilica

1. Introduction

Among many heat-trapping gases, including CO_2, SO_2, NO, N_2O, NO_2, and CH_4, carbon dioxide (CO_2) is considered one of the major contributors to global warming. According to the report released by EPA in 2011, the contribution of CO_2 is about 84% of the total discharge of gases into the atmosphere. Currently, the CO_2 concentration exceeded 400 ppm, and the average global temperature has already increased by 1–2 °C, as compared to the values reported over the last few decades. Although this temperature fluctuation seems to be small, it has a significant effect on the global climate patterns, leading to the melting of glaciers in north and south poles, eventually causing the sea level to rise. Thus, reducing the concentration level of CO_2 is an urgent requirement to keep the atmosphere safe and diminish the effects of global warming.

Nowadays, the preferred technology for industrial CO_2 capture involves chemical absorption in aqueous solutions of amines. Alkanolamines, including monoethanolamine (MEA) and diethanolamine (DEA), are commonly used in this process [1]. However, the amine-based scrubbing techniques possess several drawbacks. For instance, the regeneration of absorbed CO_2 from liquid amine is energy intensive, and replacement of amine after several cycles is needed due to its degradation under oxidizing and reducing chemical environments.

Solid sorbents have attracted a great attention as potential alternatives to the liquid amine scrubbing techniques, due to their high stability, tunability, and reusability. Unlike aqueous liquid amines, porous solid sorbents adsorb CO_2 via physisorption (physical adsorption) [2] or chemisorption (chemical adsorption) [2], and avoid many drawbacks associated with liquid amines [3,4]. Chemisorption on solid sorbents is achieved by modifying their surface with basic sites such as alkaline carbonates and different amine functional groups, which show high affinity toward acidic CO_2. Physical solid sorbents for CO_2 capture include activated carbons [5,6], carbon molecular sieves [7,8], carbon nanotube-based sorbents [9,10], zeolites [11,12], and metal-organic frameworks [13,14]. Materials like alkali metal-based sorbents [15–18], amine-functionalized solid sorbents such as activated carbons [19,20], carbon nanotubes [21,22], solid resins [23], modified zeolite-based sorbents [24,25], polymer-based sorbents [26,27], amine-functionalized silica [28,29], grafted silica sorbents [30–32], and amine-impregnated alumina sorbents [33], have been investigated as chemical solid sorbents.

Mesoporous silica materials, such as SBA-15 [34] and MCM-41 [35], have been investigated for CO_2 capture. However, these materials show low CO_2 sorption capacities due to weak interactions between silanol groups and CO_2. Therefore, MCM-41 and SBA-15 silica materials modified with different amine groups, including diethylenetriamine, tetraethylenepentamine, polyethyleneimine, and ethylenediamine, have been studied for CO_2 capture [30–32,36–40]. The chemical incorporation of amine groups onto silica matrix can be achieved either by co-condensation [41,42] or post-synthesis grafting [31,43,44]. Typically, the higher loading of amine groups leads to higher CO_2 capacity depending on the properties of mesoporous silica. Thus, the selection of proper silica support with high porosity and high surface area before amine grafting is essential. During grafting of amine groups onto silica surface, the surface silanols react with aminosilanes. Therefore, maintaining high surface silanol density is beneficial for achieving higher loading of amine groups and larger CO_2 sorption capacity.

Over the past few years, the amine-grafted mesoporous silica and impregnated alumina materials have been investigated as solid sorbents for CO_2 capture [45,46]. Mesoporous silica inherits high surface area, tunable, and well-defined porosity, which are important for uniform grafting of amine functional groups. Similarly, alumina materials show favorable properties, including high surface area, high porosity, and thermal and mechanical stability, which are suitable for amine modification. The adsorption properties of alumina strongly depend on its structure and thermal treatment temperature (calcination temperature). For example, mesoporous alumina materials feature large pore volume, three-dimensional (3D) interconnected mesoporosity, and thermal stability. Due to these unique characteristics, alumina-based materials have been investigated for gas adsorption applications. Examples of alumina-based materials for CO_2 capture include amine-impregnated Al_2O_3 [33], basic Al_2O_3 [46], amine-modified mesoporous Al_2O_3 [47,48], MgO/Al_2O_3 [49], and Al_2O_3–organosilica [50] materials.

Interestingly, the adsorption properties of alumina–mesoporous silica composites can be further improved by introducing silica precursors with nitrogen-containing functional groups, such as amines and organic bridging groups [50,51]. We previously reported the synthesis of alumina–mesoporous organosilica hybrid materials with isocyanurate bridging groups and their CO_2 adsorption properties at elevated temperatures [50]. Incorporation of organic bridging groups facilitated the formation of crosslinked mesostructures with well-ordered pores and enhanced surface properties [51]. In addition, the organic moieties of bridging groups increase hydrophobicity of the mesoporous silica matrix, which is beneficial for CO_2 sorption.

In this study, we report the synthesis and CO_2 adsorption properties of mesoporous Al–MO hybrid materials at 25 °C. Here, we also present the effects of different alumina precursors, silica precursors with varying organic bridging groups, amine-based silica hanging groups, and calcination process on the CO_2 capture at 25 °C. Al–MO hybrid materials were prepared in the presence of triblock copolymer Pluronic P123 using alumina and organosilica precursors. Two alumina precursors used in this study include aluminum nitrate nanahydrate (Al–NN) and aluminum isopropoxide (Al–IP), see Scheme 1. Silanes with three different bridging groups, including nitrogen-containing tris[3-(trimethoxysilyl)propyl]isocyanurate (ICS), and non-nitrogen-containing 1,4-bis(triethoxysilyl)benzene (BTEB), and bis(triethoxysilyl)ethane (BTEE)), were used as organosilica precursors. The grafting of amine groups was achieved by using 3-aminopropyltriethoxysilane (APTS), see Scheme 1. To our knowledge, this is the first report on the Al–MO hybrid materials prepared by using different alumina and silica precursors with various organic bridging groups and amine hanging groups for CO_2 capture at ambient conditions.

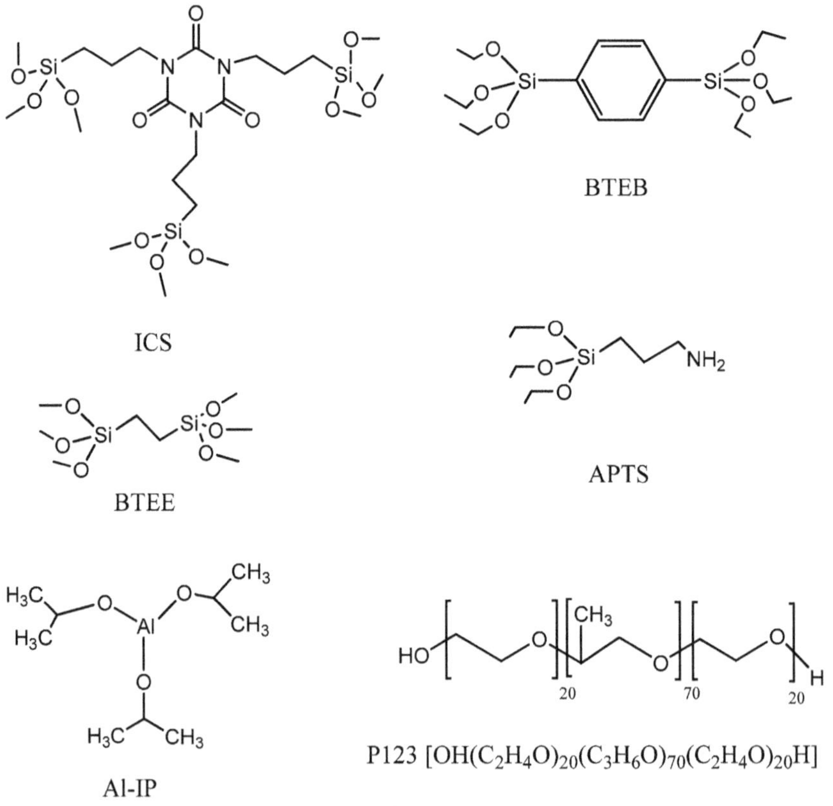

Scheme 1. Structures of the chemicals used in the synthesis of different Al–MO hybrid materials.

2. Materials and Methods

2.1. Materials

Tris[3-(trimethoxysilyl)propyl]isocyanurate (ICS), 1,4-bis(triethoxysilyl)benzene (BTEB), bis(triethoxysilyl)ethane (BTEE), and 3-aminopropyltriethoxysilane (APTS) were purchased from

Gelest Inc (Morrisville, PA, USA). Pluronic P123 ($EO_{20}PO_{70}EO_{20}$) triblock copolymer was purchased from BASF Corporation (Florham Park, NJ, USA). Absolute ethanol was obtained from AAPER Alcohol and Chemical Company (Shelbyville, KY, USA). Ethanol (95%) was purchased from Fisher Scientific (Pittsburgh, PA, USA). Aluminum isopropoxide (Al–IP) and aluminum nitrate nanahydrate (Al–NN) were purchased from Sigma Aldrich (St. Louis, MO, USA). All reagents were in analytical grade, and used without further purification.

2.2. Synthesis Method

Alumina–mesoporous organosilica (Al–MO) hybrid samples were synthesized using a slightly modified literature method [50,52]. The procedure used for the preparation of the first series of Al–MO hybrid materials was as follows: 1.0 g of Pluronic P123 ($EO_{20}PO_{70}EO_{20}$) and known amounts of Al–NN (aluminum nitrate nanahydrate) or Al–IP (aluminum isopropoxide) were added to 20 mL of absolute ethanol in a 250 mL polypropylene bottle. Then, the reaction mixture was stirred at 25 °C for 70 min. After that, a predetermined amount of silica precursor with different organic bridging groups (ICS, BTEE or BTEB) was added to the resultant mixture. The same steps were used in the preparation of the second series of Al–MO hybrid materials with APTS hanging groups. A known amount of APTS was added to the reaction mixture after adding the silica precursor (ICS, BTEE, or BTEB). In both cases, the final mixture was stirred at room temperature for 4 h and kept in an oven at 60 °C for 48 h. Then, extraction of the P123 block copolymer template from the resulting white/yellow solid was performed using 55 mL of 95 % ethanol solution in a 250 mL polypropylene bottle under vigorous stirring for a few minutes. The resulting dispersion was kept in the oven (Sheldon Manufacturing Inc., Cornelius, OR, USA) at 100 °C for another 24 h. Next, the product was filtered and rinsed thoroughly with 95 % ethanol and dried in an oven at 70 °C for an additional 20 h. Finally, calcination of the solid obtained after extraction was conducted in a horizontal quartz tube furnace (Lindberg Blue M manufactured by Thermal Product Solutions, White Deer, PA, USA) at 350 °C for 2 h in flowing N_2 at a heating rate of 2 °C/min. The samples obtained prior to the extraction step (as-synthesized), after extraction (extracted), and after extraction followed by calcination (extracted and calcined), were tested for their adsorption properties. Scheme 2 shows the route used for the synthesis of alumina–organosilica (Al-IC10-AP10) material using two organosilanes, the first with an isocyanurate bridging group (ICS) and the second with 3-aminopropyl (APTS) hanging group.

The aluminum content (0.011 mol) was kept constant for all samples studied. The number of moles of Si used in the synthesis was a percentage of the total number of Al moles. Based on our previous study, the optimum Si% for tris[3-(trimethoxysilyl)propyl]isocyanurate (ICS) and 3-aminopropyltriethoxysilane (APTS) was found to be 10% [50]. The Al–MO materials with higher Si% than 10 showed smaller pore widths due to the geometrical constraints created by bulky silane bridging groups during self-assembly process [50]. Therefore, Si percentage, with respect to the total number of Al moles, was maintained at 10% for all Al–MO samples. The samples were denoted based on the precursors used in the initial reaction mixture. For instance, the short abbreviation AN refers to aluminum nitrate nanahydrate (Al–NN), and AI to aluminum isopropoxide (Al–IP). Short abbreviations IC, B, E, and AP, were used for tris[3-(trimethoxysilyl)propyl]isocyanurate (ICS), 1,4-bis(triethoxysilyl)benzene (BTEB), bis(triethoxysilyl)ethane (BTEE), and 3-aminopropyltriethoxysilane (APTS), respectively. For instance, AN-B10 refers to the samples synthesized using Al–NN and BTEB (10% Si), whereas AN-B10-AP10 refers to the samples synthesized using Al–NN, and BTEB (10% Si) and APTS (10% Si). In both cases, Al-NN acts as an alumina precursor. Similarly, AI-B10 refers to the samples prepared using Al–IP and BTEB (10% Si), while AI-B10-AP10 refers to the samples obtained by using Al–IP, BTEB (10% Si), and APTS (10% Si)). The as-synthesized samples are denoted with a # mark. Labels for all extracted samples do not have an asterisk (*), whereas the samples subjected to the extraction, followed by calcination, are marked with an asterisk (*).

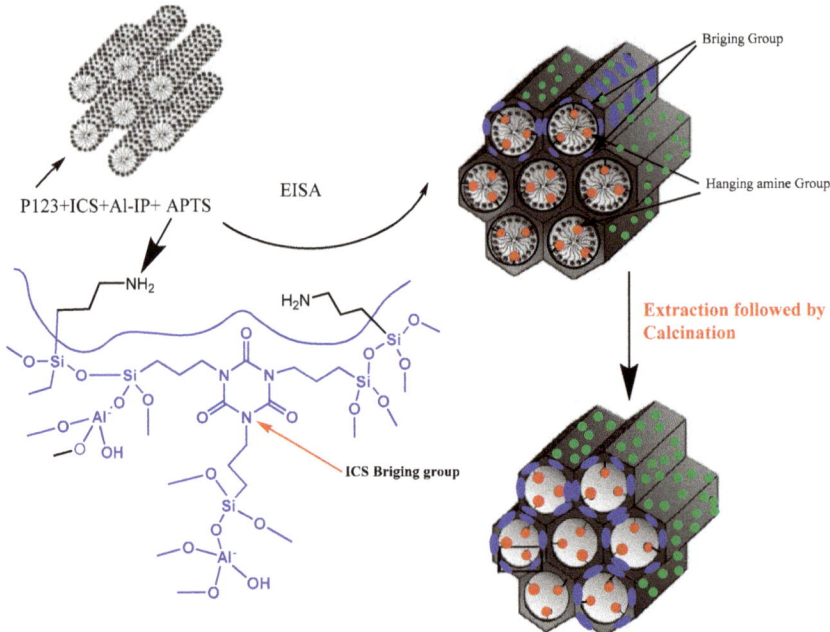

Scheme 2. Illustration of the synthesis route for alumina–organosilica material (Al-IC10-AP10) using isocyanurate bridging silane (ICS) and 3-aminopropyltriethoxysilane (APTS).

Four analogous samples to AN-IC10, AN-IC10*, AN-IC10-AP10, and AN-IC10-AP10* were studied in a previous work [50], where a series of materials was prepared by varying the amount of ICS. The synthesis and calcination conditions used in [50] and this work are slightly different, which results in different adsorption parameters. For instance, AN-IC10 in the current paper was calcined at 350 °C with heating rate of 2 °C/min, while an analogous sample reported in [50] was calcined at 300 °C with heating rate of 3 °C/min. In previous work [50], alumina precursors were mixed with one bridging silane only, isocyanurate (IC), by varying its percentage from 10 to 80%, while, in this work, silanes (10% only) with different bridging were explored. Importantly, the previous work [50] was focused on the CO_2 capture at elevated temperatures, while the current work reports the newly synthesized samples with different functional groups for CO_2 adsorption at 25 °C.

2.3. Characterization of Materials

Nitrogen (N_2) adsorption isotherms were collected at −196 °C (77 K) using an ASAP 2010 volumetric analyzer (Micromeritics, Inc., Norcross, GA, USA). All materials were outgassed under vacuum at 110 °C for 2 h before recording the adsorption measurements.

High-resolution thermogravimetric measurements were conducted using a TGA Q-500 analyzer (TA Instruments, Inc., New Castle, DE, USA). Thermogravimetric (TG) studies were performed from 25 °C to 700 °C in flowing N_2 gas, using a heating rate of 10 °C/min. The TG profiles were used to study the thermal stability of the hybrid materials and the removal of P123 template.

2.4. NMR Analysis

All NMR spectra were collected on Bruker Avance (III) 400WB NMR spectrometer (Bruker Biospin Corporation, Billerica, MA, USA) equipped with a magic-angle spinning (MAS) triple resonance probe head having zirconia rotors of 4 mm diameter by using the procedure reported in [50].

2.5. CO$_2$ Adsorption Measurements

CO$_2$ adsorption studies were conducted using an ASAP 2020 volumetric adsorption analyzer (Micromeritics, Inc., Norcross, GA, USA) at the pressures up to ~1.2 atm and 25 °C using ultrahigh pure CO$_2$ gas. All samples were outgassed at 110 °C for 2 h under vacuum before the analysis. The sample reusability was tested over a number of adsorption/desorption cycles for the sample with the highest CO$_2$ adsorption capacity. Six consecutive adsorption/desorption runs were recorded on ASAP 2020. For instance, the first adsorption isotherm measurement was carried out up to 1.2 atm followed by desorption cycle. After the completion of the first run, the second adsorption/desorption run was started manually on the same instrument, and so on.

2.6. Calculation of Surface Properties

The Brunauer–Emmett–Teller (BET) specific surface area (S_{BET}) of each sample was calculated by following its N$_2$ adsorption isotherm in the relative pressure range of 0.05 to 0.20. The single-point pore volume (V_{sp}) was determined based on the amount N$_2$ gas adsorbed at a relative pressure of ~0.98. The pore size distributions (PSD) were calculated using adsorption branches of N$_2$ adsorption/desorption isotherms and applying the improved Kruk–Jaroniec–Sayari (KJS) method developed for cylindrical pores [53]. The total pore volume (V_t) was calculated by integrating the entire PSD curve and the volume of micropores (V_{mi}) was calculated by taking the difference between V_t and V_{meso} ($V_t - V_{meso}$). The mesopore volume (V_{meso}) was determined by integrating the PSD curve from 2 nm to the end of the PSD curve. In addition to the PSD curves that display the overall distribution of pores, the pore widths (W_{max}) corresponding to the maxima of these distributions are also tabulated to show the characteristic pore sizes of the samples studied.

3. Results and Discussion

3.1. Properties of Al–MO Hybrid Composites

High-resolution thermogravimetry (HR-TG) and differential thermogravimetry (DTG) studies were conducted to identify the thermal stability of silica precursors with ethylene, benzene, isocyanurate bridging groups, and amine hanging groups. Figure 1 shows the DTG profiles recorded in flowing N$_2$ for AN-IC10#, AN-IC10*, and AN-IC10-AP10* samples. As shown in Figure 1, all samples exhibit a sharp peak approximately at 100 °C, assigned to the loss of physically adsorbed water. The as-synthesized (AN-IC10#) sample displays two additional peaks on its DTG profile in the temperature ranges of 150–300 °C and 350–480 °C, corresponding to the decomposition of P123 block copolymer template and isocyanurate bridging groups, respectively, see Figure 1.

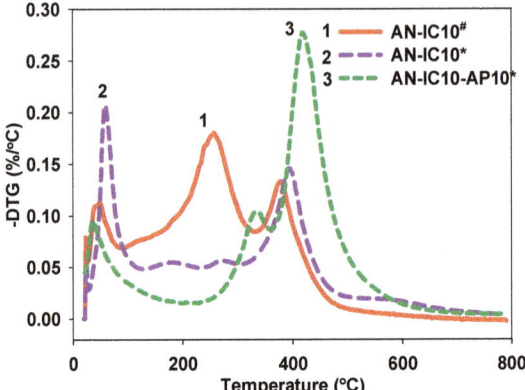

Figure 1. DTG profiles of the as-synthesized (#) alumina–silica composite with isocyanurate bridging groups and alumina–silica composites with isocyanurate bridging groups in the presence and absence of aminopropyl groups after extraction followed by calcination (*). Note: AN and AP refer to the aluminum nitrate and aminopropyl, respectively.

The DTG profiles obtained for the extracted and calcined alumina–silica samples (AN-IC10-AP10* and AN-IC10*) with isocyanurate (ICS) bridging groups in the presence and absence of aminopropyl group show a second major peak centered at 430 °C, related to the decomposition of isocyanurate bridging groups, see Figure 1. The small peak observed at 330 °C for AN-IC10-AP10* corresponds to the degradation of aminopropyl groups. The DTG profiles for AN-IC10-AP10* and AN-IC10* do not have any peak in the temperature range from 150 to 300 °C, indicating the complete removal of block copolymer template via extraction followed by calcination, see Figure 1. The corresponding TG curves of AN-IC10#, AN-IC10*, and AN-IC10-AP10* samples are shown in Figure 2. Figure 3 displays the DTG profiles of three extracted and calcined alumina–silica samples (AN-IC10*, AN-E10*, and AN-B10*) with ICS, BTEE, and BTEB bridging groups. As can be seen from Figure 3, peaks at ~400, 550, and 600 °C correspond to the thermal degradation of isocyanurate, benzene, and ethylene bridging groups, respectively.

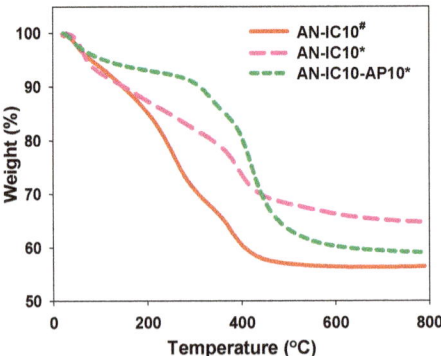

Figure 2. TG profiles of the as-synthesized (#) alumina–silica sample with isocyanurate bridging groups and alumina–silica samples with isocyanurate bridging groups in the presence and absence of aminopropyl groups after the extraction and calcination (*). Note that AN and AP refer to the aluminum nitrate and aminopropyl, respectively.

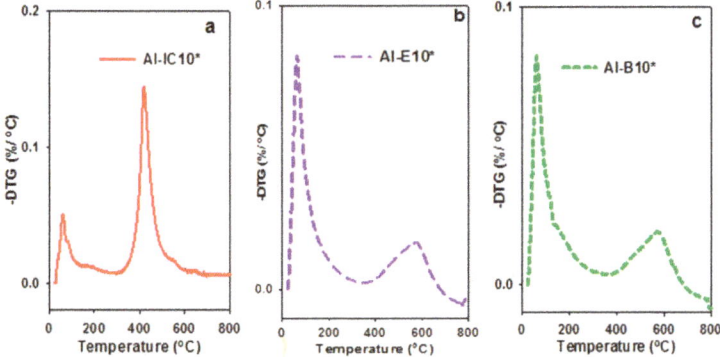

Figure 3. DTG profiles for the extracted and calcined (*) alumina–silica composites containing isocyanurate (IC; panel **a**), ethylene (E; panel **b**), and benzene (B; panel **c**) bridging groups. Note that Al refers to aluminum isopropoxide.

Figures 4 and 5 show comparisons of N_2 adsorption/desorption isotherms for the Al–MO samples synthesized using Al–NN as an alumina precursor with three different bridged organosilanes in the absence and presence of 3-aminopropyltriethoxysilane (AP) at −196 °C (77 K), respectively. All extracted (AN-B10, AN-E10, AN-IC10, AN-B10-AP10, AN-E10-AP10, AN-IC10-AP10), and extracted and calcined (AN-B10*, AN-E10*, AN-IC10*, AN-B10-AP10*, AN-E10-AP10*, AN-IC10-AP10*) samples exhibit type IV nitrogen adsorption isotherms with steep capillary condensation–evaporation steps beginning at a relative pressure of around 0.85 (see adsorption isotherms in Figures 4 and 5). Insets in Figures 4 and 5 show the pore size distribution (PSD) curves corresponding to the aforementioned adsorption isotherms obtained by the KJS method [53]. Our N_2 adsorption isotherms reveal that the extracted samples derived from Al–NN (AN-IC10, AN-B10, AN-E10, AN-B10-AP10, AN-E10-AP10) have better surface properties, as evidenced by higher surface area, total pore volume (V_t), and pore width (W_{max}), as compared to the extracted and calcined counterparts (AN-IC10*, AN-B10*, AN-E10*, AN-B10-AP10*, AN-E10-AP10*). This may be attributed to the shrinkage of organosilica structure due to the calcination process at a relatively high temperature (350 °C) [54].

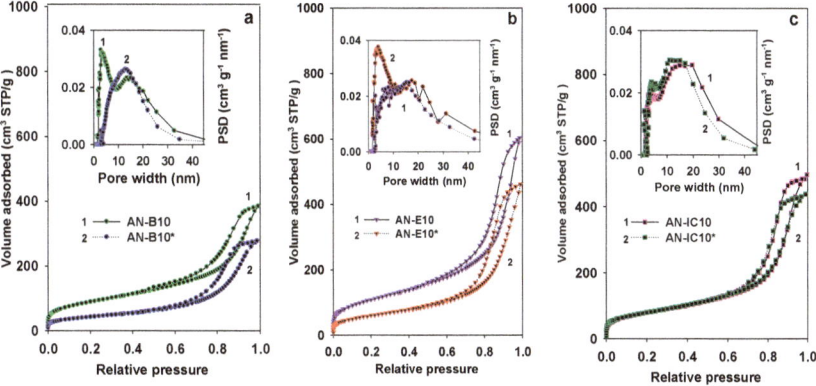

Figure 4. N_2 adsorption isotherms and their corresponding PSD curves (insets) for the extracted (AN-B10 (left panel **a**), AN-E10 (middle panel **b**), and AN-IC10 (right panel **c**)), and extracted and calcined (AN-B10* (left panel **a**), AN-E10* (middle panel **b**), and AN-IC10* (right panel **c**)) samples. AN refers to aluminum nitrate.

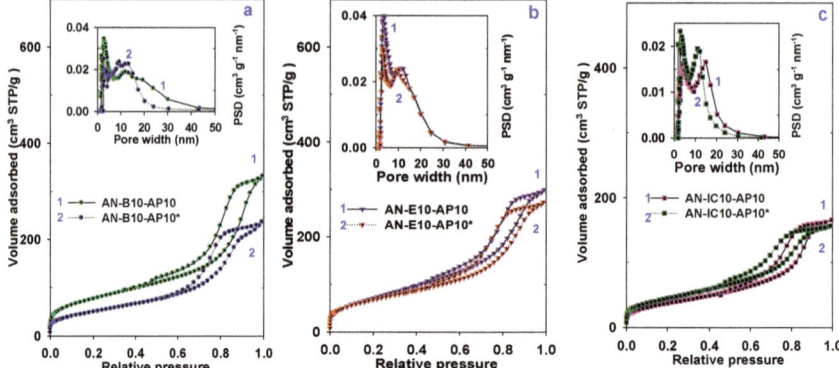

Figure 5. N$_2$ adsorption isotherms and their corresponding PSD curves (insets) for extracted (AN-B10-AP10 (left panel **a**), AN-E10-AP10 (middle panel **b**), and AN-IC10-AP10 (right panel **c**)), and extracted calcined (AN-B10-AP10*(left panel **a**), AN-E10-AP10*(middle panel **b**), and AN-IC10-AP10*(right panel **c**)) samples. Note that AN and AP refer to aluminum nitrate and aminopropyl, respectively.

Similar to the Al–NN-derived samples, all Al–IP-derived samples also show type IV isotherm with relatively broad capillary condensation–evaporation steps pertained to H1 type hysteresis loop (see Figures 6 and 7) and quite narrow PSDs (see insets in Figures 6 and 7). As shown in Table 1, all extracted and calcined samples derived from Al–IP (AI-B10*, AI-E10*, AI-IC10*, AI-B10-AP10*, AI-E10-AP10*, and AI-IC10-AP10*) exhibit lower surface parameters, including surface area, microporosity, and pore width, as compared to the corresponding extracted samples (AI-B10, AI-E10, AI-IC10, AI-B10-AP10, AI-E10-AP10, and AI-IC10-AP10). For example, the specific surface area of AI-IC10 decreases from 655 m$^2 \cdot$g^{-1} to 618 m$^2 \cdot$g^{-1} after grafting aminopropyl hanging groups (AI-IC10-AP10*). In addition, the pore width of AI-IC10 sample reduces from 5.6 nm to 5.4 nm after introducing aminopropyl hanging groups (AI-IC10-AP10*), see Table 1. The reduction in the surface properties of the extracted and calcined samples is due to the shrinkage of the organosilica structure upon calcination at 350 °C [54].

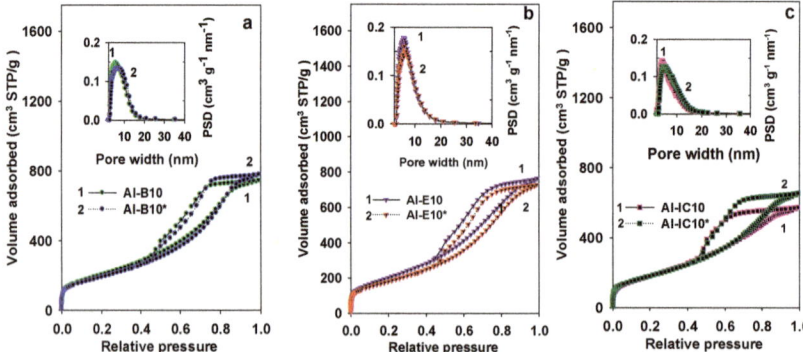

Figure 6. N$_2$ adsorption isotherms and their respective PSD curves (insets) for the extracted (AI-B10 (left panel **a**), AI-E10 (middle panel **b**), and AI-IC10 (right panel **c**)), and extracted calcined (AI-B10* (left panel **a**), AI-E10* (middle panel **b**), AI-IC10* (right panel **c**)) samples, where B, E, and IC refer to benzene, ethylene, and isocyanurate bridging groups, respectively; AI refers to aluminum isopropoxide.

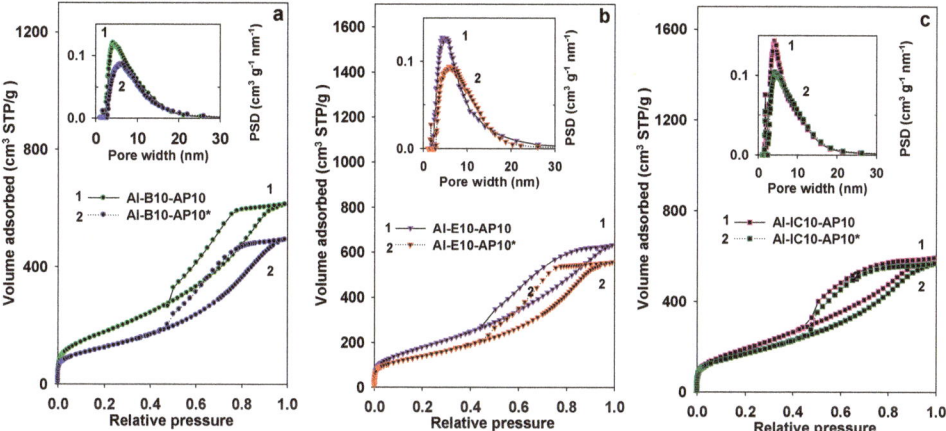

Figure 7. N_2 adsorption isotherms (left panel) and the corresponding PSD curves (insets) for the extracted (AI-B10-AP10 (left panel **a**), AI-E10-AP10 (middle panel **b**), and AI-IC10-AP10 (right panel **c**)), and extracted calcined (AI-B10-AP10* (left panel **a**), AI-E10-AP10* (middle panel **b**), AI-IC10-AP10* (right panel **c**)) samples. AI refers to aluminum isopropoxide, and AP represents aminopropyl.

Table 1. Adsorption parameters obtained for the alumina–silica samples studied.

Samples	S_{BET} (m²/g)	V_{sp} (cc/g)	V_t (cc/g)	V_{mi} (cc/g)	W_{max} (nm)	n_{CO2} (mmol/g)	n^*_{CO2} (μmol/m²)
AN-B10	319	0.60	0.63	0.02	2.9/13.6	0.78	2.45
AN-B10*	150	0.42	0.44	<0.01	12.0#	0.73	4.87
AN-E10	393	0.91	0.94	<0.01	4.1/17.0	1.09	2.77
AN-E10*	219	0.67	0.72	<0.01	15.9#	0.85	3.88
AN-IC10	289	0.74	0.77	0.01	4.0/16.4#	0.92	3.18
AN-IC10*	286	0.68	0.68	<0.01	4.0/13.2#	1.72	6.01
AN-B10-AP10	284	0.50	0.52	0.02	2.9/12.3	0.96	3.38
AN-B10-AP10*	186	0.36	0.38	<0.01	9.4/13.3	1.22	6.56
AN-E10-AP10	265	0.46	0.48	<0.01	3.4/11.6#	0.74	2.79
AN-E10-AP10*	247	0.42	0.43	<0.01	3.2/9.8#	1.51	6.11
AN-IC10-AP10	138	0.26	0.26	<0.01	4.1/14.8	0.72	5.21
AN-IC10-AP10*	163	0.24	0.24	<0.01	3.6/11.1	1.40	8.59
AI-B10	742	1.14	1.24	0.05	5.1	2.60	3.50
AI-B10*	684	1.20	1.28	0.03	6.1	1.99	2.91
AI-E10	740	1.18	1.27	0.03	5.7	2.64	3.57
AI-E10*	652	1.12	1.19	0.02	6.1	2.35	3.60
AI-IC10	664	0.89	0.95	0.04	4.1	1.66	2.50
AI-IC10*	655	1.01	1.06	0.02	4.7	2.39	3.65
AI-B10-AP10	654	0.94	1.00	0.02	4.2	1.83	2.80
AI-B10-AP10*	452	0.76	0.78	<0.01	5.6	1.74	3.85
AI-E10-AP10	664	0.97	1.03	0.02	4.0	1.86	2.80
AI-E10-AP10*	509	0.86	0.90	<0.01	5.1	1.70	3.34
AI-IC10-AP10	713	0.91	0.98	0.03	4.0	2.33	3.27
AI-IC10-AP10*	618	0.87	0.90	0.02	5.4	2.21	3.58

S_{BET}—specific surface area, V_{sp}—single point pore volume, V_t—total pore volume, V_{mi}—volume of micropores, W_{max}—pore width (# refers to the middle value of the "flat" PSD peak), n_{CO2}—number of moles of CO_2 adsorbed at 1 atm per gram of the sample, and n^*_{CO2}—number of moles of CO_2 adsorbed at 1 atm per unit surface area of the sample. B, E, or IC refer to benzene, ethylene, and isocyanurate bridging groups, respectively; AN and AP refer to aluminum nitrate and aminopropyl, respectively. Extracted, and extracted and calcined samples, are denoted without* and with*, respectively.

It is noteworthy that the samples prepared from Al–IP precursor show much higher surface area and total pore volume (V_t) as compared to the analogous samples prepared from Al–NN precursor, see Table 1 and Figure 7. For instance, the specific surface area of AI-B10 is 742 m$^2 \cdot$g^{-1}, as compared to 319 m$^2 \cdot$g^{-1} of AN-B10, and the total pore volume (V$_t$) of AI-B10 is 1.24 cm$^3 \cdot$g^{-1} as compared to 0.63 cm$^3 \cdot$g^{-1} of AN-B10, see Table 1. However, Al–NN-derived samples exhibit higher pore widths as compared to Al–IP-derived samples. For example, the pore width of AN-B10 sample is 13.6 nm as compared to 5.6 nm of AI-B10 sample.

^{13}C, ^{27}Al, and ^{29}Si solid-state NMR spectra provide a comprehensive description of the chemical environment of carbon, aluminum, and silicon atoms present in the Al–MO materials studied. For instance, ^{1}H–^{29}Si cross polarization (CP) MAS NMR spectra reflect the degree of condensation of the siloxane bonds of isocyanurate, benzene, ethylene bridging groups, and aminopropyl hanging groups present in the hybrid organosilica framework. ^{27}Al magic-angle spinning (MAS) NMR spectra reveal the coordination of the incorporated aluminum units in the Al–MO composites. ^{1}H–^{13}C CP/MAS NMR spectra confirm the presence of organic bridging (B, E, IC) and surface aminopropyl groups. The ^{27}Al MAS NMR spectra exhibit three peaks for the Al–MO samples studied, see Figure 8. As shown in Figure 8, both AI-IC10* and AN-IC10* samples display an intense resonance peak at 3.7 ppm attributed to the aluminum species in an octahedral environment. The additional weak resonance peaks, visible at about 66.8 and 34.5 ppm, can be attributed to the aluminum in tetrahedral and pentahedral symmetries, respectively. The ^{1}H–^{29}Si CP/MAS NMR spectra of the AI-B10*, AI-E10*, and AI-IC10* samples show one prominent resonance peak at −82.2 ppm (Figure 9), corresponding to Q^2 (Si–(OSi)$_2$(OH)$_2$)/(Si–(OSi)$_2$(OH)(OAl)) groups. The spectrum of AI-E10* displays an additional small peak at ~−47 ppm, corresponding to the T^1 (R–Si–(OSi)$_1$(OH)$_2$) structure. There are no visible peaks above 90 ppm, indicating the absence of Q^n (n = 3,4) structures. Figure 10 shows the ^{1}H–^{13}C CP/MAS spectra of the AI-IC10-AP10*, AI-B10-AP10*, and AI-E10-AP10* samples. The AI-IC10-AP10* and AI-B10-AP10* samples show three characteristic resonance peaks at 9, 21, and 44 ppm, referring to the carbon atoms bonded to Si atoms, carbon atoms present in the middle of propyl chain, and carbon atoms directly attached to N atoms, respectively. The characteristic sharp peaks at 150 and 138 ppm reflect carbon atoms in isocyanurate and benzene bridging groups, respectively. The spectrum of the AI-E10-AP10* sample exhibits four distinct peaks at 5.5, 9.0, 21.0, and 28.9 ppm. The first peak at 5.5 ppm refers to ethylene bridging groups. The second peak at 9 ppm refers to the carbon atoms present in the aminopropyl chain directly bonded to silicon. Note that the carbon atom directly bonded to Si also appears at 9 ppm. The third and fourth peaks at 21 and 28 ppm, of AI-E10-AP10* sample, correspond to the carbon atoms in the middle of the propyl chain and carbon atoms directly attached to N atoms in the aminopropyl chain, see Figure 10.

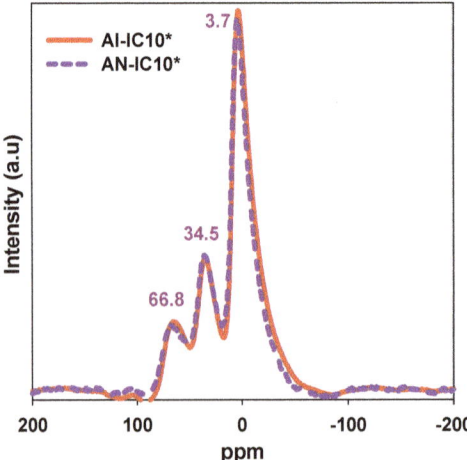

Figure 8. ^{27}Al-MAS NMR spectra of the AI-IC10* and AN-IC10* for extracted and calcined samples.

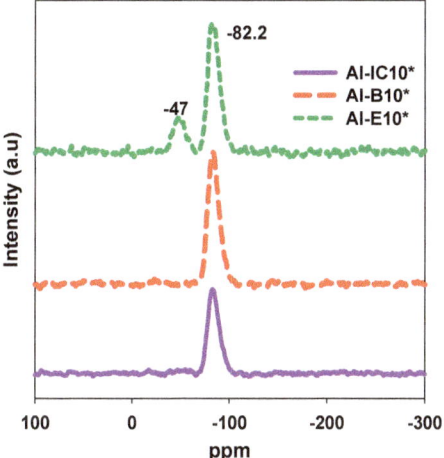

Figure 9. ^1H–^{29}Si-MAS NMR spectra of the Al-IC10*, Al-B10*, and Al-E10* samples.

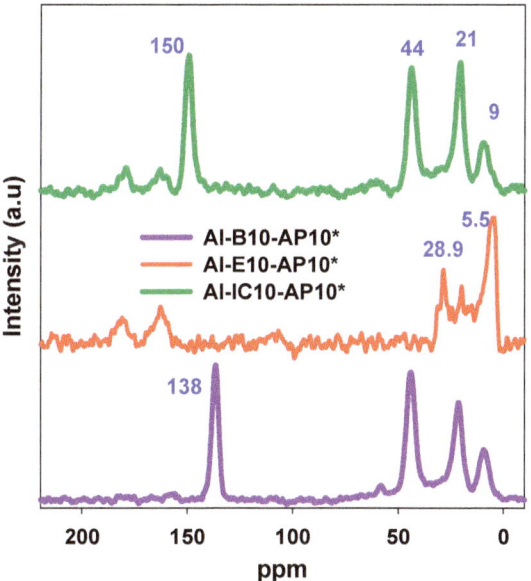

Figure 10. ^1H–^{13}C CP/MAS NMR spectra of the Al-IC10-AP10*, Al-E10-AP10*, and Al-B10-AP10* samples.

3.2. CO$_2$ Sorption

Four series of Al–MO samples were examined for CO$_2$ adsorption at pressures up to ~1.2 atm and 25 °C. Figures 11 and 12 show a comparison of CO$_2$ adsorption isotherms measured for all Al–MO samples studied. AN-B10, AN-E10, and AN-IC10 samples, and their extracted-followed-by-calcination counterparts (AN-B10*, AN-E10*, and AN-IC10*) show CO$_2$ uptakes at 1 atm in the range from 0.72 to 1.72 mmol/g (Table 1 and Figure 11a). However, Al-B10, Al-E10, and Al-IC10 samples, and their extracted-followed-by-calcination counterparts (Al-B10*, Al-E10*, and Al-IC10*) exhibit higher CO$_2$ uptakes at 1 atm, varying from 1.66 to 2.64 mmol/g (Table 1 and Figure 12a).

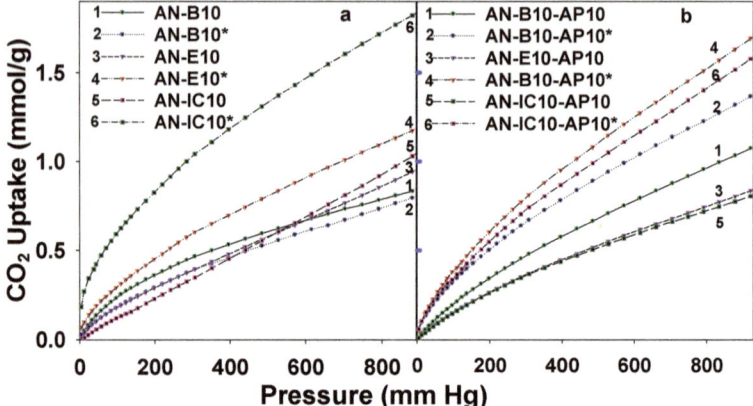

Figure 11. CO$_2$ adsorption profiles at 25 °C, measured on the extracted, and extracted and calcined a) AN-Z10 and b) AN-Z10-AP10 samples, where Z = B, E, or IC.

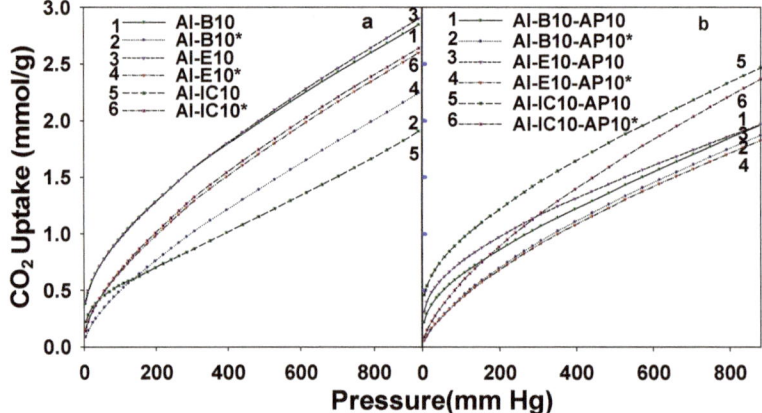

Figure 12. CO$_2$ adsorption profiles at 25 °C measured on the extracted, and extracted and calcined (*) (a) AN-Z10 and (b) AN-Z10-AP10 samples, where Z = B, E, or IC.

It has been previously reported that the CO$_2$ adsorption at ambient temperature is governed by a physisorption mechanism, which is dependent on the volume of micropores (below 2 nm), in particular, ultramicropores (below 0.7 nm) and surface area [55,56]. However, the samples studied show very low micropore volumes, suggesting that, in this case, the surface area and surface properties are major factors determining the CO$_2$ adsorption (see Table 1). Moreover, the samples synthesized using Al–IP precursor show an approximately two-fold increase in the specific surface area and microporosity, in comparison to the corresponding samples synthesized from Al–NN precursor (see Table 1) and, consequently, their CO$_2$ uptakes at 1 atm are higher. Our results also reveal that the structure of alumina precursor plays a vital role in determining the surface properties of the Al–MO hybrid materials. Therefore, aluminum isopropoxide (Al–IP) precursor works better than aluminum nitrate nanahydrate (Al–NN) precursor, due to its ability to form 3D extended mesoporous structures with block copolymers during the co-condensation process [54,57]. Among three silica precursors, organosilica composite with ethylene groups (AI-E10) exhibits the highest CO$_2$ uptake of 2.64 mmol/g at 1 atm, see Table 1. At the same pressure, the samples with benzene (AI-B10) and isocyanurate (AI-IC10*) bridging groups exhibit the CO$_2$ uptakes of 2.60 and 2.39 mmol/g, respectively. Our data suggest that the samples subjected to extraction and calcination have lower surface parameters, including microporosity, mesoporosity, surface area, and pore size (see Table 1), as compared to the samples subjected to extraction only, which is due to the shrinkage of the Al–MO materials at high temperatures. For example, the extracted and calcined samples with benzene and ethylene bridging groups show slightly lower CO_2 adsorption capacities as compared to the samples subjected to extraction only. The CO_2 uptakes measured at 1 atm for AI-B10 and AI-E10 change from 2.60 and 2.64 mmol/g, to 1.99 and 2.35 mmol/g, when calcination is performed (see Table 1). By contrast, the CO_2 uptakes measured for the extracted and calcined samples with nitrogen-containing isocyanurate group (AN-IC10* and AI-IC10*) are larger than those for the extracted samples (AN-IC10 and AI-IC10), see Table 1. Higher CO_2 adsorption on the AN-IC10* and AI-IC10* samples could be assigned to the unique structure of isocyanurate group and the presence of nitrogen species favoring interactions with CO_2. Grafting aminopropyl hanging groups on the surface of the Al–MO composites did not cause a significant effect on the CO_2 uptake (see Table 1). Al–MO samples with aminopropyl hanging groups also display lower microporosity and surface area (see Table 1 and Figures 11 and 12). Our results demonstrate that the CO_2 adsorption on mesoporous alumina–organosilica materials is mainly determined by the surface area of the sample, alumina precursor, and structure and functionality of the organic bridging groups. Calcination of the alumina–silica samples at 350 °C did not improve their adsorption properties. Amine-containing materials are commonly used as chemical sorbents at higher temperatures; however, grafting amine groups on the Al–MO materials did not affect physisorption of CO_2 at the conditions studied.

Cycle stability of the Al–MO material with the highest CO_2 adsorption capacity was also tested as described previously [58,59]. Cycle stability measurements for AI-E10 were conducted for six consecutive cycles, and CO_2 uptake was measured. The CO_2 uptake remained stable after six adsorption/desorption cycles with approximately 2.64 mmol/g CO_2 uptake at 1 atm, suggesting good stability of AI-E10 materials. Note that the alumina samples (extracted and calcined at 350 °C) do not show a long-range order on the XRD profiles; thus, they are amorphous in nature (data not shown).

Alumina-based composite materials have been investigated for CO_2 capture. Table 2 summarizes a comparison of CO_2 uptakes for various alumina-based composite materials reported in the literature at low and elevated temperatures. However, studies of alumina-based materials for CO_2 capture at ambient conditions are rare. The sorbents studied in this work show much higher CO_2 uptake at 25°C, as compared to the values reported previously for alumina-based materials under the same conditions.

Table 2. Comparison of the CO_2 sorption data for different alumina-based composite materials.

Material	Temperature (°C)	Pressure (atm)	CO_2 Uptake (mmol/g)	Ref.
Al–Mg oxide	60 (13%H_2O)	0.99	1.36	[49]
Al_2O_3–silica	120	0.99	2.20	[50]
Al–Zr-mixed oxide–silica	0	1	1.83	[58]
	25	1	1.39	
	60	1	2.60	
	120	1	2.37	
Al–Mg mixed oxide–silica	300	0.99	0.46	[60]
γ-Al_2O_3	60	9.9	1.94	[61]
N-doped γ-Al_2O_3	55	NG	0.67	[62]
Al–Mg mixed oxide–nitrate–graphene oxide	60	1	1.00	[63]
Al_2O_3–silica	25	1	2.64	This work

4. Conclusions

Mesoporous alumina–organosilica (Al–MO) materials exhibited high CO_2 adsorption capacity at 25 °C. CO_2 adsorption properties of Al–MO materials depend on the surface area of the sample, alumina precursor, and structure and functionality of the organosilica bridging group. The Al–NN- and Al–IP-derived samples exhibited the highest CO_2 uptakes of 1.72 and 2.64 mmol/g at 1 atm, respectively. High CO_2 uptake, reusability, selectivity, and good thermal and mechanical stability, make Al–MO materials potential adsorbents for larger-scale CO_2 capture at 25 °C.

Author Contributions: Writing—Review & Editing, C.G, R.S.D, C.S.K, M.J.

Funding: This research received no external funding.

Acknowledgments: Authors thanks to Mahinda Gangoda for technical support with NMR measurements.

Conflicts of Interest: The authors declare no conflict of interest.

References

1. Dassanayake, R.S.; Gunathilake, C.; Dassanayake, A.C.; Abidi, N.; Jaroniec, M. Amidoxime-functionalized nanocrystalline cellulose–mesoporous silica composites for carbon dioxide sorption at ambient and elevated temperatures. *J. Mater. Chem. A* **2017**, *5*, 7462–7473. [CrossRef]
2. Burwell, R.L. Section 1—Definitions and Terminology. In *Manual of Symbols and Terminology for Physicochemical Quantities and Units–Appendix II*; Burwell, R.L., Ed.; Pergamon Press: Oxford, UK, 1976; pp. 74–86.
3. Oschatz, M.; Antonietti, M. A search for selectivity to enable CO_2 capture with porous adsorbents. *Energy Environ. Sci.* **2018**, *11*, 57–70. [CrossRef]
4. Nugent, P.; Belmabkhout, Y.; Burd, S.D.; Cairns, A.J.; Luebke, R.; Forrest, K.; Pham, T.; Ma, S.; Space, B.; Wojtas, L.; et al. Porous materials with optimal adsorption thermodynamics and kinetics for CO_2 separation. *Nature* **2013**, *495*, 80–84. [CrossRef] [PubMed]
5. Dong, F.; Lou, H.; Goto, M.; Hirose, T. A new PSA process as an extension of the Petlyuk distillation concept. *Sep. Purif. Technol.* **1999**, *15*, 31–40. [CrossRef]
6. Kikkinides, E.S.; Yang, R.T.; Cho, S.H. Concentration and recovery of carbon dioxide from flue gas by pressure swing adsorption. *Ind. Eng. Chem. Res.* **1993**, *32*, 2714–2720. [CrossRef]
7. Burchell, T.D.; Judkins, R.R.; Rogers, M.R.; Williams, A.M. A novel process and material for the separation of carbon dioxide and hydrogen sulfide gas mixtures. *Carbon* **1997**, *35*, 1279–1294. [CrossRef]
8. Rutherford, S.W.; Do, D.D. Adsorption dynamics of carbon dioxide on a carbon molecular sieve 5A. *Carbon* **2000**, *38*, 1339–1350. [CrossRef]
9. Cinke, M.; Li, J.; Bauschlicher, C.W.; Ricca, A.; Meyyappan, M. CO_2 adsorption in single-walled carbon nanotubes. *Chem. Phys. Lett.* **2003**, *376*, 761–766. [CrossRef]
10. Huang, L.; Zhang, L.; Shao, Q.; Lu, L.; Lu, X.; Jiang, S.; Shen, W. Simulations of Binary Mixture Adsorption of Carbon Dioxide and Methane in Carbon Nanotubes: Temperature, Pressure, and Pore Size Effects. *J. Phys. Chem. C* **2007**, *111*, 11912–11920. [CrossRef]
11. Inui, T.; Okugawa, Y.; Yasuda, M. Relationship between properties of various zeolites and their carbon dioxide adsorption behaviors in pressure swing adsorption operation. *Ind. Eng. Chem. Res.* **1988**, *27*, 1103–1109. [CrossRef]
12. Siriwardane, R.V.; Shen, M.-S.; Fisher, E.P. Adsorption of CO_2, N_2, and O_2 on Natural Zeolites. *Energy Fuels* **2003**, *17*, 571–576. [CrossRef]
13. Demessence, A.; D'Alessandro, D.M.; Foo, M.L.; Long, J.R. Strong CO_2 Binding in a Water-Stable, Triazolate-Bridged Metal−Organic Framework Functionalized with Ethylenediamine. *J. Am. Chem. Soc.* **2009**, *131*, 8784–8786. [CrossRef] [PubMed]
14. Yazaydın, A.Ö.; Snurr, R.Q.; Park, T.-H.; Koh, K.; Liu, J.; LeVan, M.D.; Benin, A.I.; Jakubczak, P.; Lanuza, M.; Galloway, D.B.; et al. Screening of Metal−Organic Frameworks for Carbon Dioxide Capture from Flue Gas Using a Combined Experimental and Modeling Approach. *J. Am. Chem. Soc.* **2009**, *131*, 18198–18199. [CrossRef] [PubMed]
15. Lee, S.C.; Chae, H.J.; Lee, S.J.; Park, Y.H.; Ryu, C.K.; Yi, C.K.; Kim, J.C. Novel regenerable potassium-based dry sorbents for CO_2 capture at low temperatures. *J. Mol. Catal. B Enzym.* **2009**, *56*, 179–184. [CrossRef]
16. Lee, S.C.; Choi, B.Y.; Lee, T.J.; Ryu, C.K.; Ahn, Y.S.; Kim, J.C. CO_2 absorption and regeneration of alkali metal-based solid sorbents. *Catal. Today* **2006**, *111*, 385–390. [CrossRef]

17. Lee, S.C.; Kim, J.C. Dry Potassium-Based Sorbents for CO_2 Capture. *Catal. Surv. Asia* **2007**, *11*, 171–185. [CrossRef]
18. Lee, S.C.; Kwon, Y.M.; Ryu, C.Y.; Chae, H.J.; Ragupathy, D.; Jung, S.Y.; Lee, J.B.; Ryu, C.K.; Kim, J.C. Development of new aluminium oxide-modified sorbents for CO_2 sorption and regeneration at temperatures below 200 °C. *Fuel* **2011**, *90*, 1465–1470. [CrossRef]
19. Plaza, M.G.; Pevida, C.; Arias, B.; Fermoso, J.; Rubiera, F.; Pis, J.J. A comparison of two methods for producing CO_2 capture adsorbents. *Energy Procedia* **2009**, *1*, 1107–1113. [CrossRef]
20. Przepiórski, J.; Skrodzewicz, M.; Morawski, A.W. High temperature ammonia treatment of activated carbon for enhancement of CO_2 adsorption. *Appl. Surf. Sci.* **2004**, *225*, 235–242. [CrossRef]
21. Lu, C.; Bai, H.; Wu, B.; Su, F.; Hwang, J.F. Comparative Study of CO_2 Capture by Carbon Nanotubes, Activated Carbons, and Zeolites. *Energy Fuels* **2008**, *22*, 3050–3056. [CrossRef]
22. Su, F.; Lu, C.; Cnen, W.; Bai, H.; Hwang, J.F. Capture of CO_2 from flue gas via multiwalled carbon nanotubes. *Sci. Total Environ.* **2009**, *407*, 3017–3023. [CrossRef] [PubMed]
23. Drage, T.C.; Arenillas, A.; Smith, K.M.; Pevida, C.; Piippo, S.; Snape, C.E. Preparation of carbon dioxide adsorbents from the chemical activation of urea–formaldehyde and melamine–formaldehyde resins. *Fuel* **2007**, *86*, 22–31. [CrossRef]
24. Jadhav, P.D.; Chatti, R.V.; Biniwale, R.B.; Labhsetwar, N.K.; Devotta, S.; Rayalu, S.S. Monoethanol Amine Modified Zeolite 13X for CO_2 Adsorption at Different Temperatures. *Energy Fuels* **2007**, *21*, 3555–3559. [CrossRef]
25. Su, F.; Lu, C.; Kuo, S.-C.; Zeng, W. Adsorption of CO_2 on Amine-Functionalized Y-Type Zeolites. *Energy Fuels* **2010**, *24*, 1441–1448. [CrossRef]
26. Lee, S.; Filburn, T.P.; Gray, M.; Park, J.-W.; Song, H.-J. Screening Test of Solid Amine Sorbents for CO_2 Capture. *Ind. Eng. Chem. Res.* **2008**, *47*, 7419–7423. [CrossRef]
27. Cho, E.-B.; Kim, D.; Mandal, M.; Gunathilake, C.A.; Jaroniec, M. Benzene-Silica with Hexagonal and Cubic Ordered Mesostructures Synthesized in the Presence of Block Copolymers and Weak Acid Catalysts. *J. Phys. Chem. C* **2012**, *116*, 16023–16029. [CrossRef]
28. Ma, X.; Wang, X.; Song, C. "Molecular Basket" Sorbents for Separation of CO_2 and H_2S from Various Gas Streams. *J. Am. Chem. Soc.* **2009**, *131*, 5777–5783. [CrossRef] [PubMed]
29. Yue, M.B.; Chun, Y.; Cao, Y.; Dong, X.; Zhu, J.H. CO_2 Capture by As-Prepared SBA-15 with an Occluded Organic Template. *Adv. Funct. Mater.* **2006**, *16*, 1717–1722. [CrossRef]
30. Hiyoshi, N.; Yogo, K.; Yashima, T. Adsorption characteristics of carbon dioxide on organically functionalized SBA-15. *Microporous Mesoporous Mater.* **2005**, *84*, 357–365. [CrossRef]
31. Knowles, G.P.; Delaney, S.W.; Chaffee, A.L. Diethylenetriamine[propyl(silyl)]-Functionalized (DT) Mesoporous Silicas as CO_2 Adsorbents. *Ind. Eng. Chem. Res.* **2006**, *45*, 2626–2633. [CrossRef]
32. Norihito, H.; Katsunori, Y.; Tatsuaki, Y. Adsorption of Carbon Dioxide on Amine Modified SBA-15 in the Presence of Water Vapor. *Chem. Lett.* **2004**, *33*, 510–511.
33. Plaza, M.G.; Pevida, C.; Arias, B.; Fermoso, J.; Arenillas, A.; Rubiera, F.; Pis, J.J. Application of thermogravimetric analysis to the evaluation of aminated solid sorbents for CO_2 capture. *J. Therm. Anal. Calorim.* **2008**, *92*, 601–606. [CrossRef]
34. Liu, X.; Li, J.; Zhou, L.; Huang, D.; Zhou, Y. Adsorption of CO_2, CH_4 and N_2 on ordered mesoporous silica molecular sieve. *Chem. Phys. Lett.* **2005**, *415*, 198–201. [CrossRef]
35. He, Y.; Seaton, N.A. Heats of Adsorption and Adsorption Heterogeneity for Methane, Ethane, and Carbon Dioxide in MCM-41. *Langmuir* **2006**, *22*, 1150–1155. [CrossRef] [PubMed]
36. Choi, S.; Drese, J.H.; Eisenberger, P.M.; Jones, C.W. Application of Amine-Tethered Solid Sorbents for Direct CO_2 Capture from the Ambient Air. *Environ. Sci. Technol.* **2011**, *45*, 2420–2427. [CrossRef] [PubMed]
37. Drese, J.H.; Choi, S.; Lively, R.P.; Koros, W.J.; Fauth, D.J.; Gray, M.L.; Jones, C.W. Synthesis–Structure–Property Relationships for Hyperbranched Aminosilica CO_2 Adsorbents. *Adv. Funct. Mater.* **2009**, *19*, 3821–3832. [CrossRef]
38. Sayari, A.; Belmabkhout, Y. Stabilization of Amine-Containing CO_2 Adsorbents: Dramatic Effect of Water Vapor. *J. Am. Chem. Soc.* **2010**, *132*, 6312–6314. [CrossRef] [PubMed]
39. Zeleňák, V.; Badaničová, M.; Halamová, D.; Čejka, J.; Zukal, A.; Murafa, N.; Goerigk, G. Amine-modified ordered mesoporous silica: Effect of pore size on carbon dioxide capture. *Chem. Eng. J.* **2008**, *144*, 336–342. [CrossRef]

40. Zelenak, V.; Halamova, D.; Gaberova, L.; Bloch, E.; Llewellyn, P. Amine-modified SBA-12 mesoporous silica for carbon dioxide capture: Effect of amine basicity on sorption properties. *Microporous Mesoporous Mater.* **2008**, *116*, 358–364. [CrossRef]
41. Burkett, S.L.; Sims, S.D.; Mann, S. Synthesis of hybrid inorganic–organic mesoporous silica by co-condensation of siloxane and organosiloxane precursors. *Chem. Commun.* **1996**, *11*, 1367–1368. [CrossRef]
42. Macquarrie, D.J. Direct preparation of organically modified MCM-type materials. Preparation and characterisation of aminopropyl–MCM and 2-cyanoethyl–MCM. *Chem. Commun.* **1996**, *16*, 1961–1962. [CrossRef]
43. Harlick, P.J.E.; Sayari, A. Applications of Pore-Expanded Mesoporous Silicas. 3. Triamine Silane Grafting for Enhanced CO_2 Adsorption. *Ind. Eng. Chem. Res.* **2006**, *45*, 3248–3255. [CrossRef]
44. Knowles, G.P.; Graham, J.V.; Delaney, S.W.; Chaffee, A.L. Aminopropyl-functionalized mesoporous silicas as CO_2 adsorbents. *Fuel Process. Technol.* **2005**, *86*, 1435–1448. [CrossRef]
45. Mao, C.-F.; snm Vannice, M.A. High surface area α-aluminium oxide. I.: Adsorption properties and heats of adsorption of carbon monoxide, carbon dioxide, and ethylene. *Appl. Catal. A General* **1994**, *111*, 151–173. [CrossRef]
46. Yong, Z.; Mata, V.; Rodrigues, A.E. Adsorption of Carbon Dioxide on Basic Aluminium oxide at High Temperatures. *J. Chem. Eng. Data* **2000**, *45*, 1093–1095. [CrossRef]
47. Chaikittisilp, W.; Kim, H.-J.; Jones, C.W. Mesoporous Aluminium Oxide-Supported Amines as Potential Steam-Stable Adsorbents for Capturing CO_2 from Simulated Flue Gas and Ambient Air. *Energy Fuels* **2011**, *25*, 5528–5537. [CrossRef]
48. Jeon, H.; Ahn, S.H.; Kim, J.H.; Min, Y.J.; Lee, K.B. Templated synthesis of mesoporous aluminium oxides by graft copolymer and their CO_2 adsorption capacities. *J. Mater. Sci.* **2011**, *46*, 4020–4025. [CrossRef]
49. Li, L.; Wen, X.; Fu, X.; Wang, F.; Zhao, N.; Xiao, F.; Wei, W.; Sun, Y. MgO/Al_2O_3 Sorbent for CO_2 Capture. *Energy Fuels* **2010**, *24*, 5773–5780. [CrossRef]
50. Gunathilake, C.; Gangoda, M.; Jaroniec, M. Mesoporous isocyanurate-containing organosilica–aluminium oxide composites and their thermal treatment in nitrogen for carbon dioxide sorption at elevated temperatures. *J. Mater. Chem. A* **2013**, *1*, 8244–8252. [CrossRef]
51. Asefa, T.; MacLachlan, M.J.; Coombs, N.; Ozin, G.A. Periodic mesoporous organosilicas with organic groups inside the channel walls. *Nature* **1999**, *402*, 867. [CrossRef]
52. Cai, W.; Yu, J.; Anand, C.; Vinu, A.; Jaroniec, M. Facile Synthesis of Ordered Mesoporous Aluminium Oxide and Aluminium Oxide-Supported Metal Oxides with Tailored Adsorption and Framework Properties. *Chem. Mater.* **2011**, *23*, 1147–1157. [CrossRef]
53. Kruk, M.; Jaroniec, M.; Sayari, A. Application of Large Pore MCM-41 Molecular Sieves to Improve Pore Size Analysis Using Nitrogen Adsorption Measurements. *Langmuir* **1997**, *13*, 6267–6273. [CrossRef]
54. Morris, S.M.; Fulvio, P.F.; Jaroniec, M. Ordered Mesoporous Alumina-Supported Metal Oxides. *J. Am. Chem. Soc.* **2008**, *130*, 15210–15216. [CrossRef] [PubMed]
55. Kumar, A.; Hua, C.; Madden, D.G.; O'Nolan, D.; Chen, K.-J.; Keane, L.-A.J.; Perry, J.J.; Zaworotko, M.J. Hybrid ultramicroporous materials (HUMs) with enhanced stability and trace carbon capture performance. *Chem. Commun.* **2017**, *53*, 5946–5949. [CrossRef] [PubMed]
56. Wickramaratne, N.P.; Jaroniec, M. Importance of small micropores in CO_2 capture by phenolic resin-based activated carbon spheres. *J. Mater. Chem. A* **2013**, *1*, 112–116. [CrossRef]
57. Bromberg, L.; Su, X.; Hatton, T.A. Aldehyde Self-Condensation Catalysis by Aluminum Aminoterephthalate Metal–Organic Frameworks Modified with Aluminum Isopropoxide. *Chem. Mater.* **2013**, *25*, 1636–1642. [CrossRef]
58. Gunathilake, C.; Jaroniec, M. Mesoporous alumina–zirconia–organosilica composites for CO_2 capture at ambient and elevated temperatures. *J. Mater. Chem. A* **2015**, *3*, 2707–2716. [CrossRef]
59. Gunathilake, C.; Dassanayake, R.S.; Abidi, N.; Jaroniec, M. Amidoxime-functionalized microcrystalline cellulose–mesoporous silica composites for carbon dioxide sorption at elevated temperatures. *J. Mater. Chem. A* **2016**, *4*, 4808–4819. [CrossRef]
60. Hanif, A.; Dasgupta, S.; Nanoti, A. High temperature CO_2 adsorption by mesoporous silica supported magnesium aluminum mixed oxide. *Chem. Eng. J.* **2015**, *280*, 703–710. [CrossRef]

61. Granados-Correa, F.; Bonifacio-Martinez, J.; Hernandez-Mendoza, H.; Bulbulian, S. Capture of CO_2 on gamma-Al_2O_3 materials prepared by solution-combustion and ball-milling processes. *J. Air Waste Manag. Assoc.* **2016**, *66*, 643–654. [CrossRef] [PubMed]
62. Thote, J.A.; Chatti, R.V.; Iyer, K.S.; Kumar, V.; Valechha, A.N.; Labhsetwar, N.K.; Biniwale, R.B.; Yenkie, M.K.; Rayalu, S.S. N-doped mesoporous alumina for adsorption of carbon dioxide. *J. Environ. Sci.* **2012**, *24*, 1979–1984. [CrossRef]
63. Wang, J.; Mei, X.; Huang, L.; Zheng, Q.; Qiao, Y.; Zang, K.; Mao, S.; Yang, R.; Zhang, Z.; Gao, Y.; et al. Synthesis of layered double hydroxides/graphene oxide nanocomposite as a novel high-temperature CO_2 adsorbent. *J. Energy Chem.* **2015**, *24*, 127–137. [CrossRef]

© 2018 by the authors. Licensee MDPI, Basel, Switzerland. This article is an open access article distributed under the terms and conditions of the Creative Commons Attribution (CC BY) license (http://creativecommons.org/licenses/by/4.0/).

Article

Self-Propagating Synthesis and Characterization Studies of Gd-Bearing Hf-Zirconolite Ceramic Waste Forms

Kuibao Zhang [1,3,*], **Dan Yin** [1], **Kai Xu** [2] **and Haibin Zhang** [4,*]

1. State Key Laboratory of Environment-friendly Energy Materials,
   Southwest University of Science and Technology, Mianyang 621010, China; yindan@swust.edu.cn
2. State Key Laboratory of Silicate Materials for Architectures, Wuhan University of Technology,
   Wuhan 430070, China; kaixu@whut.edu.cn
3. Sichuan Civil-Military Integration Institute, Mianyang 621010, China
4. Institute of Nuclear Physics and Chemistry, China Academy of Engineering Physics,
   Mianyang 621900, China
* Correspondence: xiaobao320@163.com (K.Z.); wsschbinzhang@163.com (H.Z.);
  Tel./Fax: +86-816-241-9492 (K.Z.)

Received: 14 December 2018; Accepted: 30 December 2018; Published: 7 January 2019

Abstract: Synroc is recognized as the second-generation waste matrice for nuclear waste disposal. Zirconolite is one of the most durable Synroc minerals. In this study, Gd and Hf were selected as the surrogates of trivalent and tetravalent actinide nuclides. Gd-bearing Hf-zirconolite ($Ca_{1-x}Hf_{1-x}Gd_{2x}Ti_2O_7$) ceramic waste forms were rapidly synthesized from a self-propagating technique using CuO as the oxidant. The results indicate that Gd can concurrently replace the Ca and Hf sites. However, Gd_2O_3 could not completely be incorporated into the lattice structure of zirconolite when the x value is higher than 0.8. The aqueous durability of selected Gd-Hf codoped sample (Hf-Gd-0.6) was tested, where the 42 days normalized leaching rates (LR_i) of Ca, Cu, Gd and Hf are measured to be 1.57, 0.13, 4.72 × 10^{-7} and 1.59 × 10^{-8} g·m^{-2}·d^{-1}.

Keywords: self-propagating; nuclear waste; zirconolite; actinide; aqueous durability

---

1. Introduction

Due to the main contribution of minor actinides (Np, Am, Cm) to the long term radiotoxicity of high-level nuclear wastes (HLW) recovered from spent fuel reprocessing, the separation of the actinide nuclides and their immobilization in durable matrices have been of prime importance [1,2]. A large number of fundamental and engineering orientated studies have been launched in several countries (France, Japan, Russia, et al.) to explore the feasibility of highly stable matrices, such as ceramics, glass-ceramics or glasses [3–15]. Among these host materials, borosilicate glass has been proved as a desirable matrix for large-scale applications [4,11]. However, the low solubility of minor actinides in glass matrix and the relatively low thermal stability of glass are the major limitations for the disposal of actinide-rich wastes [10,16]. Alternatively, Synroc has been proposed as a potential matrice for HLW immobilization by Ringwood et al. [17]. Synroc is mainly composed of multiple titanate mineral phases, such as zirconolite ($CaZrTi_2O_7$), pyrochlore ($A_2B_2O_6X$), perovskite ($CaTiO_3$), hollandite ($BaAl_2Ti_6O_{16}$), rutile (TiO_2), spinel (AB_2O_4), et al. These mineral phases have accommodated actinide elements in the natural environment for over tens of millions of years. According to the theory of isomorphism substitution, radioactive nuclides can be included into the lattice structure of above-mentioned mineral phases, which can significantly promote the waste loading and long-term stability [17–23].

Zirconolite, which is one of the most durable phases among Synroc minerals, has been extensively investigated as a ceramic matrice [24–27]. Zirconolite exhibits a layered structure, which is formed by the stacking layers of edge shared Ti-O polyhedra (TiO$_6$ and TiO$_5$) and layers of Ca^{2+} and Zr^{4+} ions [25–28]. Due to the composition and nature of substitution, zirconolite can transform into different polytypes like zirconolite-2M (monoclinic), zirconolite-3T (trigonal), zirconolite-3O (orthorhombic). The different forms of coordination make zirconolite structure capable of accommodating large cations like rare-earth, actinide and alkaline earth ions, as well as small cations like transition metal ions [28–30]. Moreover, zirconolite-base waste forms exhibit excellent performances in waste loading, aqueous durability, chemical flexibility, radiation resistance and existence of nature analogues, which make it a potential host phase for the immobilization of separated minor actinides [4,7,8].

In general, zirconolite-rich Synroc waste forms were mainly synthesized by traditional methods, such as liquid phase synthesis (hydroxide and sol-gel methods) and solid state reaction [31–33]. These approaches usually require a long-time sintering process under high-temperature and high pressure, which is time consuming and evokes the risk of nuclide volatilization. Muthuraman et al. have been proposed an alternative synthesis approach, self-propagating high-temperature synthesis (SHS), for the immobilization of nuclear waste [34]. Because of its special advantages [35], SHS technique has been considered as a candidate approach for environment protection, such as stabilization of radioactive and toxic wastes. In recent years, we have explored the rapid synthesis of zirconolite and pyrochlore based waste forms using SHS [36–41]. Quick pressing (QP) was also introduced to obtain highly densified samples. The results demonstrate that highly densified ceramic-based waste forms can be synthesized within several minutes using this SHS/QP technique.

As real actinides contained HLW is not available in laboratory, simulated actinide nuclides are widely employed in fundamental research. From the consideration of crystal chemistry and ionic radius [42], Gd and Ce were usually employed as the surrogates of trivalent and tetravalent actinide elements. From previous studies [43,44], the charge state of Ce is not stable as Ce^{4+} usually transforms to Ce^{3+} under high temperature sintering. Actually, Hf is a better surrogate of tetravalent actinides (especially Pu) over Ce as the charge state of Hf^{4+} is extremely stable. Hf exhibits similar density and solubility as Pu in vitreous waste forms. Hf can also partially or totally replace the Zr site of zirconolite (Hf-zirconolite, CaHfTi$_2$O$_7$) [45,46]. Gd and Hf are considered as a neutron poison for fission reactions because they have extreme high capture cross-sections of thermal neutron [47]. Thus, the Gd-bearing Hf-zirconolite waste forms possess high critical safety when loaded with fissile actinide isotopes of ^{239}Pu and ^{235}U. In this study, Hf-zirconolite was rapidly prepared from an SHS/QP technique using CuO as the oxidant. The Zr site was totally replaced by Hf with chemical composition of CaHfTi$_2$O$_7$. On this basis, Gd$_2$O$_3$ was introduced as the surrogate of trivalent actinides, which was designed to concurrently occupy the Ca and Hf sites of Hf-zriconolite. The phase composition, crystal structure, site occupancy and microstructure of the Gd-bearing samples were investigated. In addition, the aqueous durability was evaluated using the standard MCC-1 leaching test [48].

2. Materials and Methods

Analytical grade CuO, CaO, Ti, TiO$_2$, ZrO$_2$, as well as high purity Gd$_2$O$_3$ and HfO$_2$ (purity ≥ 99.9 wt. %), were purchased as the raw materials. Firstly, the Hf-zirconolite was prepared according to the following chemical equation:

$$6CuO + 2CaO + 3Ti + TiO_2 + 2HfO_2 = 2CaHfTi_2O_7 + 6Cu \quad (1)$$

After that, a series of compositions with stoichiometry as Ca$_{1-x}$Hf$_{1-x}$Gd$_{2x}$Ti$_2$O$_7$ (x = 0.2, 0.4, 0.6, 0.8 and 1.0, named as Hf-Gd-0.2, Hf-Gd-0.4, Hf-Gd-0.6, Hf-Gd-0.8 and Hf-Gd-1.0) were synthesized from this SHS technique. The designed SHS reactions were conducted as follows:

$$6CuO + 2(1-x)CaO + 3Ti + TiO_2 + 2(1-x)HfO_2 + 2xGd_2O_3 = 2Ca_{1-x}Hf_{1-x}Gd_{2x}Ti_2O_7 + 6Cu \quad (2)$$

The weight percentages of raw materials are listed in Table 1. About 20 g reactants were completely homogenized using planetary ball milling. The mixed powders were then preformed into cylindrical pellets with dimension of Φ25 × 12 mm. The pressed pellets were then ignited and densified similarly to in our previous report [36]. Before pressure exertion, the reaction temperatures of all samples were measured by a W/Re 5/26 thermocouple located at the sample center.

Table 1. Weight percentage of the raw reactants for Gd-doped Hf-zirconolite samples.

Sample No.	Addictive Amount of Raw Materials (g)					
	CuO	CaO	HfO$_2$	Ti	TiO$_2$	Gd$_2$O$_3$
Hf-Gd-0.2	7.502	1.410	5.294	2.257	1.255	2.279
Hf-Gd-0.4	7.282	1.026	3.854	2.191	1.218	4.426
Hf-Gd-0.6	7.075	0.665	2.496	2.128	1.184	6.450
Hf-Gd-0.8	6.879	0.323	1.213	2.069	1.151	8.362
Hf-Gd-1.0	6.694	-	-	2.014	1.120	10.171

The as-synthesized specimens were pulverized into fine powders, which were characterized by X-ray diffractometer (XRD; D/MAX-RB, Rigaku Corporation, Tokyo, Japan) with Cu Kα radiation to obtain the phase composition. The ignited samples were compressed by a quick pressing of 45 MPa with 60 s holding time after about 25–30 s delay of combustion. The obtained samples were then sliced and polished using different grades of emery paper and 0.5 μm diamond pastes. After cleaning and drying, the samples were subjected to further characterizations. Microstructure of the selected Hf-Gd-0.6 sample was typically observed using field-emission scanning electron microscopy (FESEM; Zeiss Ultra-55, Oberkochen, Germany) under 15 KV energy. The phase composition and elemental distribution were analyzed from the results of energy-dispersive X-ray spectrometer (EDX, ULTRA 55, ZEISS, Oberkochen, Germany) attached with the FESEM equipment. The chemical durability of Hf-Gd-0.6 sample was evaluated using standard MCC-1 leaching test. The specimen was sliced and grinded into dimension of 5.28 mm × 5.30 mm × 5.24 mm, which was suspended by a copper wire and immersed in 80 mL deionized water. Completely cleaned polytetra-fluoroethylene (PTFE) was utilized the leaching container. The leaching tests were carried out at 90 °C with durations of 1, 3, 7, 14, 21, 28, 35 and 42 days. The elemental concentrations of Ca and Cu in the leachates were obtained by inductively coupled plasma (ICP) analysis (iCPA 6500, ThermoFisher, Waltham, MA, USA), while Hf and Gd were collected by inductively coupled plasma-mass spectrometry (ICP-MS) analysis using an Agilent 7700× spectrometer (Santa Clara, CA, USA).

3. Results and Discussion

3.1. Combustion Temperature and XRD Analysis of the Hf-Zirconolite Sample

According to the previous research [45], Hf can totally replace the Zr site of zirconolite. In this experiment, we firstly testify the feasibility for the SHS preparation of Hf-zirconolite. The combustion experiment of the above-mentioned Equation (1) was conducted. The result demonstrates that the green body can be successfully ignited with self-sustaining reaction. The combustion lasts for about 10 s after ignition, which leads to a reaction speed of about 2–3 mm/s. The center temperature of this sample was measured as depicted in Figure 1a. The maximum temperature is 1177 °C and the temperature duration ($\geq$1000 °C) is longer than 30 s. As there is heat dissipation during the combustion reaction and subsequent testing, the real temperature should be much higher than the measured one. This temperature is adequate and beneficial for subsequent compression as it is higher than the melting point of Cu (1083 °C). Figure 1b shows the XRD pattern of the obtained Hf-zirconolite sample, which indicates the phase composition mostly conforms to the original design. Hf-zirconolite and Cu demonstrate are the main phases with a trace of CaTiO$_3$ phase. As no peaks correspond to HfO$_2$, we can confirm that HfO$_2$ has been completely incorporated into the Zr site of zirconolite. Because the Zr^{4+} and Hf^{4+} cations are in the same charge state and close ionic radius (0.72 Å for Zr^{4+}

and 0.71 Å for Hf^{4+}), they can mutually substituted under random proportion. This result testifies that Hf-zirconolite can be readily synthesized using the SHS method.

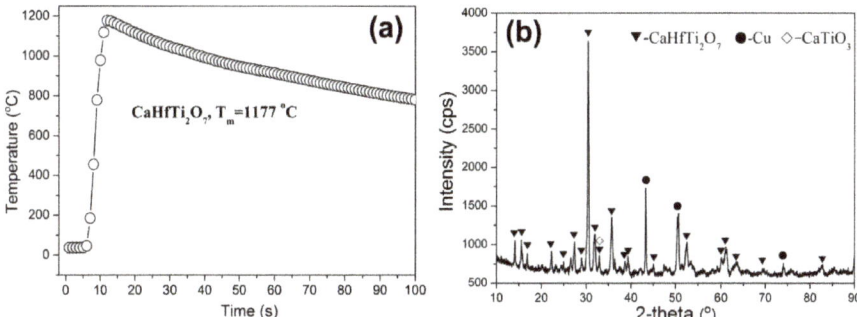

Figure 1. (a) Reaction temperature of the Hf-zirconolite sample, (b) XRD pattern of the Hf-zirconolite sample.

3.2. Reaction Temperature and Phase Composition of Gd-Bearing Hf-Zirconolite Samples

The Gd-bearing Hf-zirconolite waste forms were subsequently synthesized. All the designed SHS reactions were successfully ignited and the combustions lasted for about 10 s after tungsten wire ignition. The center temperatures were collected and depicted in Figure 2a. There is not a trend of regularity for the temperature of Gd-bearing samples. The maximum temperatures of these five samples reach to 1392 °C, 1120 °C, 1403 °C, 1458 °C and 1298 °C as the x value is elevated from 0.2 to 1.0. Compared with the original Hf-zirconolite (1177 °C), the Gd_2O_3 doped samples exhibit much higher temperatures (except for the Hf-Gd-0.4 sample). This result reveals that the reactivity of Gd_2O_3 is higher than CaO or/and HfO_2. Although the temperatures are not high, they are adequate and facilitate the subsequent densification process to get highly densified samples.

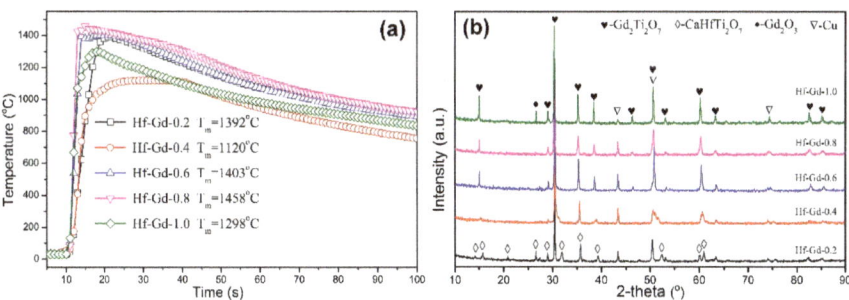

Figure 2. (a) Reaction temperatures, (b) XRD patterns of the Gd-bearing Hf-zirconolite samples with x values of 0.2–1.0.

According to previous studies [29,31], the Ca and Zr site of zirconolite could be concurrently occupied by trivalent actinides. The phase compositions of Gd-bearing Hf-zirconolite samples were characterized with the XRD patterns presented in Figure 2b. It is distinctly demonstrated that there is a phase transformation from 2M-zirconolite to cubic pyrochlore as the x value is elevated. There are only Hf-zirconolite ($CaHfTi_2O_7$, PDF No. 84-0163) and Cu phases in the Hf-Gd-0.2 sample. Minor pyrochlore appears when the x value is 0.4. The pyrochlore phase demonstrates as the main phase when the x value is 0.6, which can be verified by the superlattice (100) diffraction peak at around 15°. This result is similar as the Nd-bearing zirconolite in the solid-state synthesized $CaZrTi_2O_7$-$Nd_2Ti_2O_7$ system [31]. However, unreacted Gd_2O_3 is detected in the Hf-Gd-1.0 sample, which indicates that Gd could not totally substitute the Ca and Hf sites. This phenomenon may be related with the highly different ionic radius between Gd^{3+} (0.938 Å) and Hf^{4+} (0.71 Å). The maximum loading capacity of Gd_2O_3 is the Hf-Gd-0.8 sample, and only $Gd_2Ti_2O_7$-based pyrochlore (PDF No. 73-1698) and Cu are demonstrated as the constituent phases in this sample.

3.3. SEM and EDX Analysis of the Gd-Doped Samples

The observed phase fields in the $Ca_{1-x}Hf_{1-x}Gd_{2x}Ti_2O_7$ system were further supported by the SEM and EDX analysis. Typical back-scattered electron (BSE) image of the selected Hf-Gd-0.6 sample is shown in Figure 3a. No obvious pores can be observed in the surface image, which indicates this sample was well densified. Meanwhile, two different phases with distinct contrasts can be detected in the polish surface. The ceramic matrix phase is labeled as "A" and the metallic Cu phase is labelled as "B". The Cu phase can be readily determined because it is segregated by a distinct boundary. The brightness of "A" district is obviously higher than "B", which is attributed to the higher atomic number over Cu for $Ca_{1-x}Hf_{1-x}Gd_{2x}Ti_2O_7$ phase. According to the XRD result, the ceramic matrix should be pyrochlore-based titanate with a small amount of zirconolite phase. Figure 3b presents the fracture surface of Hf-Gd-0.6 sample, which exhibits a dense microstructure with tightly contacted submicron sized grains. The grain boundary is not very clear in the polishing surface and fracture surface, which reveals the feature of combustion synthesis as the reaction speed is high and soaking time is short. There is no time for the formation of grain boundary and grain growth.

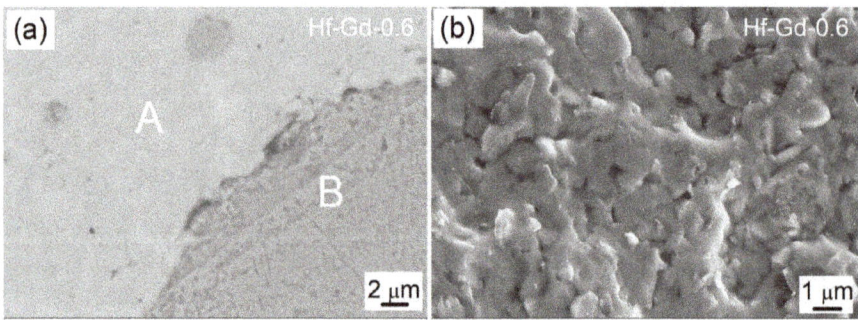

Figure 3. SEM images of the Hf-Gd-0.6 sample: (**a**) the polished surface, (**b**) the fracture surface.

Elemental EDX characterization was further conducted to determine the phase composition and elemental distribution of the typical Hf-Gd-0.6 specimen. The BSE and EDX mapping images are presented in Figure 4, where all the metallic elements of Ca, Ti, Hf, Gd and Cu are listed. The representative BSE image of Figure 4a supports the coexistence of "A" and "B" phases. Obviously, the "B" area must be Cu phase, which is testified by the EDX mapping image of Figure 4d. The "A" phase should be $Ca_{1-x}Hf_{1-x}Gd_{2x}Ti_2O_7$ phase as the Ca, Ti, Gd elements are enriched in this area. This result conforms to the phase composition of XRD analysis. It's worth noting that Hf not only appears in the matrix A area but also in the Cu phase. The enrichment of Cu and Hf elements is slightly overlapping in the "B" area. This phenomenon is strange as there is no peak corresponding to Hf or HfO_2 in the XRD pattern. It may be attributed to the adjacent energy characteristic peaks of Cu and Hf in the EDX spectra (Hf: K_α = 8.040, K_β = 8.903, Cu: K_α = 7.898, K_β = 9.021).

The EDX spotting analysis was further conducted to determine the chemical composition of the constituent ceramic phase, where the results are demonstrated in Figure 5. The EDX spotting analysis demonstrates that Hf has not been detected in the Cu phase. The EDX spotting image of "A" phase in Figure 5a is presented in Figure 5b. Similar as the EDX mapping results, the existence of Ca, Ti, Zr, Hf and O in the EDX spotting spectra indicates that the "A" phase is Gd and Hf doped pyrochlore phase. At least five points of "A" area were calculated to obtain the average elemental quantities as listed in Figure 5b. Based on this data, the chemical formulation of ceramic phase is calculated as $Ca_{0.39}Hf_{0.37}Gd_{1.38}Ti_{1.80}O_7$. Compared with the designed formulation of Hf-Gd-0.6 sample ($Ca_{0.4}Hf_{0.4}Gd_{1.2}Ti_2O_7$), the obtained ceramic phase is slightly deficient in Ti while rich in Gd. The Ca and Hf elements are very close to the designed values. This result testifies that the ceramic phase is in pyrochlore structure, where the Ca and Hf elements occupy the A site (Gd site in this study) of $A_2B_2O_7$ pyrochlore.

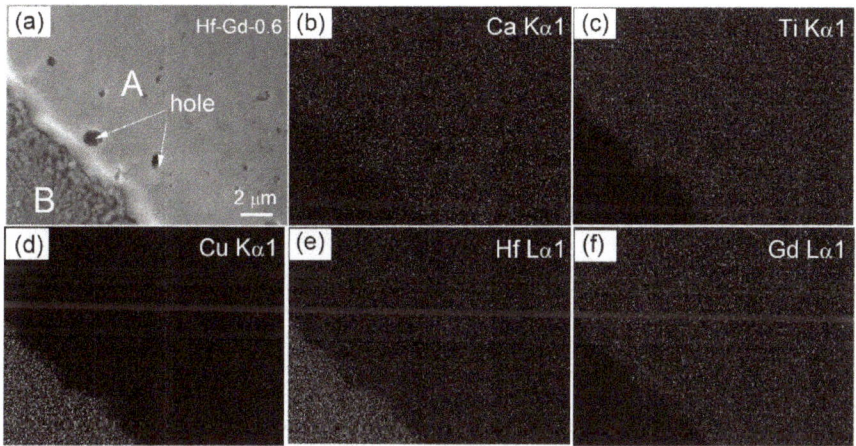

Figure 4. SEM-EDX mapping images of the Hf-Gd-0.6 sample: (**a**) representative BSE image, (**b**–**f**) elemental distribution of Ca, Ti, Cu, Hf and Gd elements.

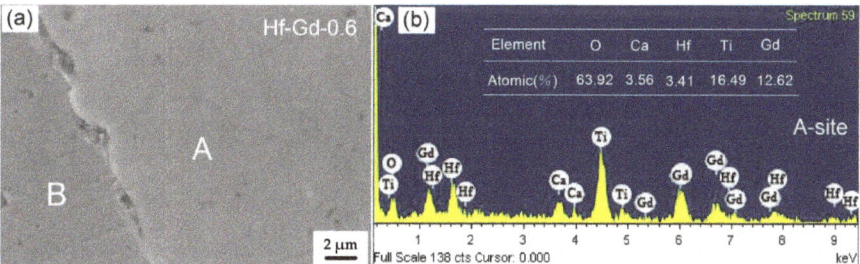

Figure 5. EDX spotting results of the Hf-Gd-0.6 sample: (**a**) representative BSE image, (**b**) elemental spotting analysis of the "A" region.

3.4. Chemical Stability of the Hf-Gd-0.6 Sample

The representative Hf-Gd-0.6 specimen was selected for the standard MCC-1 leaching test. The 1–42 days normalized elemental leaching rate of Ca, Cu, Gd and Hf are computed and depicted in Figure 6a–d. With the increase of soaking duration, all the normalized leaching rates firstly decrease in 1–7 days. However, the LR_{Cu} and LR_{Gd} exhibit slight ascension when the leaching time is prolonged (7 days for Cu and 21 days for Gd). Anyhow, the LR_{Ca} and LR_{Cu} values are 1.57 $g·m^{-2}·d^{-1}$ and 0.13 $g·m^{-2}·d^{-1}$ after 42 days. Gd and Hf are highly durable elements as shown in Figure 6c,d. Although there is a slight increase, the 42 days LR_{Gd} value is as low as 4.72×10^{-7} $g·m^{-2}·d^{-1}$. The LR_{Hf} value exhibits a congruent decrease tendency during 1–42 days leaching, where the leaching rate is 1.11×10^{-8} $g·m^{-2}·d^{-1}$ after 42 days. In this experiment, the leaching rate of Ca and Cu is comparable while Gd and Hf are even lower than Synroc waste forms prepared by hot pressing (HP) or hot isostatic pressing (HIP) [32,33]. The leaching rates are also significantly lower than borosilicate glass (about 1 g·m^{-2}·d^{-1}, 90 °C) [2,3,49].

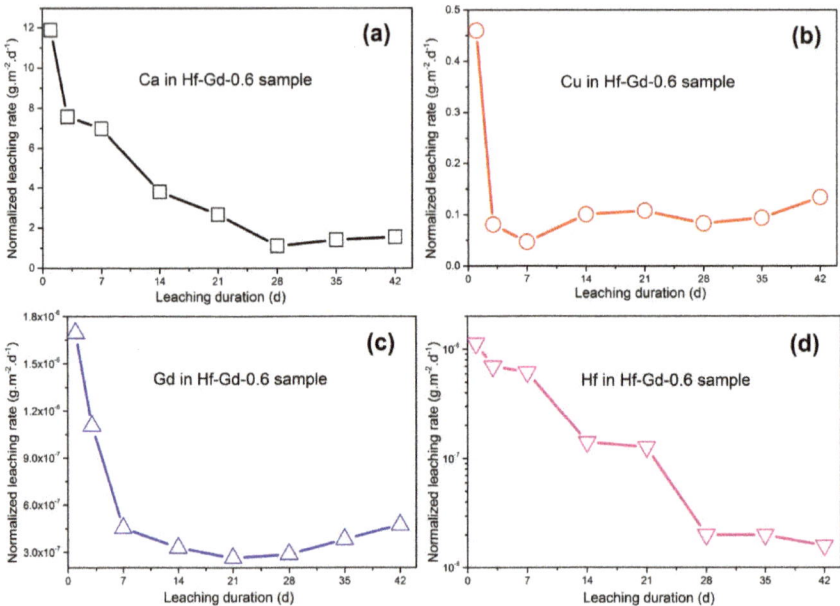

Figure 6. 1–42 days normalized leaching rates of the Hf-Gd-0.6 sample: (**a**) element Ca, (**b**) element Cu, (**c**) element Gd, (**d**) element Hf.

4. Conclusions

In this study, Gd-bearing Hf-zirconolite ($Ca_{1-x}Hf_{1-x}Gd_{2x}Ti_2O_7$) waste forms were rapidly synthesized from the SHS/QP method using CuO as the oxidant. Gd and Hf were employed as the simulates of trivalent and tetravalent actinides. The results indicate that Hf can totally replace the Zr site using this SHS process, and Gd can concurrently replace the Ca and Hf sites (Gd preferentially substitutes the Ca site). Gd_2O_3 could not completely be incorporated into the lattice structure of zirconolite when the x value is higher than 0.8. The aqueous durability of selected Hf-Gd-0.6 sample was tested, where the 42 days normalized leaching rates (LR_i) of Ca, Cu, Gd and Hf are measured to be 1.57, 0.13, 4.72 × 10^{-7} and 1.59 × 10^{-8} g·m^{-2}·d^{-1}. These results demonstrate that the SHS/QP route is suitable for the preparation of zirconolite and pyrochlore based waste forms for HLW immobilization.

Author Contributions: Conceptualization, K.Z. and H.Z.; Methodology, K.Z. and D.Y.; Formal Analysis, D.Y. and K.X.; Writing—Original Draft Preparation, D.Y.; Writing—Review & Editing, K.Z.

Funding: This research was funded by the National Natural Science Foundation of China (No. 51202203, 51672228), the Project of State Key Laboratory of Environment-friendly Energy Materials (Southwest University of Science and Technology, No. 16kffk05, 17FKSY0104) and Science Development Foundation of China Academy of Engineering Physics.

Conflicts of Interest: The authors declare no conflicts of interest.

References

1. International Atomic Energy Agency. *Design and Operation of High Level Waste, Vitrification and Storage Facility*; Technical Report Series No. 176; IAEA: Vienna, Austria, 1977.
2. Ojovan, M.I.; Lee, W.E. *An Introduction to Nuclear Waste Immobilization*; Elsevier Ltd.: Oxford, UK, 2005; pp. 213–267.

3. Caurant, D.; Loiseau, P.; Majérus, O.; Aubin-Chevaldonnet, V.; Bardez, I.; Quintas, A. *Glass, Glass-Ceramics and Ceramics for Immobilization of Highly Radioactive Nuclear Wastes*; Nova Science Publishers: New York, NY, USA, 2009.
4. Ewing, R.C. Nuclear waste forms for actinides. *Proc. Natl. Acad. Sci. USA* **1999**, *96*, 3432–3439. [CrossRef] [PubMed]
5. Lee, W.E.; Ojovan, M.I.; Stennett, M.C.; Hyatt, N.C. Immobilisation of radioactive waste in glasses, glass composite materials and ceramics. *Adv. Appl. Ceram.* **2006**, *105*, 3–12. [CrossRef]
6. Donald, I.W.; Metcalfe, B.L.; Taylor, R.N.J. The immobilization of high level radioactive wastes using ceramics and glasses. *J. Mater. Sci.* **1997**, *32*, 5851–5887. [CrossRef]
7. Vance, E.R.; Lumpkin, G.R.; Carter, M.L.; Cassidy, D.J.; Ball, C.J.; Day, R.A.; Begg, B.D. Incorporation of uranium in zirconolite ($CaZrTi_2O_7$). *J. Am. Ceram. Soc.* **2002**, *85*, 1853–1859. [CrossRef]
8. Ewing, R.C.; Weber, W.J.; Lian, J. Nuclear waste disposal-pyrochlore $A_2B_2O_7$: Nuclear waste form for the immobilization of plutonium and "minor" actinides. *J. Appl. Phys.* **2004**, *95*, 5929–5971. [CrossRef]
9. Loiseau, P.; Caurant, D.; Baffier, N.; Mazerolles, L.; Fillet, C. Glass-ceramic nuclear waste forms obtained from SiO_2-Al_2O_3-CaO-ZrO_2-TiO_2 glasses containing lanthanides (Ce, Nd, Eu, Gd, Yb) and actinides (Th): Study of internal crystallization. *J. Nucl. Mater.* **2004**, *335*, 827–837. [CrossRef]
10. Caurant, D.; Majérus, O.; Losieau, P.; Bardez, I.; Baffier, N.; Dussossoy, J.L. Crystallization of neodymium-rich phases in silicate glasses developed for nuclear waste immobilization. *J. Nucl. Mater.* **2006**, *354*, 143–162. [CrossRef]
11. Weber, W.J.; Navrotsky, A.; Stefanovsky, S.; Vance, E.R.; Vernaz, E. Materials science of high-level nuclear waste immobilization. *MRS Bull.* **2009**, *34*, 46–53. [CrossRef]
12. Vance, E.R. Synroc: A suitable waste form for actinides. *MRS Bull.* **1994**, *19*, 28–32. [CrossRef]
13. Ojovan, M.I.; Lee, W.E. *New Developments in Glassy Nuclear Wasteforms*; Nova Science Publishers: New York, NY, USA, 2007.
14. Donald, I.W. *Waste Immobilization in Glass and Ceramic Based Hosts. Radioactive, Toxic and Hazardous Waste*; Wiley: Chichester, UK, 2010.
15. Stefanovsky, S.V.; Yudintsev, S.V.; Giere, R.; Lumpkin, G.R. Nuclear waste forms. In *Energy, Waste and the Environment: A Geochemical Perspective*; Special Publications; Giere, R., Stille, P., Eds.; Geochemical Society: London, UK, 2004; Volume 236, pp. 37–63.
16. Loiseau, P.; Caurant, D.; Majerus, O.; Baffier, N.; Fillet, C. Crystallization study of (TiO_2, ZrO_2)-rich SiO_2-Al_2O_3-CaO glasses. Part II. Surface and internal crystallization processes investigated by differential thermal analysis (DTA). *J. Mater. Sci.* **2003**, *38*, 843–852. [CrossRef]
17. Ringwood, A.E.; Kesson, S.E.; Ware, N.G.; Hibberson, W. Immobilization of high level nuclear reactor wastes in SYNROC. *Nature* **1979**, *278*, 219–223. [CrossRef]
18. Vance, E.R.; Ball, C.J.; Day, R.A.; Smith, K.L.; Blackford, M.G.; Begg, B.D.; Angel, P.J. Actinide and rare earth incorporation into zirconolite. *J. Alloy. Compd.* **1994**, *213–214*, 406–409. [CrossRef]
19. Begg, B.D.; Vance, E.R.; Conradson, S.D. The incorporation of plutonium and neptunium in zirconolite and perovskite. *J. Alloy. Compd.* **1998**, *271–273*, 221–226. [CrossRef]
20. Peng, L.; Zhang, K.B.; Yin, D.; Wu, J.J. Self-propagating synthesis, mechanical property and aqueous durability of $Gd_2Ti_2O_7$ pyrochlore. *Ceram. Int.* **2016**, *42*, 18907–18913. [CrossRef]
21. Zhang, K.B.; He, Z.S.; Peng, L.; Zhang, H.; Lu, X. Self-propagating synthesis of $Y_{2-x}Nd_xTi_2O_7$ pyrochlore and its aqueous durability as nuclear waste form. *Scr. Mater.* **2018**, *146*, 300–303. [CrossRef]
22. He, Z.S.; Zhang, K.B.; Yin, D.; Peng, L.; Zhao, W.W. Self-propagating plus quick pressing synthesis and characterizations of $Gd_{2-x}Nd_xTi_{1.3}Zr_{0.7}O_7$ ($0 \leq x \leq 1.4$) pyrochlores. *J. Nucl. Mater.* **2018**, *504*, 61–67. [CrossRef]
23. Zhang, K.B.; Wen, G.J.; Zhang, H.B.; Teng, Y.C. Self-Propagating high-temperature synthesis of $Y_2Ti_2O_7$ pyrochlore and its aqueous durability. *J. Nucl. Mater.* **2015**, *465*, 1–5. [CrossRef]
24. Rossell, H.J. Zirconolite—A fluorite-related superstructure. *Nature* **1980**, *283*, 282–283. [CrossRef]
25. Gatehouse, B.M.; Grey, I.E.; Hill, R.J.; Rossell, H.J. Zironolite, $CaZr_xTi_{3-x}O_7$; structure refinement for near-end-member composition with x = 0.85 and 1.30. *Acta Cryst. B* **1981**, *37*, 306–312. [CrossRef]
26. Rossell, H.J. Solid solution of metal oxides in the zirconolite phase $CaZrTi_2O_7$. I. Heterotype solid solution. *J. Solid State Chem.* **1992**, *99*, 38–51. [CrossRef]
27. Cheray, R.W. Zirconolite $CaZr_{0.92}Ti_{2.08}O_7$ at 295 K to 1173 K. *J. Solid State Chem.* **1992**, *98*, 323–329. [CrossRef]

28. White, T.J. The microstructure and microchemistry of synthetic zirconolite, zirkelite and related phases. *Am. Mineral.* **1984**, *69*, 1156–1172.
29. Coelho, A.; Cheary, R.W.; Smith, K.L. Analusis and structural determination of Nd-substituted zirconolite-4M. *J. Solid State Chem.* **1997**, *129*, 346–359. [CrossRef]
30. Subramanian, M.A.; Aravamudan, G.; Subbba Rao, G.V. Oxide pyrochlores—A review. *Prog. Solid State Chem.* **1983**, *15*, 55–143. [CrossRef]
31. Jafar, M.; Sengupta, P.; Achary, S.N.; Tyagi, A.K. Phase evolution and microstructural studies in $CaZrTi_2O_7$-$Nd_2Ti_2O_7$ system. *J. Am. Ceram. Soc.* **2014**, *97*, 609–616. [CrossRef]
32. Teng, Y.C.; Wang, S.L.; Huang, Y.; Zhang, K.B. Low-temperature reaction hot-pressing of cerium-doped titanate composite ceramics and their aqueous stability. *J. Eur. Ceram. Soc.* **2014**, *34*, 985–990. [CrossRef]
33. Zhang, Y.; Stewart, M.W.A.; Li, H.; Carter, M.I.; Vance, E.R.; Moricca, S. Zirconolite-rich titanate ceramics for immobilization of actinides-Waste form/HIP can interaction and chemical durability. *J. Nucl. Mater.* **2009**, *395*, 67–74. [CrossRef]
34. Muthuraman, M.; Patil, K.C.; Senbagaraman, S.; Umarji, A.M. Sintering, microstructure and dilatometric studies of combustion synthesized Synroc phases. *Mater. Res. Bull.* **1996**, *32*, 1375–1381. [CrossRef]
35. Glagovskii, É.M.; Kuprin, A.V.; Pelevin, L.P.; Konovalov, E.E.; Starkov, O.V.; Levakov, E.V.; Postnikov, A.Y.; Lisitsa, F.D. Immobilization of high-level wastes in stable mineral-like materials in a self-propagating high-temperature synthesis regime. *Atom. Energy* **1999**, *87*, 514–518. [CrossRef]
36. Zhang, K.B.; Wen, G.J.; Yin, D.; Zhang, H.B. Self-propagating high-temperature synthesis of Ce-bearing zirconolite-rich minerals using $Ca(NO_3)_2$ as the oxidant. *J. Nucl. Mater.* **2015**, *467*, 214–223. [CrossRef]
37. Zhang, K.B.; Wen, G.J.; Zhang, H.B.; Teng, Y.C. Self-propagating high-temperature synthesis of CeO_2 incorporated zirconolite-rich waste forms and the aqueous durability. *J. Eur. Ceram. Soc.* **2015**, *35*, 3085–3093. [CrossRef]
38. He, Z.S.; Zhang, K.B.; Xue, J.L.; Zhao, W.W.; Zhang, H.B. Self-propagating chemical furnace synthesis of nanograin $Gd_2Zr_2O_7$ ceramic and its aqueous durability. *J. Nucl. Mater.* **2018**, *512*, 385–390. [CrossRef]
39. Peng, L.; Zhang, K.B.; He, Z.S.; Yin, D.; Xue, J.; Xu, C.; Zhang, H. Self-propagating high-temperature synthesis of ZrO_2 incorporated $Gd_2Ti_2O_7$ pyrochlore. *J. Adv. Ceram.* **2018**, *7*, 41–49. [CrossRef]
40. Zhang, K.B.; He, Z.S.; Xue, J.L.; Zhao, W.W.; Zhang, H.B. Self-propagating synthesis of $Y_{2-x}Nd_xTi_2O_7$ pyrochlores using CuO as the oxidant and its characterizations as waste form. *J. Nucl. Mater.* **2018**, *507*, 93–100. [CrossRef]
41. Zhang, K.B.; Yin, D.; Han, P.W.; Zhang, H.B. Two-step synthesis of zirconolite-rich ceramic waste matrice and its physicochemical properties. *Int. J. Appl. Ceram. Technol.* **2018**, *15*, 171–178. [CrossRef]
42. Seaburg, G.T. Overview of actinide and lanthanide (the f) elements. *Radiochim. Acta* **1993**, *61*, 115–122. [CrossRef]
43. Holgado, J.P.; Alvarez, R.; Munuera, G. Study of CeO_2 XPS spectra by factor analysis: Reduction of CeO_2. *Appl. Surf. Sci.* **2000**, *161*, 301–315. [CrossRef]
44. Zhang, K.B.; Yin, D.; Peng, L.; Wu, J.J. Self-propagating synthesis and CeO_2 immobilization of zirconolite-rich composites using CuO as the oxidant. *Ceram. Int.* **2017**, *43*, 1415–1423. [CrossRef]
45. Caurant, D.; Loiseau, P.; Bardez, I. Structural characterization of Nd-doped Hf-zirconolite $Ca_{1-x}Nd_x$ $HfTi_{2-x}Al_xO_7$ ceramics. *J. Nucl. Mater.* **2017**, *407*, 88–99. [CrossRef]
46. Perera, D.S.; Stewart, M.W.A.; Li, H.; Day, R.A.; Vance, E.R. Tentative Phase Relationships in the System $CaHfTi_2O_7$-$Gd_2Ti_2O_7$ with up to 15 mol% Additions of Al_2TiO_5 and $MgTi_2O_5$. *J. Am. Ceram. Soc.* **2002**, *85*, 2919–2924. [CrossRef]
47. Emsley, J. *The Elements*; Clarendon Press: Oxford, UK, 1992.
48. ASTM C1220-98. *Standard Test Method for Static Leaching of Monolithic Wasteforms for Disposal of Radioactive Waste*; ASTM International: West Conshohocken, PA, USA, 1998.
49. Smith, K.L.; Lumpkin, G.R.; Blackford, M.G.; Day, R.A.; Hart, K.P. The durability of Synroc. *J. Nucl. Mater.* **1992**, *190*, 287–294. [CrossRef]

© 2019 by the authors. Licensee MDPI, Basel, Switzerland. This article is an open access article distributed under the terms and conditions of the Creative Commons Attribution (CC BY) license (http://creativecommons.org/licenses/by/4.0/).

Review

Removal of Hazardous Oxyanions from the Environment Using Metal-Oxide-Based Materials

Ewelina Weidner and Filip Ciesielczyk *

Institute of Chemical Technology and Engineering, Faculty of Chemical Technology, Poznan University of Technology, Berdychowo 4, PL-60965 Poznan, Poland; ewelina.a.weidner@doctorate.put.poznan.pl
* Correspondence: filip.ciesielczyk@put.poznan.pl; Tel.: +48-61-6653626

Received: 11 February 2019; Accepted: 14 March 2019; Published: 20 March 2019

Abstract: Scientific development has increased the awareness of water pollutant forms and has reawakened the need for its effective purification. Oxyanions are created by a variety of redox-sensitive metals and metalloids. These species are harmful to living matter due to their toxicity, nondegradibility, and mobility in aquatic environments. Among a variety of water treatment techniques, adsorption is one of the simplest, cheapest, and most effective. Since metal-oxide-based adsorbents poses a variety of functional groups onto their surface, they were widely applied in ions sorption. In this paper adsorption of harmful oxyanions by metal oxide-based materials according to literature survey was studied. Characteristic of oxyanions originating from As, V, B, W and Mo, their probable adsorption mechanisms and comparison of their sorption affinity for metal-oxide-based materials such as iron oxides, aluminum oxides, titanium dioxide, manganese dioxide, and various oxide minerals and their combinations are presented in this paper.

Keywords: oxyanions; sorption; metal oxides; environment pollution; water purification; adsorbents; hazardous metals

1. Introduction

Intensive industrial development has contributed to the increased pollution of the environment with hazardous metals and metalloids. The majority of these elements are redox sensitive and some of their oxidation states can form oxyanions in solution. Oxyanions (or oxoanions) are polyatomic negatively charged ions containing oxygen with the generic formula $A_xO_y^{z-}$ (where A represents a chemical element and O represents an oxygen atom) [1]. Those compounds represent a range of different species depending both on pH and redox potential [2]. Oxyanions of As, V, B, W, and Mo are commonly found trace pollutants in various waste streams [1,3]. Metal(loid) oxyanions are characterized by toxicity [4–6], nonbiodegradability [5,6], and high solubility in water [7], which makes them extremely mobile harmful species which easily bioaccumulate in the environment and in the food chain [8]. Thus these species are very dangerous even at low concentrations. They can be transferred into living organisms via inhalation, ingestion, and skin adsorption, causing irreversible effects [5]. The main sources of hazardous oxyanions are alkaline wastes originating from high temperature processes with the thermal treatment of waste, fossil fuel combustion, and ferrous or non-ferrous metal smelting [2], nonetheless they are also produced in microelectronics, electroplating, metal finishing, battery manufacturing, tannery, metallurgy, and fertilizer industries [5]. However industrial activities are significantly increasing the concentration of oxyanions, and there are environments where geological formations promote dissolution of weak-acid oxyanion species, like arsenic, vanadium, or antimony, which pollute ground water sources used by local communities [9]. Considering the harmful properties of metal(loid) oxyanions, their effective elimination from water and wastewater is becoming a key issue for environment and public health protection. Nonetheless in comparison with research concerning cationic pollutants, number of works pertained to oxyanions removal is still dramatically low, what can be seen on the chart presented in Figure 1.

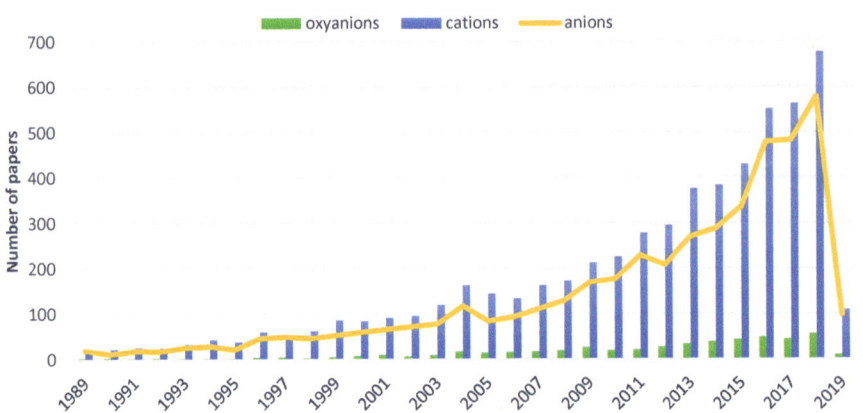

Figure 1. Bar chart of the number of articles per year concerning cations, anions and oxyanions adsorption for the 1989–28 January 2019. The statistical data were obtained by searching "adsorption metal oxide cations/anions/oxyanions" phrases in the Scopus data base as title and keywords.

The removal of those hazardous species from wastewater and water sources have been subjected to a variety of techniques e.g., ion exchange, filtration, adsorption, reverse osmosis, solvent extraction, chemical precipitation, evaporation and concentration, electrodialysis, and biomethods [10–13]. The main methods used for wastewater treatment with particular regard to adsorption has been shown in Figure 2.

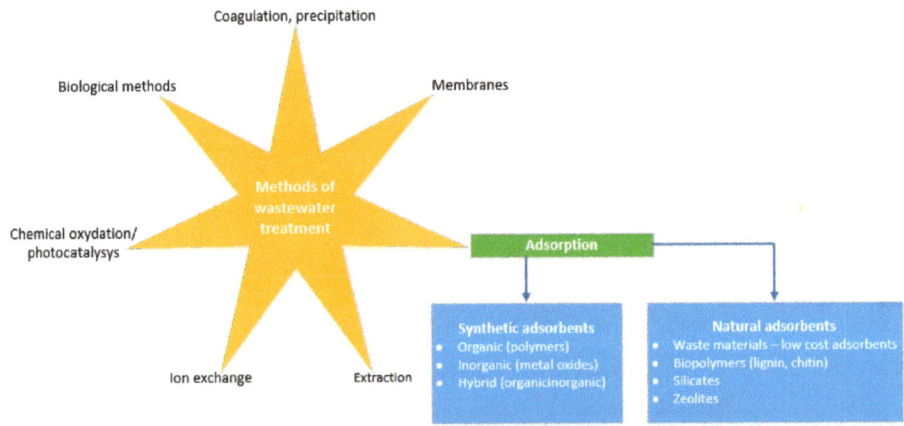

Figure 2. Methods of wastewater treatment.

The main problem with the majority of those methods is their high cost and need for advanced equipment. Adsorption is in the advantage over other techniques in water treatment due to its low cost, simple design, easy operation, insensitivity to toxic species, no secondary pollutants production and high efficiency [14]. Many adsorbents have been proposed in literature for water purification from harmful oxyanion species, including activated carbons [15,16], lignocellulosic materials [8], gold [17,18], silver [19], zero valent iron [20], natural minerals like zeolites [21–24], calcite [25], bentonite [14], goethite [5,26–30], kaolinite [31], graphene oxide [32], metal hydroxides, or metal oxides. Metal oxides are characterized by the presence of different types of surface groups [33]. Thus metal-oxide-based materials reveal a capability to adsorb ions [34] and because of that one of their major uses is for adsorbing metal or nonmetal ions from aqueous wastewater streams [35]. In this work adsorption of harmful oxyanions by metal oxide-based materials according to a literature survey have been studied.

2. Arsenic Oxyanions

2.1. Arsenic Pollution

Arsenic is a metalloid that occurs in in the +III and +V oxidation states. It is a building element of the Earth's crust and is naturally occurring in the environment in the air, soils and rocks, natural water, and organisms [36,37]. In its inorganic form it is strongly carcinogenic and highly toxic. Nowadays arsenic pollution has been recognized to be one of the world's greatest environmental hazards. The World Health Organization consider it one of the top ten major public health concern chemicals. The United States Environmental Protection Agency have located inorganic compounds of arsenic like arsenic acid, arsenic(III) oxide, and arsenic(V) oxide on the hazardous waste list [38]. The International Agency for Research on Cancer has classified arsenic as a human carcinogenic substance, group one [39]. Besides cancer, other negative health effects that may be associated with long-term ingestion of arsenic include developmental effects, diabetes, pulmonary disease, skin lesions, and cardiovascular disease can occur in living organisms. The current WHO recommendation of arsenic concentration in drinking water is 10 μg/L, remarking that this is only a provisional guideline arising from practical difficulties in removing arsenic from drinking water and those concentration should be as low as possible to eliminate all of its negative results. Inorganic arsenic is the most significant chemical contaminant in drinking water all over the globe. In natural waters, concentration of arsenic range from less than 0.5 μg/L to more than 5000 μg/L [37]. It is naturally present at high levels in the groundwater of a number of countries, including Argentina, Bangladesh, Chile, China, India, Mexico, and the United States of America [40]. Such high arsenic concentrations were connected to geothermal influence, mineral dissolution (e.g., pyrite oxidation), desorption in the oxidizing environment, and reductive desorption and dissolution [37]. Moreover arsenic is widely used industrially in the smelters, coal-burning, electric plants, processing of glass, pigments, textiles, paper, metal adhesives, wood preservatives, ammunition pesticides, natural weathering processes, runoff from mining operations, feed additives, and pharmaceuticals [41]. The outflow of arsenic-contaminated industrial wastewater to an aquatic system can cause deleterious effects on human health, animals, and plants. Thus elimination of arsenic from the environment is a task of high interest in research communities all over the world. Currently available techniques for arsenic removal include coagulation/precipitation, ion exchange, lime softening, reverse osmosis, electrodialysis, and adsorption. Conventionally, coagulation/precipitation with ferric and aluminum salts were used to remove arsenic from aqueous systems, but the waste sludge resulting from this process is creating problems associated with its treatment and disposal [42].

2.2. Characteristic of Arsenic Oxyanions in Aquatic Environment

In the aquatic environment arsenic is able to create inorganic oxyanions—an oxidized form arsenate [As(V)] and a reduced form arsenite [As(III)] [15]. The percentage content of arsenic species in dependence with pH conditions of water is demonstrated in Figure 3. Under oxidizing conditions, arsenic usually exists in the pentavalent (arsenate) form such as H_3AsO_4 (dominates in the pH < 9.2), $H_2AsO_4^-$, $HAsO_4^{2-}$, or AsO_4^{3-} depending on the activity of electrons (Eh) and activity of hydrogen ions (pH). Under reducing conditions arsenic mainly exist in the trivalent form (arsenite)—$H_2AsO_3^-$ and $HAsO_3^{2-}$. In the typical pH for majority of natural waters (6.5–8.5) $H_2AsO_4^-$ and $HAsO_4^{2-}$ are predicted to appear. The behavior of arsenic ions in the groundwater and water treatments systems is determined by the electrical charge. The strength of sorption of anions onto metal oxides is strongly dependent on the pH of the environment [36].

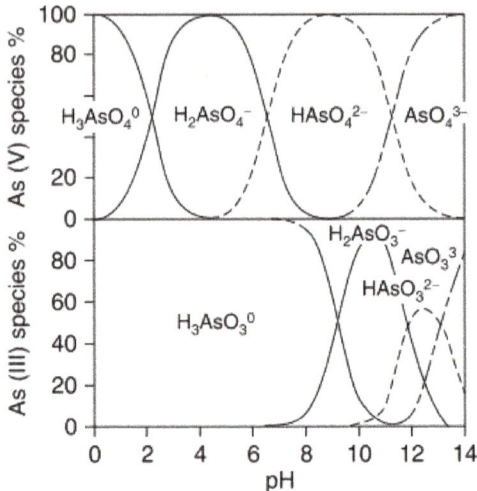

Figure 3. Distribution of As(V) and As(III) species as a function of pH, ionic strength = 0.04 M [43].

2.3. Adsorbents for Arsenic Removal from Water Sources

Adsorption technologies represent an innovative and economic approach to the arsenic removal problem. Metal oxide-based materials were successfully applied for arsenic adsorption from environment since its pollution problem was noticed. In the Table 1 sorption conditions and capacities for different arsenic species adsorbents were collected.

Aluminum oxide (Al_2O_3), also called activated alumina (AA), is produced by thermal dehydration of aluminum hydroxide, so that the surface can exchange contaminants for hydroxyl groups. It is characterized by a relatively high surface area (about 200 m^2/g) and diverse pore distribution of macro and micropores. It can be regenerated with sodium hydroxide, followed by neutralization with sulphuric acid [36]. Currently it is classified by the USEPA as among the best available technology for arsenic removal in drinking water [44,45]. Activated alumina has strong selectivity to arsenate ion, is nonhazardous and can be safely disposed on landfills. Among the treatment processes for the arsenic elimination, Al_2O_3 adsorption is less expensive than the membrane separation, and more versatile than the ion exchange process [46]. Aluminum oxide has been widely used in West Bengal (India) [47]. Its main drawbacks are its pH sensitivity and low regeneration range of about 50–70% (must be replaced after four to five regeneration cycles) [48]. The surface of activated alumina is positively charged until the pH is lower than point of zero charge (pH_{pzc}), which for different type of alumina is around 8.4–9.1 [46]. Dambies [44] in his review reported that the optimum pH value of oxyanions adsorption onto activated alumina is located in the range of 6–8, where it is predominantly positively charged, and with the increase of pH, the positive charge of Al_2O_3 increases, decreasing the sorption of arsenic oxyanions.

Iron oxide materials are characterized by their low cost and environmental friendliness [49]. They reveal a high affinity towards arsenic oxyanions, which makes it possible to apply them as adsorbents in water purification. An important mechanism for As(V) and As(III) removal by iron oxides is surface complexation [41]. However the most popular iron adsorbent used for arsenic removal is granular ferric hydroxide, and several iron(III) oxide materials (i.e., amorphous hydrous ferric oxide, goethite, and poorly crystalline hydrous ferric oxide) proved to be promising adsorptive materials for arsenic removal as well. Oscarson et al. [50] in 1982 investigated amounts of As(III) and As(V) adsorbed by pure Fe oxide and Al oxide. They used 0.1 g of each oxide in 70 mL of arsenic solution (adsorbent concentration 7 g/L) per 0.5–12 h. The adsorption capacities of Al_2O_3 were confirmed as to 16 mg/L for arsenite and 24.5 mg/L for arsenate, while for Fe_2O_3 60.9 mg/g and 21.3 mg/g respectively. In their previous work they revealed that iron and aluminum oxides do not oxidize arsenite to arsenate [50]. In 2007 Jeong et al. [45] studied the adsorption of arsenate [As(V)] onto Fe_2O_3 and Al_2O_3 and they obtained significantly lower adsorption capacities using similar adsorbates concentrations—0.05–1 g/L of Fe_2O_3 and 0.5–6 g/L of Al_2O_3. The maximum adsorption capacities of Fe_2O_3 and Al_2O_3 at pH 6 were estimated from the Langmuir isotherm, and found to be 0.66 mg/g and 0.17 mg/g, respectively. However, though adsorption capacities for Al_2O_3 and Fe_2O_3 are significantly low, they are still one of the most popular adsorbents used for arsenic removal from water environment.

Lin et al. [46] investigated commercially available amorphous granular activated alumina (Macherey-Nagel, Düren, Germany) as arsenic oxyanions sorbents. Before sorption, study samples were dried in the oven at 105 ± 5 °C for 24 h and stored in a desiccator for further analysis and experiments. For better results narrow size ranges of samples were analyzed. Surface area of granular activated alumina varied from 115 to 118 m^2/g. Arsenic sorption studies were conducted from model concentrations obtained from $Na_2HAsO_4 \cdot 7H_2O$ (KR Grade, Sigma-Aldrich, St. Louis, MO, USA) and $NaAsO_2$ (GR Grade, Sigma-Aldrich). Adsorption equilibrium was established within 40 h for arsenite and 170 h for arsenate and were studied for different pH and concentration of arsenic species. Obtained data fitted with both Freundlich and Langmuir isotherm equations and all the nonlinear regression coefficients were larger than 0.93 which indicated that both models successfully describe the partition behavior between water and the granular activated alumina surface for arsenite and arsenate. Davis and Misra [35] investigated the influence lanthanum oxide presence on Al_2O_3 adsorption properties regarding to As(V) oxyanions. An obtained hybrid oxide system containing of 10% lanthanum oxide and 90% of aluminum oxide (activated γ-alumina) revealed adsorption capacities of 0.050 mg/g for $H_2AsO_4^-$ and 0.029 mg/g for $HAsO_4^{2-}$, which is a noticeable decrease in comparison with results of pure aluminum oxide. Researchers recognized optimal pH of the process close to and above the pH of the point zero charge of activated alumina, where its surface is neutral or negatively charged. Repulsion of negative ions from negative surface translates to a very low adsorption capacity obtained for their material in comparison with pure Al_2O_3 in slightly acidic pH. Perhaps authors could obtain better results if they had obeyed the basic laws of electrochemistry and set beneficial sorption conditions.

To take heed of cautionary notice on the use of aluminum-based compounds for water treatment published by World Health Organization in 1997 and problems with granular ferric hydroxide Manna et al. [47] synthesized crystalline hydrous ferric oxide (CHFO) and investigated its sorption properties for arsenic removal. Tests were run onto model solutions prepared from sodium metaarsenite and disodium hydrogen arsenate of A.R. grade (British Drug Houses). CHFO were prepared by hydrolysis of 0.1 M $FeCl_3$ in 0.01 M HCl with 0.1 M NH_3 solution to obtain a pH in the range of 4–5. The precipitate was aged with the mother liquor for five days, then the acid was removed and material was dried in 40 °C in an air oven. Experiments revealed that adsorption follows a first-order Lagergren kinetic model and the data fit the Langmuir isotherm. For maximum As(III) and As(V) adsorption CHO needed 3 and 5 h respectively. The increase of adsorbent drying temperature onto As(V) sorption from 25 to 300 °C resulted in an increase in the number of active sites and porosity due to removal of physically adsorbed water molecules. The optimum drying temperature for adsorption of inorganic arsenic species from natural water samples is 200 to 300 °C. The surface area of CHO had not been investigated. The regeneration of arsenic(III)-rich CHFO (As content: 66.6 mg/g) conducted by the authors revealed that a 1.0 M solution of NaOH or KOH is able to desorb ~60 ± 1% of initial arsenic content. About 15–20% of adsorbed arsenic does not desorb even in these harsh conditions, which may be the result of chemisorption or fouling of the adsorbent. Thus after regeneration CHFO will be 15–20% less effective in arsenic adsorption. A total of 99 ± 0.5% of the arsenic content was recovered from arsenic-rich regenerate, thus the solution obtained after its recovery can be discharged safely onto surface soil, which prevents its further disposal in the environment.

Arsenate and arsenite can be successfully removed by zerovalent iron (ZVI), which corrodes in water environment creating magnetite, a permeable reactive barrier of ZVI in the subsurface. This fact pushed Su et al. [51] to investigate magnetite as a sorbent for inorganic arsenic. In their research eight different magnetite types with different surface area and purity of the material were used. Reagent grade Na_2HAsO_4 (Aldrich) and $NaAsO_2$ (Baker) were used as the inorganic arsenic source. Below pH 5.6–6.8 As(V) sorption were favored, while in a pH value above 7, As(III) was strongly attracted to the minerals. Magnetites revealed an ability to oxidize As(III) to As(V) and the oxidation range increased with increase of the pH from 2 to 12. The authors suggested the preparation of an engineered system where magnetite will be a favorable corrosion product of ZVI, which would be effective for arsenic removal.

However, though adsorption of hazardous metals is an effective removal technique, it does not cause their annihilation. Some researchers have worked on combining adsorption with other techniques to completely destroy or remold heavy metals into harmless compounds. One of the most popular transformation processes is oxidation. Photocatalytic activity of titanium dioxide was previously used in the oxidation of arsenite to arsenate. Bissen et al. [52] effectively photooxidized As(III) to As(V) using suspension of TiO_2 in water with simulated or natural sunlight as irradiation sources. However, TiO_2 has a low surface area and low adsorption capability, batch experiments proved that in the natural sunlight part of the arsenic was adsorbed onto it. Moreover it is very hard to remove the arsenate from contaminated water at the same time as oxidation occurs. Those facts pushed Zhang and Itoh [53] to combine titanium dioxide with iron oxide and slag (SIOT) to obtain device for arsenic removal from high-concentration arsenic contaminated wastewaters (100–20,000 mg/L). They used slag obtained from a municipal solid waste incinerator (Resource Recovery Center of Toyohashi, Aichi, Japan), TiO_2 in an anatase form with purity 99.9% (High Purity Chemicals) and analytical grade $FeCl_3$ (Wako or Aldrich). Slag (50 g) was aged for 2 days in a NaOH solution to obtain nearly neutral pH, then $FeCl_3$ solution was added and aged for 12 h on magnetic stirrer and finally 5 g TiO_2 was added. After 2 h the slurry was filtrated and dried at 105 °C for 2 h and then at 550 °C for another 1 h. Finally, the composite material has been grinded into separate grains and dried at 105 °C for 24 h under the vacuum. The obtained material's surface area was investigated by the BET method (Quantachrome Monosorb MS-21, Boynton Beach, FL, USA) and was equal to 163 m^2/g which was lower than material without TiO_2 (196 m^2/g) which was previously synthesized by the research group. The addition of 10% of TiO_2 reduced surface area by about 20%. However, though the oxidation of As(III) to As(V) was rapid and effective, the adsorption of produced As^{5+} ions was slow. Adsorption capacity for pure TiO_2 was 0.0001 mg/g, while for SIOT increased to 0.0047 mg/g and remained still very low. Analogical system without the slag (NHITO) was investigated by Gupta et al. [54] and they obtained slightly different results. Material was prepared in a low temperature process of slow injection of 10 g $TiCl_4$ into a 0.5 M $FeCl_3$ in hot 0.1 M HCl solution (~60 °C) with mechanical stirring. The pH of the mixture was regulated to 5.0–6.0 with 1 M NaOH. The formed precipitate was aged for 6 days in mother liquor, washed with deionized water till the alkali were free, dried in air oven at 60–70 °C, cooled with ice cold water, broken into agglomerated particles, and sieved for use. However, though the BET surface area of the obtained bimetal oxide was equal to 77.8 (±0.2) m^2/g, which is much lower than for material obtained by Zhang and Itoh [53], it was characterized with a much higher adsorption capacity equal to 85.0 mg of As(III) per g of adsorbent and 14.3 mg of As(V) per g. Adsorption of arsenic species followed the Langmuir model and was favorable in pH = 7 at a temperature of around 30 °C. Adsorption of the As(III) species is not one of the key factors in arsenic removal, because those species can be easily oxidized to As(V) ones. The main problem is arsenate elimination, and for As(V) removal, NHITO is no competition for other adsorbents which are cheap and available, like crystalline hydrous ferric oxide.

Another metal oxide with the ability to convert arsenite to arsenate is manganese dioxide (MnO_2). Manganese dioxide in its mineral form has been used in water treatment for more than 75 years, efficiently removing iron, manganese, and arsenic at pH between 5–9 [55]. Manganese oxides are very active components of natural environments, able to strongly sorb ions and participate in redox reactions. They have been identified as the primary electron acceptor in the oxidation of As(III) to As(V) by freshwater lake sediments [56]. Oscarson et al. [56] investigated arsenic oxidation by manganese dioxide and proposed mechanism of it, which can be described by five equations:

$$HAsO_2 + MnO_2 = (MnO_2) \cdot HAsO_2 \quad (1)$$

$$(MnO_2) \cdot HAsO_2 + HAsO_2 + H_2O = H_3AsO_4 + MnO \quad (2)$$

$$H_3AsO_4 = H_2AsO_4^- + H^+ \quad (3)$$

$$H_2AsO_4^- = HAsO_4^{2-} + H^+ \quad (4)$$

$$(MnO_2) \cdot HAsO_2 + 2H^+ = H_3AsO_4 + Mn^{2+} \quad (5)$$

The first step of the oxidation process is adsorption of arsenic species creating a layer onto the MnO_2 surface (Equation (1)). Next $HAsO_2$ is oxidized to H_3AsO_4 with subsequent oxygen transfer. However when the pH is equal to 7 or less, the predominant trivalent arsenic form is arsenious acid ($HAsO_2$), the oxidation products dissociate forming almost equal amounts of $H_2AsO_4^-$ and $HAsO_4^{2-}$ with little H_3AsO_4 presence at equilibrium (Equations (3) and (4)). During dissociation each mole of As(III) release 1.5 moles of hydrogen ions, which should significantly lower the pH of the system when no other reactions occur. However, after the reaction the solution pH stays close to neutral, which indicates a reaction of hydrogen ions with adsorbed $HAsO_2$ on MnO_2 surface. In such a reaction the H_3AsO_4 is formed and manganese is reduced and dissoluted (Equation (5)).

Despite MnO_2 being a common and effective oxidizing agent for As(III), it is characterized by a significantly low surface area, which limits the arsenic sorption capacity of this adsorbent. To overcome this disadvantage Lei et al. [57] combined MnO_2 with iron oxide, which is known as efficient arsenic adsorbent. Using a hydrothermal method researchers prepared a flower-like three-dimensional nanostructure Fe–Mn binary oxide and compared its arsenic adsorption properties with pure manganium oxide and iron oxides. The preparation procedure of Fe–Mn binary oxide was as follows: In 76 mL of deionized water $MnSO_4 \cdot H_2O$ (0.6830 g), $Fe(NO_3)_3 \cdot 9H_2O$ (1.6406 g), $K_2S_2O_8$ (1.0868 g), and 4 mL of concentrated sulfuric acid were mixed and stirred at room temperature. The homogeneous solution was autoclaved and preheated to 110 °C for 6 h. Impurities were removed using deionized water and ethanol and then precipitates were dried at 60 °C for 8 h. Pure MnO_2 and iron oxides were prepared in the same way except that the $Fe(NO_3)_3 \cdot 9H_2O$ and $MnSO_4 \cdot H_2O$, respectively, were absent. The adsorption onto prepared materials had fitted well to the Freundlich isotherm, which suggested that the adsorption mechanism is a multilayer physisorption. The surface area was 123 m^2/g for Fe–Mn binary oxide, 77 m^2/g for MnO_2, and 43 m^2/g for iron oxides measured by an unspecified technique. The highest sorption capacities occurred to be 26.50, 23.40, and 11.22 mg/g respectively. Despite quite a high increase of Fe–Mn binary oxide's surface area in comparison with iron or manganium oxide, arsenic sorption capacity increased insignificantly compared to iron oxides. This fact throws into question the sense of using such hybrid systems in commercial arsenic removal.

Considering the harsh conditions of waste streams containing arsenic pollution, Ren et al. [49] combined iron oxide with hydrous zirconium oxide, which is characterized by high resistance to acids, alkalis, oxidants, and reductants. The Fe–Zr binary oxide was prepared by a simple coprecipitation method at ambient temperature. Ferric chloride hexahydrate (0.05 mol) and zirconyl chloride octahydrate (0.0125 mol) were dissolved in deionized water (400 mL). During stirring pH was established to the level of 7.5 by adding sodium hydroxide (2 mol/L). The formed suspension was stirred for 1 h, aged at room temperature for 4 h, washed with deionized water, filtered, dried at 65 °C for 4 h, and crushed. The surface area of obtained material examined via BET analysis was equal to 339 m^2/g with pore volume of 0.21 cm^3/g. The SEM images revealed amorphous structure of obtained binary oxide. The adsorption of arsenic onto Fe-Zr binary oxide was described well by the Freundlich model. The Langmuir isotherm failed to describe the adsorption behavior, despite the adsorption capacities having been calculated from

Langmuir equations and found to be equal to 46.1 mg/g for As(V) and 120.0 mg/g for As (III) at pH 7.0. Similar research was carried out by Gupta et al. [58,59]. Researchers obtained nanostructure of Fe-Zr binary oxide (NHIZO) by the hydrolysis of hot (60 °C) ferric chloride (0.18 M $FeCl_3$ in 0.1 M HCl) and zirconium oxychloride (0.02 M $ZrOCl_2$ in 0.1 M HCl). Then the precipitate was aged, filtered, washed with deionized water, dried at 80 °C, and treated with cold water to obtain agglomerated particles ranging in size from 140 to 290 µm. As opposed to Ren et al. [49], experimental adsorption data fit well to the Langmuir isotherm. Adsorption capacities were determined by Langmuir model and were equal 65.5 mg/g for As(II) and 9.4 mg/g for As(III). Revealed data suggests that the As(III) sorption by NHIZO is physisorption in nature, while As(V) sorption reaction with NHIZO is a chemisorption phenomenon. However the hydrolysis method is environmentally friendly compared to Fe-Zr binary oxide obtained by the coprecipitation method, and NHIZO shows much lower adsorption capacities for arsenic species. Erdoğan et al. [60] obtained a nano ZrO_2/B_2O_3 oxide system which was used for arsenate ion determination in tap and underground waters. Unfortunately, the authors had not explained the advantages resulting from use of such an oxide combination. Sorption capacity was determined using the batch method (pH = 3, room temperature, Langmuir model), and it was equal to 98.04 mg/g. However, while the obtained material had revealed promising sorption properties in optimal conditions, its application in the real system had not been tested. The normal range for pH for groundwater systems is 6 to 8.5 [61], while drinking water must have a pH value of 6.5–8.5, so results obtained by the authors cannot be related to real conditions.

Kwon et al. [62] immobilized zirconium oxide on alginate beads obtaining a composite adsorbent for arsenite and arsenate removal, reaching adsorption capacity of 32.3 mg/g for arsenite and 28.5 mg/g for arsenate. They used alginate as a matrix due to its strong affinity for metal ions. Immobilization using bead encapsulation is an effective way to prevent ZrO_2 to environment. However, though the obtained material is characterized by a satisfactory adsorption capacity for arsenic ions, the system reached an equilibrium state within 240 h and the pH_{pzc} of the obtained material was 4.3, which is much lower in comparison with conventional sorbents like activated alumina.

The adsorption of arsenic oxyanions onto metal oxide materials has been widely studied by scientists all over the world. Due to the presence of a positive surface charge onto the majority of adsorbents in low pH, arsenic sorption is favorable in acidic conditions. As can be seen from the data gathered in Table 1, a lot of experiments were performed in neutral pH, which was estimated as an optimal value, which suggest that besides electrostatic attraction, another bonding takes place in arsenic adsorption. Arsenic species are effectively adsorbed in room temperature.

Table 1. Sorption properties of metal oxide-based adsorbents for arsenic oxyanions removal.

Adsorbent	Surface Area (m²/g)	As Concentration (mg/L)	Adsorption Capacity (mg/g)		Temperature (°C)	Contact Time (h)	pH	Ref.
			As^{3+}	As^{5+}				
Al_2O_3	- 0.55 0.55	100 0.6 0.2	16.0 ± 0.9 - -	24.5 ± 1.6 0.14 0.098	25 23 ± 0.5 25 ± 0.5	12 2 1–2	7.0 7.0 ± 0.1 6 ± 0.1	[56] [45] [63]
Al_2O_3 (granular)	115–118 115–118	0.79–4.90 2.85–11.50	1.69 -	- 15.90	25 ± 0.5	40 170	6.1 (±0.1) 5.2 (±0.1)	[46] [46]
Al_2O_3-La_2O_3	- -	0.51 3.62	- -	0.050 0.029	21 21	48 48	7.8–9.3 7.8–9.3	[35] [35]
Fe_2O_3	- 5.05 5.05	100 0.6 0.2	60.9 ± 1.1 - -	21.3 ± 0.1 0.56 0.616	25 23 ± 0.5 25 ± 0.5	12 1 1–2	7.0 7 ± 0.1 6 ± 0.1	[56] [45] [63]
Crystalline hydrous ferric oxide	-	50	66–68	55–58	30 ± 2	4	7.0	[47]
Fe_3O_4 (magnetite)	2.43–16.5 2.43–16.5	2 2	0.65 -	- 0.7	- -	24 24	7.0 2.5–4.0	[51] [51]
TiO_2	-	100	0.0001	-	40	10	3.0	[53]
Slag-Fe_2O_3-TiO_2	163	100	0.0047	-	40	10	3.0	[53]
Fe_2O_3-TiO_2	77.8 ± 0.2	5–10	85.0	14.3	30 ± 2	3.5/6	7.0 ± 0.1	[54]
MnO_2 Fe_2O_3-MnO_2	77 123	60 60	2.55 (As^{3+} + As^{5+}) 9.89 (As^{3+} + As^{5+})		22 22	1/6 1/6	4.0 4.0	[57] [57]
Fe_2O_3-ZrO_2	339 - 263	5–40 10 10	120.0 66.5 ± 1.8 -	46.1 - 9.36	25 ± 1 30 ± 1.6 30 ± 1.6	36 2 1.6	7.0 ± 0.1 7.0 ± 0.2 7.0 ± 0.2	[49] [58] [59]
Nano ZrO_2-B_2O_3	-	5–300	-	98.04	room	2	3.0	[60]
ZrO_2-alginate beads (ZOAB)	13.2 13.2	32.9 35.2	32.3 -	- 28.5	25 25	240 240	~5.0 ~5.0	[62] [62]

135

3. Vanadium Oxyanions

3.1. Vanadium Pollution

Vanadium is a transition metal able to create variety of compounds on oxidation states ranging from −III to V. This element is characterized by very high solubility, which causes it to distribute widely in water, soil, crude oil and air [64]. Vanadium is a redox-sensitive element that exists mainly in oxidation states: V^{5+}, V^{4+}, and V^{3+} [65]. Vanadium is widely applied in industries like photography, glass, rubber, ceramic, textile, mining, metallurgy, oil refiling, automobile, and in the production of pigments and inorganic chemicals [66,67]. Such multiplicitous applications results in the production of huge amounts of vanadium polluted wastes, which are discharged into environmental waters. Vanadium has been recognized as a potentially dangerous pollutant in the same class as lead, mercury, and arsenic [68]. In cases of large accidental spills or dumping of contaminated ash, there may be major toxic effects on fauna and flora. Vanadium binds strongly to soil particles and sediments, which makes an immobile element. In European soils vanadium concentration varies between 1.28 and 537 mg/kg [69]. This element can accumulate in some plants, but not in animals. Vanadium can affect organisms via inhalation of air, ingestion of food or water, or by dermal contact. Vanadium(V) (vanadate) and vanadium(IV) (vanadyl) oxyanions can have a large effect on the function of a variety of enzymes either as an activator or inhibitor of the enzyme function [70]. Pentavalent vanadium is especially harmful to human health—it can cause damage to the respiratory, gastrointestinal, and central nervous systems and disturbs metabolism [71]. The International Agency for Research on Cancer had classified vanadium pentoxide as a possible carcinogen. Currently vanadium is on the USEPA (United States Environmental Protection Agency) Drinking Water Contaminant Candidate List (CCL3) due to its potential carcinogenic effects [68,72]. Maximum concentrations of vanadium in drinking water range from about 0.2 to 100 µg/L, with typical values ranging from 1 to 6 µg/L [73,74].

3.2. Characteristic of Vanadium Oxyanions in Aquatic Environment

Baes and Mesmer [75] revealed that vanadium exists in different hydrolyzed forms depending upon its concentration and the pH of the environment. The pentavalent form is a favored state of soluble vanadium, due to V(III) and V(IV) easily undergoing rapid oxidation by a variety of oxidizing agents including air [76]. However, simple reducing agents which are frequently present in waters, i.e., oxalates, can reduce V(V) to V(IV) [71]. In an aquatic environment pentavalent vanadium occurs mostly in the presence of oxygen, creating oxyanions [72]. Twelve vanadium species can coexist in the solution [68]. Under acidic conditions (pH < 3) vanadium(V) exists in cationic form as VO_2^+, while in the neutral-alkaline (pH = 4–11) they occur in neutral ($VO(OH)_3$) and anionic forms including decavanadate species ($V_{10}O_{26}(OH)_2^{4-}$, $V_{10}O_{27}(OH)^{5-}$ $V_{10}O_{28}^{6-}$) and mono- or polyvanadate species (e.g., $VO_2(OH)_2^-$, $VO_3(OH)^{2-}$, VO_4^{3-}, and $V_2O_6(OH)^{3-}$, $V_2O_7^{4-}$, $V_3O_9^{3-}$, $V_4O_{12}^{4-}$) [72]. The specification of pentavalent vanadium forms in dependence of the environment's pH as shown in Figure 4.

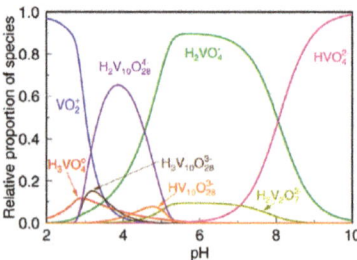

Figure 4. Distribution of vanadium(V) species in function of pH (initial vanadium concentration = 0.5 mM, T = 25 °C, 1 atm, ionic strength 0.15 M NaCl) taken from [74].

3.3. Oxide-Based Material for Vanadium Oxyanions Adsorption

Vanadium(IV) in contrast to vanadium(V) can be adsorbed onto various oxides and form complexes with organic matter, which makes it easily removable from the water phase into the sediment phase. Therefore, the removal of vanadium(V) from industrial wastes is of great importance for environmental protection. Nowadays vanadium is removed from water using biological, physical, and chemical techniques [76]. Adsorption, as an environmentally friendly and economic method, is one of the possible ways to do it.

Naeem et al. [68] examined vanadium adsorption onto commercially available metal oxide adsorbents currently used for arsenic removal—GTO from Dow, which is adsorbent-based on TiO_2, E-33 from Seven Trents, and GFH (Granular Ferric Hydroxide) from US Filter, which are iron-based ones. Experimental data revealed that pH in the range of 3.0–3.5 is favorable for vanadium sorption, which is in agreement with the electrostatic attraction between protonated sites and strongly anionic metal species. Competition between hydroxide and aqueous vanadium ions for available surface sites may cause a decrease in vanadate adsorption in high pH values. At a high pH the vanadate anion can specifically adsorb in the form of HVO_4^{2-} via a ligand exchange process, which is also characteristic for phosphate, which may suggest similar adsorption behavior of these two oxyanions.

The adsorption capacities for vanadium removal increase in the following order: GTO < E-22 < GFH. GFH revealed almost four times higher effectiveness than the other iron-based adsorbent E-22, which may be a result of its higher porosity and surface area, and less crystalline, more amorphous mineralogy. However, adsorption efficiency differs significantly between tested samples, and all of them gave similar adsorption isotherm shapes (Figure 5), which indirectly confirms that the same vanadate sorption mechanism is correct for the different metal oxides/hydroxides. The uptake of vanadate on oxides/hydroxides occurs through an anion exchange mechanism including the formation of binuclear-bridged complexes. The change of initial and final pH of the system results from substituting surface OH^- groups with HVO_4^{2-} ions from the solution. The hydroxide ion desorbs from the metal oxide surface being neutralized by H^+ ions, coming from the deprotonation of the $H_2VO_4^{2-}$, forming a water molecule, which is shown in Figure 6.

Leiviskä et al. [72] examined six commercial iron sorbents in vanadium removal from real industrial wastewater—commercial iron sorbent (CFH-12), commercial mineral sorbent (AQM), blast furnace sludge (BFS), steel converter sludge (SCS), ferrochrome slag (FeCr) and slag from a steel foundry (OKTO). Composition of sorbents was determined via XRF and XRD analyses and are shown in the Table 2. Experiments were carried out in batch and continuous flow pilot systems.

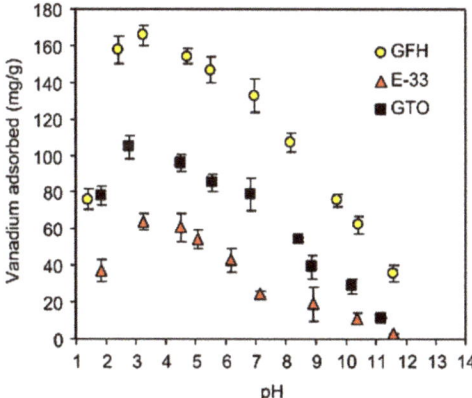

Figure 5. Vanadium adsorption isotherms obtained for 0.35 g/L dry mass of GFH and E-33, and 0.50 g/L of GTO (25 °C, ionic strength 0.01 M $NaClO_4$, initial vanadium concentration 50 mg/L), taken from [68].

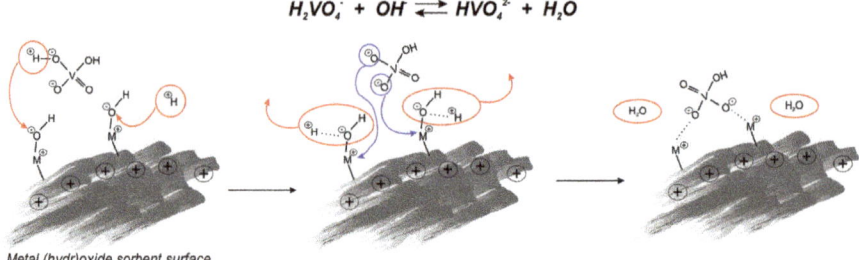

Figure 6. Mechanism of vanadate bonding to the surface of metal (hydr)oxide adsorbents proposed by Naeem et al. [68].

Table 2. Iron sorbents used for vanadium removal by Leiviskä et al., reproduced from [72]. Commercial iron sorbent (CFH-12), commercial mineral sorbent (AQM), blast furnace sludge (BFS), steel converter sludge (SCS), ferrochrome slag (FeCr), and slag from a steel foundry (OKTO).

Material	XRF Results	XRD Results
CFH-12	83% FeO, 6.1% S, 4.2% MgO, 1.4% SiO$_2$, 1.1% CaO	Gypsum (CaSO$_4$·2H$_2$O) mostly amorphous iron material
AQM	40.1% SiO$_2$, 24.8% Al$_2$O$_3$, 18.3% FeO, 3.4% MgO, 2.9% K$_2$O	Quartz (SiO$_2$) Muscovite (KAl$_2$(Si$_3$Al)O$_{10}$(OH, F)$_2$) Kaolinite (Al$_2$Si$_2$O$_5$(OH)$_4$)
BFS	63.2% FeO, 12.5% CaO, 11.0% SiO$_2$, 2.9% Al$_2$O$_3$, 2.2% MgO, 1.0% K$_2$O	Hematite (Fe$_2$O$_3$) Calcite (CaCO$_3$) Quartz (SiO$_2$)
SCS	90.3% FeO, 5.0% CaO, 1.4% SiO$_2$ 56.3%	Magnetite (Fe$_3$O$_4$) Hematite (Fe$_2$O$_3$) Cuspidine (Ca$_4$(F$_{1.5}$(OH)$_{0.5}$)Si$_2$O$_7$)
OKTO	56.3% CaO, 26.6% SiO$_2$, 6.6% MgO, 3.1% F, 2.3% Al$_2$O$_3$, 1.3% Cr$_2$O$_3$	Periclase (MgO) Calcium hydroxide (Ca(OH)$_2$) Enstatite (Fe$_{0.3}$Mg$_{0.7}$SiO$_3$) Calcium silicate (Ca$_2$SiO$_4$)
FeCr	32.5% SiO$_2$, 25.8% Al$_2$O$_3$, 24.1% MgO, 11.2% Cr$_2$O$_3$, 4.3% FeO, 1.4% CaO	Spinel magnesioferrite Aluminum iron oxide (AlFe$_2$O$_4$) Iron silicon oxide Chromium iron (Cr$_{0.7}$Fe$_{0.3}$) Magnesium aluminum chromium oxide (Mg(Al$_{1.5}$Cr$_{0.5}$)O$_4$)

Firstly, all sorbents (5 g/L) were tested in the original pH (5.8) of wastewater in room temperature. Commercial iron sorbent achieved the highest vanadium reduction (73%). BFS, SCS, and AQM reached removal efficiency around 20% (27, 22, and 16%, respectively). OKTO and FeCr reveled vanadium removal efficiency below 10%. The effect of pH on vanadium sorption with sorbents CFH-12, AQM, BFS, and SCS were investigated at a fixed sorbent dosage of 10 g/L. CFH-12 was stable in a whole pH range reaching removal rates at the level of 91–94% and adsorption capacities in the range 4.7–5.1 mg/g. BFS exhibited the highest vanadium removal of 93% at low pH (4.2–5.0). For AQM and SCS sorbents the effect of pH was less pronounced, but the lowest efficiencies were observed in high pH. The CFH-12 was proven to be the most effective in vanadium removal, so the authors examined deeply only that sorbent. CFH-12 was able to reach equilibrium at the level of 10 g/L and adsorption data were fitted to the Langmuir model, which refers to monolayer sorption phenomenon [77]. The highest efficiency of that sorbent could be explained by its amorphous structure and high iron content. Amorphous iron oxides have a large surface area and hence a greater amount of sorption sites, which results in a higher adsorption capacity compared with crystalline iron materials. In comparison with other iron sorbents (Table 3) it can be seen that CFH-12 obtained a much lower sorption capacity, which might be caused by higher particle size. For iron sorbents, the pH of the environment had a significant effect on the surface hydroxyl groups' protonation as well as for the vanadium form. In the studied pH range, vanadium exists mainly as anionic forms. The increase of efficiency in acidic pH, especially visible for the BFS sorbent, probably occurs due to the presence of higher amounts of positively charged surface groups (>Fe-OH$_2^+$) and thus electrostatic attraction between the vanadates and the surface increases. The surfaces of iron oxides generally have a net positive surface charge at acidic-neutral pH conditions (pH below the pH of point zero charge of a sorbent). Desorption of vanadium from CFH-12 was investigated to be successful—2 M sodium hydroxide was able to desorb vanadium efficiently. The recovery and renewed usage of vanadium removed from wastewaters is possible with commercial iron sorbent. The possibility to reduce its particle sizes could be tested to optimize sorption abilities.

Considering the ability of vanadium to adsorb onto other metal oxides/hydroxides and its behavior in water, Su et al. [78] established that vanadium should have similar adsorption characteristics to arsenic and selenium. As activated alumina is an inexpensive and efficient material for arsenic and selenium removal, researchers decided to test the adsorption of vanadium on its surface. Activated alumina used in this research was purchased from Tramfloc, Inc. (Tempe, AZ, USA) and was characterized with a BET surface area equal to 363 m^2/g and a pH of point zero charge 8.8. The authors tested five different initial concentrations of vanadium to test the adsorption capacity of activated alumina. In this work authors tested also adsorption of arsenic and selenium, and they noticed that all ions act similarly, and environment pH influences maximum adsorption—it decreases in more acidic conditions and increases in more basic ones. Previous literature research indicates that arsenic adsorption is favorable in acidic pH which is in contrast to results presented in this work. Experimental data is presented only vaguely in the form of graphs, on which it can be seen that adsorbed amount varies from about 1 to about 45 mg/g in the initial vanadium concentration range of 10–493 mg/L, which is not an impressive result. Such low adsorption efficiency could be forecasted via a literature survey: in 1971 Golob et al. [79] reported that vanadium(V) can be only poorly adsorbed on activated aluminum oxide [63]. Unfortunately, data presented did not provide all derivatives to calculate adsorption capacity, which makes this research hard to follow and incomparable with other works.

Titanium dioxide is widely applied in water treatment, owing mainly to its photocatalytic properties. In the case of vanadium, its pentavalent form is not able to be further oxidized. Between 1990 and 2010 some researchers investigated adsorption of vanadium onto TiO$_2$ surface in order to obtain vanadia-titania catalysts [80–82], which are still used for the catalytic reduction of nitrogen oxide [83]. Those researchers proceeded onto model solutions and adsorption did not occur as an effective method to do so. Nevertheless there is still a place for research on vanadium recovery from real wastewaters by sorbing it onto TiO$_2$ in order to obtain V-Ti catalysts.

Activated carbon derived from various natural materials is one of the most widely used adsorbents i.e., for removal of organic pollutants [84]. It is known for its extended surface area, microporous structure, and great sorption abilities [85], however, it reveals a poor ion adsorption capacity. Thus, Sharififard et al. [66] decided to impregnate it with iron-oxide-hydroxide to increase its affinity to vanadium oxyanions and create new adsorbent for its removal from water. Commercially available activated carbon, manufactured by Norit (Amersfoort, the Netherlands) with the trade name Norit ROY 0.8, was modified via a permanganate/ferrous iron synthesis method. Optimum synthesis conditions were as follows: concentration of FeSO$_4$ = 0.4 M, contact time = 24 h, and temperature = 55 °C. The obtained hybrid material was characterized by a lower surface area (777 m^2/g) but higher vanadium adsorption capacity (119.01 mg/g) in comparison with pure activated carbon (surface area 1062 m^2/g., adsorption capacity 37.87 mg/g). The adsorption equilibrium data fitted well with Freundlich isotherm, which suggested heterogeneous adsorption, what might be caused by the coexistence of different sorption sites, and/or different sorption mechanisms, or sorption of different vanadium species.

Carbon nanotubes (CNTs) doped with metal oxides are one of the most effective adsorbents used by researchers. Gupta et al. [12] prepared multi-walled carbon nanotubes doped with a palladium oxide (PdO-MWCNTs) adsorbent for vanadium removal. Multi-walled carbon nanotubes (MWCNTs) were synthesized by CVD method (purity N95%) from raw materials. Pure MWCNTs were treated with a H_2SO_4/HNO_3 (3:1 v/v) mixture for 8 h at 10 °C at ultrasonic conditions. Then product was washed by deionized water and dried. The obtained acidic-MWCNTs powder was dispersed in water and combined with $Pd(NO_3)_2 \cdot 2H_2O$ and NaOH to pH 10. The product was sonificated in an ultrasonic homogenizer and then stirred at 80 °C for 6 h and dried. Vanadium sorption was investigated by batch experiments. The optimal conditions for vanadium removal were as follows: pH = 3.0, initial vanadium(V) concentration = 60.0 mg/L, adsorbent dosage = 1.0 g/L, temperature 25 °C, and contact time 30 min. Adsorption isotherms and reaction kinetics imply that adsorption by PdO-MWCNTs could follow the Langmuir model and is a pseudo-second reaction. The highest vanadium removal efficiency of 93.7% was reached for initial vanadium concentration 60 mg/L and pH 3. Adsorbent recovery tests had not been proceeded. However, though the PdO-MWCNTs obtained high adsorption capacity for vanadium removal, their effectiveness had not been enormously high. Tests were carried only for model solutions. Thus complicated preparation and lack of information of adsorbent/adsorbate recovery and competitive adsorption of other ions strongly limit their usage.

Raw materials are gaining more and more attention from the research community in the field of adsorption. They are cost effective and highly available materials [86], which may be able to replace activated carbon adsorbents. Chitosan is characterized by the high capacity for the sorption of oxyanions, which are efficiently sorbed in acidic solutions by ionic interactions. This material was investigated in pentavalent vanadium sorption by many scientists [87–91]. Omidinasab et al. [67] decided to connect chitosan with magnetite to makes easier the separation after adsorption process. Besides its magnetic properties, Fe_3O_4 is characterized by chemical inertness, biocompatibility, non-toxicity, good thermal stability, and high surface area [92,93], which makes it a perfect material to create environmental friendly adsorbent. Chitosan was chemically extracted from chitin originating from shrimp shell wastes and then dissoluted in distilled water. Ferrous and ferric salts were co-precipitated by an ammonia solution at room temperature while the chitosan solution was slowly dripped into the mixture. Such prepared nanoparticles (Fe_3O_4-CSN) were collected using an external magnetic field and washed with distilled water. For testing the ability of vanadium removal the real wastewater samples originating from oil refinery were used. Fe_3O_4-CSN composite occurred to be very efficient—99.9% of vanadium was removed from the solution. The system reached equilibrium in the very short time of 10 min. Vanadium sorption is favorable in low temperatures and acidic pH. Kinetic data implies that the reaction was pseudo-second order, while equilibrium data fit better with the Freundlich isotherm model. Thus the adsorption onto chitosan-magnetite composite was a combination of physi- and chemisorption. Thermodynamic data revealed that the process was exothermic and spontaneous. However authors checked adsorbent only on two solutions in case of vanadium and palladium recovery, they claim that the Fe_3O_4-CSN composite can be used effectively for the removal of metal ions and for the treatment of real wastewaters without remarkable matrix effect. Nevertheless these conclusions seem to be too far-reaching and their confirmation requires further investigation.

A variety of metal oxides were used for purification of vanadium-contaminated water and wastewater. Similar to arsenic, vanadium is better adsorbed in acidic conditions, however, satisfactory results are also obtained in neutral conditions. A comparison of some adsorbents used by researchers for vanadate removal is shown in Table 3.

Table 3. Sorption properties of metal oxide-based adsorbents for vanadium oxyanions removal.

Adsorbent	Surface Area (m^2/g)	V Concentration (mg/L)	Adsorption Capacity (mg/g)	Temperature (°C)	Contact Time (h)	pH	Ref.
GFH (584 mg Fe/g GFH)	231	1–250	111.11	25	24	7.0 ± 0.1	[68]
E-33 (574 mg Fe/g E-33)	128	1–250	25.06	25	24	7.0 ± 0.1	[68]
GTO (650 mg Ti/g TiO$_2$)	150	1–250	45.66	25	24	7.0 ± 0.1	[68]
CFH-12	173	58.2	5.71	room	24	5.8	[72]
AQM	-	58.2	1.72	room	24	5.8	[72]
BFS	-	58.2	1.93	room	24	5.8	[72]
SCS	-	58.2	2.62	room	24	5.8	[72]
Fe-AC	777	25–200	119.01	25	24	4.5	[66]
CeO$_2$/CuFe$_2$O$_4$	190.2	30–250	798.6	25	3	6.0	[94]
Fe$_3$O$_4$-CSN	35.6	16.37	186.6	19.85	1/6	5.0	[67]
PdO-MWCNTs nanocomposites	209.59	60	245.05	25	0.5	3.0	[12]

4. Boron Oxyanions

4.1. Boron Pollution and Its Behavior in Aquatic Media

Boron is a metalloid that creates various compounds in five different oxidation states: −V, −I, +I, +II, and +III. Except for small amount in meteoroids, uncombined boron is never found in the elemental form in nature. Boron is a naturally found mainly as oxygen compounds (e.g., borate minerals) in oceans, sedimentary rocks, coal, shale, and some soils [95,96]. Its formation in aquatic environments is highly dependent on the hydrogen ion concentration—in pH above 8 it exists mainly as boric acid H_3BO_3, and below as a borate oxyanion $B(OH)_4^-$ [97]. Both of them have solution chemistries quite different from most other oxyanions. Borate is formed by the addition of a hydroxyl group to the trigonal planar boric acid molecule, creating a tetrahedral anion. In low concentrations (below 25 mmol/L) boric acid and borate exist as monomers, but with increasing concentration formation of poly-borate polymers is possible [98]. Boron species distribution is hardly dependent on the pH of the environment, as shown in Figure 7. Borates are widely used in glass production, as flame retardants, in leather production, in photographic materials, in wood preservatives and pesticides, as a high energy fuel, and in soaps and cleaners. Wastewaters polluted with boron are created mostly by glass producers and facilitate the burning of wood and coal. Boron concentration in water depends on the geochemical nature of the drainage area, proximity to marine coastal regions, and inputs from industrial and municipal effluents. In Europe, boron concentration varies from 0.001 to 2 mg/L in fresh drinking water, and similar values were reported for Russia, Turkey, Pakistan (0.01–7 mg/L), Japan (0.001 mg/L), and South Africa (0.03 mg/L). The highest concentrations were investigated in the Americas. In South America, the highest boron concentrations in boron-rich regions varied in range from 4–26 mg/L, while in other regions it was equal to 0.3 mg/L. In surface waters of North America boron concentration ranged from 0.02 mg/L to 360 mg/L in boron-rich regions, while the majority of samples were less than 0.1 mg/L [99]. The guideline value of boron concentration in drinking water was estimated as 2.4 mg/L by the World Health Organization (WHO) [100].

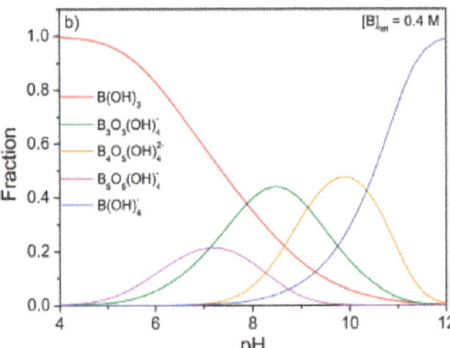

Figure 7. Distribution of boron species as a function of the solution pH (total boron concentration 0.4 M), taken from [101].

Waterborne boron may be adsorbed by soils and sediments and can accumulate in plants. Ingestion of high levels of boron can cause nausea, vomiting, abnormally low blood pressure, convulsions, and red lesions on the skin. Extremely low levels of boron in humans cause an increased heart rate and change of skin color to blue. High level exposure can affect the central nervous system, kidneys, and liver, and may be a leading cause of death. Borate in wastewaters is difficult to treat because it does not generate insoluble compounds with hazardous metal ions or alkaline earth metals [102]. Conventional means of water treatment (coagulation, sedimentation, filtration) are not able to remove boron completely, so more advanced and specific methods are needed to eliminate it from highly boron-polluted waters [100]. Between those methods, the most effective one is adsorption technique, due to its simplicity and the possibility to apply it in aqueous media with low concentrations of boron. Boron adsorption can be conducted on various sorbents, e.g., chelating resins, polysaccharides, synthesized clay, fly ash, and oxides [103].

4.2. Materials for Effective Adsorption of Boron Oxyanions

According to the literature survey made by Demetriou and Pashalidis [95], aluminum and iron oxides are the primary boron adsorption surfaces in soils, which encourage them to proceed adsorption tests using iron oxide (FeO(OH)). The adsorbent, iron-oxide (Fe(O)OH, mesh-325, Aldrich Co) was used without any further purification or other pre-treatment, while boron solutions were made from standard boron solution (99.99%, Aldrich Co) by addition of distilled water. Iron oxide's point zero charge was reached at pH = 8. The authors investigated the optimal pH in the range of 4–12, temperature from range 20–70 °C, initial boron concentration range from 0.1–7.0 mg/L, and amount of the adsorbent range from 0.05–2.5. Optimum pH oscillates from 7 to 9 with a maximum at about 8 and it is close to the pH_{pzc} of iron oxide, and slightly lower than pKa (9.2) of the boric acid, which indicates that the best adsorption occurs when the surface has no charge and boric acid is predominant in solution. Adsorption capacity was investigated using the Langmuir model at 22 °C, and the initial boron concentration, 55 mg/L, was equal to 0.324 mg/g. The authors underline the importance of iron oxide as a boric acid adsorbent, because it affects the chemical behavior and migration of boron in the natural environment and in the purification of industrial wastewaters. It is known that iron oxide as well as activated alumina are popularly-used adsorbents for industrial wastewater treatment, not specially for boron removal. In highly boron-contaminated water, iron oxide would not be able to purify the water because of its low boron sorption capacity.

Peak et al. [104] examined hydrous ferric oxide (HFO) in boron removal obtained even worse results than Demetriou and Pashalidis [95]. The authors were not able to determine maximum adsorption capacity of HFO but the adsorbed boron amount tested for three different pHs (6.5, 9.4, and 10.4) varied from almost 0 to 160 μmol/g, which is extremely low. Moreover, adsorption isotherms did not display a particularly high affinity of boric acid for the HFO surface—to achieve a high surface loading, a high solution concentration is necessary. Except for the obtained results, the authors suggested that boric acid adsorbs via both physical adsorption (outer-sphere) and ligand exchange (inner-sphere) reactions.

Due to the need to develop alternative and cost-effective adsorbents, Irawan et al. [95] decided to test aluminum-based water treatment residuals (Al-WTRs) in boron removal. Al-WTRs consist mainly of aluminum, iron, and silica oxides with some organic compounds and they are generated from drinking water treatment facilities. Previously the Al-WTRs were investigated as anionic contaminants adsorbents i.e., fluoride (F^-), phosphate (PO_4^{3-}), perchlorate (ClO_4^-), arsenate (AsO_4^{3-}), and selenium (as SeO_3^{2-} and SeO_4^{2-}). A boron solution was prepared from analytical grade H_3BO_3 and adsorbents were obtained from three different water treatment plants in Taiwan. Coagulants used in water treatment plants were aluminum sulfate (sample Al-WTR1) and polyaluminum chloride (samples Al-WTR2 and Al-WTR3). Before adsorption experiments Al-WTRs were washed with deionized water to remove impurities, dried overnight in 150 °C, crushed and sieved. The chemical composition of used aluminum water treatment residuals was investigated using aqua regia-HF procedure in a Teflon closed vessel. Al-WTR1 was characterized by highest Al_2O_3 content (408 ± 1 mg/g), followed by Al-WTR2 (227 ± 4 mg/g) ad Al-WTR3 (150 ± 2 mg/g). Iron oxide content was similar for all samples and was equal to 195 ± 3 mg/g for Al-WTR1, 194 ± 3 mg/g for Al-WTR2 and 197 ± 2 mg/g for Al-WTR3. Silica content was highest in Al-WTR3 (528 ± 2 mg/g), followed by Al-WTR1 (376 ± 1 mg/g) and Al-WTR2 (322 ± 1 mg/g). Al-WTR1 occurred to be the best sorbent for boron probably due to its highest alumina content and highest surface area. Silica does not adsorb boron, so its role in boron removal is negligible, and surprisingly the Al-WTR3 with the highest silica content is characterized by lowest surface area. The data obtained during experiments fitted the Langmuir adsorption isotherm and the reaction rate was described as pseudo-second order model. Adsorption capacities for boron were equal to 0.98, 0.70, and 0.19 mg/g for Al-WTR1, Al-WTR2, and Al-WTR3, respectively. Authors concluded that aluminum-based water treatment residuals (Al-WTRs) can be used as alternative adsorbent for boron removal. However, it should be remembered that Al-WTRs will not be as effective as adsorbents designed specifically for boron removal because they have been designed for general water purification. Thus, Al-WTRs will not be effective adsorbents for highly boron contaminated water. To make great use from Al-WTRs in boron removal from industrial wastewaters further research to extend sorption capacity and affinity for boron compounds must be proceeded.

Adsorbents with magnetic properties are a promising technology for the future of water treatment. Such sorbents reveal the great potential of functionalization and the effective sorption of different substances [105]. Magnetic properties enable effective elimination of adsorbent from water treatment systems which simplifies its regeneration. Fe_3O_4-based adsorbents are characterized by large surface area and magnetic recoverability [106]. Considering the facts presented above, Liu et al. [98] tested Fe_3O_4 and its two composites derived from magnetite and bis(trimethoxysilylpropyl)amine (TSPA), and from magnetite and a flocculating agent 1010f (a copolymer of acrylamide, sodium acrylate, and [2-(acryloyloxy)ethyl]trimethylammonium chloride) in boron removal by means of adsorption. Pure magnetite was obtained by coprecipitation of Fe(II) and Fe(III) ions in aqueous solution with ammonia. For Fe_3O_4-TSPA composite, to 5.0 g of wet magnetite dispersed in water 2.5 mL of TSPA (Gelest) was added and stirred within 30 min. Composite material was removed from reaction environment using magnet and washed to inert pH. Second composite was prepared analogically, the flocculating agent was 0.5 g/L 1010f (Zibo Zhisheng Industrial Co., Ltd., PR China). For adsorption experiments 3.0 g of the wet particles were dispersed in 50 mL solution at the desired initial boron concentration, pH, and ionic strength. All of them were carried out using a SHA-C shaking water bath (Changzhou Guohua Co., Ltd., Changzhou, China) with a shaking speed of 80 rpm at 22 °C. Experimental data revealed that boron adsorption is the most favorable in pH = 6 (three initial pH were tested and adsorption decreased in order 6.0 > 2.2 > 11.7). From the boron speciation chart (Figure 6) it can be seen that researchers missed the pH range in which boron oxyanions are formed. Probably if more careful research or pH influence study had been done, the authors could have obtained much better results for their adsorbents. For all adsorbents the amount of boron adsorbed is highest in neutral solution, which may be caused by the creation of hydrogen bonding, and electrostatic and hydrophobic attractions. In alkaline conditions adsorption is the lowest, what may be the result of the electrostatic repulsion. Increasing ionic strength decreases adsorption efficiency. The highest efficiency in boron removal was observed for Fe_3O_4–TSPA, followed by Fe_3O_4–1010f, and the pure Fe_3O_4. The adsorption decreases with the increase in ionic strength. Adsorbents' surfaces have the suitable functional groups and atoms for the formation of hydrogen bonds (including ionic hydrogen bonds) with boric acid and borate, which can promote the adsorption. The authors determined that three types of interactions determine boron adsorption onto Fe_3O_4–TSPA: (i) electrostatic interaction; (ii) hydrogen bonding; and (iii) hydrophobic interaction and proposed specific mechanism of it.

Öztürk and Kavak [107] investigated boron removal from aqueous solutions by batch adsorption onto cerium oxide. Maximum boron adsorption was obtained at original pH value of boron solution (6.18) and 40 °C by powdered cerium oxide, but the authors did not give an exact value. Experimental data were neither fitted to Langmuir nor Freundlich isotherm, which confirmed that boron adsorption onto cerium oxide is unfavorable.

In their work de la Fuente García-Soto et al. [97] investigated ability of magnesium oxide for boron removal by means of adsorption from water environment. Researchers were working on commercially available magnesium oxide produced by Panreac Chemical (Castellar del Vallès, Spain) and solutions of boric acid supplied by Merck, Darmstadt (Germany) in distilled water. Adsorption isotherms were prepared by combining growing amount of adsorbent in constant adsorbate solution in time necessary to reach equilibrium. Then the remaining boron concentration was measured. Conditions of the process were: Mg/B mol ratio, 20; stirring speed, 200 rpm; stirring time, 2 h; repose time,

48 h; room temperature; pH 9.50–10.50. Experimental data were compared using the Langmuir isotherm model. The authors studied the mechanism of boron adsorption onto magnesium oxide surface and proposed it as a three step chemisorption: (i) hydration reaction of MgO with water creating a magnesium hydroxide gel possessing active centra over the surface; (ii) alkalization due to acid–base reaction between magnesium oxide and water; and (iii) stereospecific chemical reaction between borate ions and active centers. Adsorption is an effective method of medium to high boron concentrations (50–500 mg/L) and its efficiency is set over 95% of boron removal, which is higher than natural adsorbent materials such as clays, boron containing minerals, or humic acids. This process seems to be relatively inexpensive—authors estimated the cost of highly boron-contaminated water (500 mg/L) purification at 0.96 € per m^3. Nevertheless, this process is effective only for medium-high concentration of boron in the water and adsorbent can be used only in 3 cycles due to irreversibility of boron adsorption onto magnesium oxide.

Considering the need to design recyclable adsorbent for boron removal Kameda et al. [102] decided to investigate usage of Mg-Al oxide through the production of Mg-Al layered double hydroxides intercalated with $B(OH)_4^-$. Maximum adsorption capacity was obtained for Mg-Al oxide when the Mg/Al = 2 and was equal to 80 mg/g. Recycling of the adsorbent was proposed in two steps: (i) borate intercalated Mg-Al LDH treatment with carbonate ions in water in order to enable anionic exchange between carbonate and borate, to produce CO_3·Mg-Al LDH, (ii) CO_3·Mg-Al LDH calcination in 400–800 °C in order to recover Mg-Al LDH. After regeneration tests the adsorbent maintained the ability to remove boron from aqueous solution, but efficiency declined significantly—boron concentration at 480 min of adsorption for the first cycle was 32.3 mg/L, and for the regenerated adsorbent 62.1 mg/L. Such a low performance of Mg-Al oxide is connected to the decreased crystallinity and remains of adsorbed boron into the Mg-Al oxide structure. To consider the commercial use of Mg-Al oxide as boron adsorbent, the regeneration procedure needs to be refined and the contact time should be significantly reduced.

The adverse effects of boron pollution for living organisms could not be ignored. Research community developed some efficient borate sorbents based on metal oxides. Borates are preferably sorbed in alkaline pH. At pH below 8 the main boron compound is orthoboric acid and due to its lack of charge adsorbed quantity is insignificant. Thus, neutral pH could not be optimal one for boron species adsorption. The data gathered during the literature survey is shown in Table 4. However many more adsorbents were discussed in this section, only a few publications included all necessary information.

Table 4. Sorption properties of metal oxide-based adsorbents for boron oxyanions removal.

Adsorbent	Surface Area (m^2/g)	B Concentration (mg/L)	Adsorption Capacity (mg/g)	T (°C)	Contact Time (h)	pH	Ref.
MgO	-	50	303.87	room	48	9.5–10.5	[97]
	-	500	542.11	room	48	9.5–10.5	[97]
FeO(OH)	-	55	0.324	22 ± 3	-	8	[95]
Al_2O_3-Fe_2O_3-SiO_2 (Al-WTR1)	40.5 ± 5	5–100	0.980	room	24	8.3 ± 0.2	[108]
Al_2O_3-Fe_2O_3-SiO_2 (Al-WTR2)	34.6 ± 3	5–100	0.700	room	24	8.3 ± 0.2	[108]
Al_2O_3-Fe_2O_3-SiO_2 (Al-WTR3)	14.5 ± 1	5–100	0.190	room	24	8.3 ± 0.2	[108]
MgO-Al_2O_3	-	108–648	80.00	30	168	10.5	[102]

5. Tungsten Oxyanions

5.1. Tungsten as an Environmental Threat and Its Performance in Water

Tungsten (W) is a heavy metal which creates compounds in oxidation states ranging from −4, to +6, while the most stable state is +6. In the environment, it occurs naturally in soils and sediments. In solution

W is oxidized to soluble WO_4 ions. Tungstate is able to occur in various complexes depending on the pH of the environment and total W concentration. In the range from neutral to alkaline conditions, oxyanion WO_4^{2-} with tetrahedral coordination is the predominant form. In acidic media, tungsten aims for the creation of polymeric compound such as paratungstate (i.e., $W_7O_{24}^{6-}$ and $H_2W_{12}O_{43}^{10-}$) with octahedral coordination [27,109,110]. Currently tungsten is classified as an "emerging contaminant" of concern by the U.S. Environmental Protection Agency (EPA) [111]. Its average concentration in the lithosphere varies in the range of 0.2–2.4 mg/kg [112]. The residence time of tungstate in the solution amounts to about 20,000 years, which is ten times longer than the time needed for ocean mixing resulting in total homogeneity of water worldwide [113]. This phenomenon causes an increase of tungsten pollution all around the world. Tungsten was believed to be inert in the environment and less-toxic than other heavy metals, so it was used in many fields of industry [27,114]. Its major uses are tungsten-cemented carbides, metal wires, turbine blades, high temperature lubricants, catalysts, incandescent lamp filaments, television sets, heat sinks, and golf clubs [110,115]. Some phosphate fertilizers may contain 100 mg of tungsten per kg. Moreover it had been treated as an alternative to lead to produce fishing weights and ammunition. After the ban on lead shot in the USA and Norway for hunting waterfowl it was used in hunting and recreational shooting. Pollution prevention program, the Green Armament Technology (GAT), developed by the US Army proposed the usage of tungsten-tin and tungsten-nylon composites as less hazardous materials for low caliber ammunition [27,114,116]. Thus in sites of firing activities, such as combat operation zones, military, commercial, and private shooting ranges, the concentration of tungsten in soil may be higher. In 2005 Strigul et al. [112] reported in probably the first paper on the treatment of tungsten as an environmental threat revealing that its levels above 1% mass basis (i.e., 10,000 mg/kg) resulted in the death of 95% of bacterial components (*Bacillus subtilis* and *Pseudomonas fluorescen*) in 3 months, and caused death of ryegrass plants and red worms. Such tungsten concentrations also inhibit the growth of bacterial colonies, what could possibly deteriorate process performance of biological wastewater treatment systems.

5.2. Adsorption of Tungsten Species

Sorption processes onto the surface of minerals are crucial for regulating the distribution and mobility of trace metals in natural aquatic environments and soils, and it is hardly dependent on environmental pH. The mobility of tungstate oxyanion WO_4 increases more in alkaline than in acidic conditions due to the occurrence of increased repulsive force between the negatively-charged mineral surface and the tungstate oxyanion [114]. Gustafsson [117] investigated tungstate and molybdate sorption onto ferrihydrite. Results indicated that both adsorptions can be described with two monodentate surface complexes in a surface complexation model, which does not exclude the existence of other surface complexes, but suggests their lower importance. Ferrihydrite exhibited a higher affinity for tungstate than molybdate and both adsorptions were strongly pH-dependent (100% efficiency of W adsorbed in pH 0–8). Iwai and Hashimoto [27] had adsorbed tungstate ions onto different clay minerals: metal oxide minerals (gibbsite, ferrihydrite, goethite, and birnessite) and montmorillonite, which is a phyllosilicate mineral. All materials were synthesized in the laboratory. To determine tungstate's affinity for prepared adsorbents batch experiments in three different pHs—3, 6, and 9—were modeled using the Freundlich equation. Results indicate that the adsorption affinity for WO_4 is higher for metal oxide minerals (especially for Al and Fe oxide minerals) than for montmorillonite. Generally it follows the order of Al hydroxide or Fe (oxyhydr)oxides (goethite, ferrihydrite, gibbsite) > Mn oxide (birnessite) > phyllosilicate (montmorillonite) in the whole pH range. The best adsorption capacities were obtained for acidic conditions (pH 4). Aside from adsorption capacities, researchers evaluated an influence of the presence of phosphate and molybdate oxyanions on tungstate sorption. Oxyanions of PO_4 and MoO_4 revealed higher affinity than tungstate in neutral-alkaline conditions. In acidic media, tungstate is more preferably adsorbed on clay minerals. Unfortunately, adsorption–desorption tests had not been conducted, so the reusability of adsorbents remains unknown. Hur and Reeder [114] investigated tungstate sorption on boehmite, which is an aluminum oxide hydroxide (γ-AlO(OH)) mineral. Boehmite occurs naturally as a common weathering product and is known as an effective sorbent for cations and anions. Adsorption was investigated during batch uptake experiments for a range of tungsten concentrations from 50 to 2000 µM, at pH 4, 6, and 8 and two different ionic strength 0.01 or 0.1 M (calibrated with NaCl). Adsorption edges exhibits the general behavior for anions, with maximum sorption in pH range 5.0–5.5. With pH increase the sorption ability decreases, with minimum values around the point of zero charge of boehmite, which is 8.6–9.1. A smaller decrease in sorption is observed at pH values below 5. Tungstate reveals a strong affinity for the boehmite surface at acidic and neutral pH. The greatest sorption of tungsten was observed at pH 4. The maximum uptake of tungstate has not been clearly determined, and adsorption capacities varied between 7.35 and 132.36 mg/g for optimal pH conditions. Lack of maxima suggest that tungstate sorption is not limited by surface site availability over the studied concentration. Desorption tests revealed that tungstate sorption is irreversible at pH 4, and slightly reversible at pH 8. However, boehmite shows good adsorption properties, but it has limited application possibilities due to its irreversible adsorption character.

Due to their large surface area and small diffusion resistance, magnetic adsorbents are of great interest. Afkhami et al. [109] investigated the effectiveness of $Ni_{0.5}Zn_{0.5}Fe_2O_4$ prepared according to the chemical co-precipitation method at room temperature for four different oxyanions' removal, including W(IV). The adsorption capacity was in the order W(VI) > Cr(VI) > Mo(VI) > V(V) and for W(IV) was significantly higher than for other investigated oxyanions—72 mg/g. Desorption efficiency exceeded 98% while process was carried using 2 mol/L NaOH, as the most effective eluent and its equilibrium was reached in 15 min. Thus recovery of such adsorbents should be possible. Prepared nanomaterial can be easily dispersed in water, and due to its magnetic properties can be easily removed from the adsorption environment. The authors predict that $Ni_{0.5}Zn_{0.5}Fe_2O_4$ nanocomposites could be candidates for the removal of trace amounts of chromium, molybdenum, vanadium, and tungsten ions.

Due to the late discovery of tungsten's harmful properties, only few works consider its removal via means of adsorption onto metal oxide-based materials. Researchers are at the beginning of the road to find effective sorbents for W contamination's removal. Specific data gathered during literature studies were presented in Table 5. From the research already proceeded it is known that tungsten is favorably sorbed in acidic conditions at room temperature. In such conditions the surface of Fe, Al, and Mn oxides is positively charged, which enables tungsten oxyanions to bond via electrical attraction. However specific mechanism of tungsten adsorption still needs to be discovered and confirmed.

Table 5. Sorption properties of metal oxide-based adsorbents for tungsten oxyanions removal.

Adsorbent	Surface Area (m²/g)	W Concentration (mg/L)	Adsorption Capacity (mg/g)	Temperature (°C)	Contact Time (h)	pH	Ref.
$Ni_{0.5}Zn_{0.5}Fe_2O_4$	-	10–250	72	25	0.5	5	[109]
Boehmite (γ-AlO(OH))	136	1000	7.35–132.36	room	24	4	[114]
Birnessite (MnO_2)	-	18–359	6.15	25	24	4	[27]
Ferrihydrite (Fe_2O_3)	-	18–359	30.24	25	24	4	[27]
Gibbsite ($Al(OH)_3$)	-	18–359	49.82	25	24	4	[27]
Goethite (α-FeO(OH))	-	18–359	43.12	25	24	4	[27]

6. Molybdenum Oxyanions

6.1. Molybdenum Pollution and Behavior of Molybdenum Species in Water

Molybdenum is a transitional metal occurring on oxidation states from −II to +VI [118], which are easily convertible into each other. Naturally it is found only in minerals such as molydbenite, wulfenite, ferrimolybdite, and jordistite, which are mostly used for its commercial production [118]. Its main uses are metallurgical applications and as an alloying element in the production of stainless steel or cast-iron alloys. It is also crucial for the flame retardant, pigment, and catalyst industries [119]. The chemical properties of molybdenum exhibit similarities to tungsten and vanadium more than to chromium [118]. In an aquatic environment its main oxidation state is +VI on which it creates oxo/hydroxospecies, including polyanions. In natural waters Mo occurs mostly in inorganic forms on +V and +VI oxidation state as oxyanions such as MoO_4^{3-}, $HMoO_4^{2-}$, $H_2MoO_4^-$, $HMoO_4^-$, and MoO_4^{2-} depending on the pH of the environment [120]. In pHs on the level of 5–6 the dominant form is the molybdate anion, MoO_4^{2-}. In more acidic conditions molybdate is protonated to the less charged anionic species ($HMoO_4^-$) and in strongly acidic media the neutral molybdic acid $MoO_3(H_2O)_3$ is created. In high molybdenum concentrations at pH below 5–6 it is able to form isopolymetalates such as $Mo_7O_{24}^{6-}$ or $Mo_8O_{26}^{4-}$ [121]. The distribution of molybdenum species in the function of the pH is shown in Figure 8.

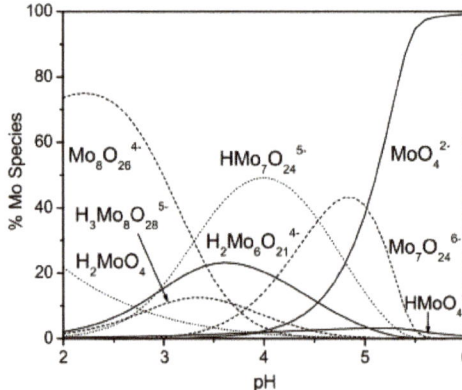

Figure 8. Distribution of Mo species in function of pH (initial molybdenum concentration = 10 mM), taken from [122].

However, though its presence is essential for life on Earth, due to its ability to form active sites for many enzymes (such as xanthine oxidase, aldehyde oxidase, and hepatic sulphite oxidase [118]), that catalyze redox reactions, in high concentrations it is toxic [121,123], and may pose health problems. Subchronic and chronic oral exposure, i.e., by drinking Mo polluted water, results in gastrointestinal disturbances, growth retardation, anemia, hypothyroidism, bone and joint deformities, liver and kidney abnormalities, sterility, and also death [119]. Most molybdate compounds are harmful and toxic when injected intraperitoneally or orally in large doses (400–800 mg per kg of body weight). The WHO established 70 µg/L as a maximum molybdenium concentration in drinking water. Total molybdenum concentrations in fresh waters range from 0.03 to 475 µg/L [121]. Near the industrial sources like molybdenum mining areas its concentration in surface water can reach 200–400 µg/L, while in groundwater to even 25,000 µg/L [119]. The molybdate ion MoO_4^{2-}, as the most common oxyform of this element, occurs in many types of industrial wastewaters i.e., wastewater from a styrene monomer plant (1000 mg/L), scrubber effluent of a municipal solid waste incinerator (0.95 mg/L) and mining water (0.1–2.2 mg/L) [119]. Thus effective methods for molybdenum removal need to be discovered.

6.2. Adsorbents for Molybdenum Removal from Water Environment

Adsorption as a relatively cheap and effective method for many oxyanions removal had also been considered for molybdenum oxyanions elimination. Iron, aluminum, and, to some extent, titanium oxides may be important adsorbent minerals for MoO_4^{2-}, as they may acquire positive charge at low pH [123].

Goldberg et al. [124] were interested in molybdenum adsorption onto Al and Fe oxide minerals and their adsorption mechanisms. Molybdenum adsorption on oxides exhibited maximum near pH 4 and iron oxides were more efficient adsorbents than aluminum oxides. Experimental data were fitted well to the constant capacitance model, which assumes an inner-sphere adsorption mechanism. Adsorption of Fe oxides based on a weight increased in the order: hematite < goethite < amorphous Fe oxide < poorly crystalline goethite, while for Al oxides: δ-Al_2O_3 < gibbsite < amorphous Al_2O_3. Adsorption of molydbdeum oxyanions onto aluminum oxide was a subject of research of many scientists [125–129] due to application of Co-MO-Al_2O_3 catalysts for hydrotreatment of petroleum fractions. Luthra and Cheng [126] examined molybdates ($Mo_7O_{24}^{6-}$ and MoO_4^{2-}) onto γ-alumina via Molybdenum-95 NMR Study. Both oxyanions were attracted to the positively charged alumina surface in pH lower than the isoelectric point of the alumina (pH 8.5). NMR studies showed that $Mo_7O_{24}^{6-}$ after the contact with alumina decomposes to MoO_4^{2-} in minutes. The decomposition was believed to be caused by the increase in the pH of the impregnation solution inside the pores of the alumina. Wu et al. [128] investigated competition adsorption for oxyanions of similar affinity to the γ-Al_2O_3 surface: molybdate, selenite, selenate, chromate, and sulfate. Specific surface area of obtained γ-Al_2O_3 was measured by BET method and was equal to 100 m^2/g. Molybdate and selenite are both strongly binding adsorbates—molybdate depresses selenite sorption at acidic pH and selenite suppresses molybdate sorption at alkaline conditions. The affinity for aluminum oxide surface was the highest for molybdate oxyanions, followed by selenite, selenate, sulfate, and chromate, respectively.

Xu et al. [33,123] investigated the influence of presence of phosphate, sulfate, silicate, and tungstate anions on the adsorption of molybdate onto goethite under anoxic conditions. Experiments were conducted in a glove bag (N_2), which maintained oxygen concentration on the 1 mg/L level. Goethite slurry was synthesized by the researchers. Results indicated that MoO_4^{2-} adsorption on goethite obeys the Langmuir model. Adsorption capacity of MoO_4^{2-} on goethite amounted 25.9 mg/g at pH 4. Competition tests revealed that an affinity for goethite follows the order phosphate > tungstate > molybdate > silicate > sulfate. Adsorption of Mo seems to be more affected by the presence of the phosphate than the tungstate anion, which is attributable to increased repulsion between Mo and more negatively charged surface sites after phosphate adsorption onto goethite surface. The goethite surface probably contains adsorption sites common for Mo, P and W anions, as well as specific for each element. Gustafsson [117] examined molybdate and tungstate adsorption to ferrihydrite synthesized by author. However, though tungstate revealed a higher affinity for ferrihydrite, Mo oxyanions were sorbed effectively. Molybdate sorption was strongly pH dependent—the best efficiencies were obtained for acidic pH, which correlates with previous research.

Iron(II, III) oxide (Fe_3O_4) were used in water purification due to its magnetic properties. The main drawback of magnetite usage is its dissolution in water. High iron concentrations are toxic for humans and other living organisms. Consider that Keskin [120] decided to use an Fe_3O_4-embedded 1,3,5-triacryloylhexahydro-1,3,5-triazine-acrylamide hybrid polymer for molybdenum oxyanions removal. Triazine compounds are able to create complexes with iron ions which connected with polymeric network prevent iron release. The triazine molecules were used as a crosslinker and the polymeric material was synthesized by free radical polymerization. Adsorption tests were performed for pure Fe_3O_4, Fe_3O_4 embedded hydrolyzed, and nonhydrolyzed polymers. The highest removal efficiency was obtained at pH 2.5 and pure magnetite was subtly more effective than both composite materials (97% for Fe_3O_4 and 96.87% and 96.36% for hydrolyzed and nonhydrolyzed polymers). The iron release between those three materials varied insignificantly—the highest value was noted for magnetite—1.16 mg/L, followed by magnetite embedded by nonhydrolyzed polymer—1.02 mg/L, and the lowest value of

0.22 mg/L characterized magnetite-embedded by hydrolyzed polymer. Kinetic studies and isotherms were investigated only for an Fe_3O_4-embedded hydrolyzed acrylamide -1,3,5-triacryloylhexahydro-1,3,5-triazine polymer. The experimental data was fitted well to the pseudo-second-order kinetic model and Langmuir isotherm. Maximum Langmuir adsorption capacity was equal to 0.213 mg/g. Author performed reusability tests of obtained material by HNO_3, $CaCl_2$, NaOH, EDTA, HCl, and HCl/HNO_3 solutions. The best results were obtained by $CaCl_2$ treatment—sorption efficiency for molybdenum removal after regeneration increased to 97%, while its mass decreased from 0.25 g to 0.21 g, which throws into question if the adsorbent surface was chemically modified by the regenerating medium.

An innovative, attractive, and economical approach for oxyanion removal is the usage of sorbent materials consisting of a matrix and a proper adsorbent. Verbinnen et al. [119] decided to test zeolite-supported magnetite for the removal of molybdenum. The material was supposed to combine good sorption affinity for oxyanions (magnetite) and high cation affinity for cations (zeolite matrix). The composite was obtained by precipitating magnetite onto zeolite surface from Fe(II) and Fe(III) salts, and the final magnetite to zeolite ratio was equal to 1:2. A molybdenum model solution was prepared by dissolving $Na_2MoO_4 \cdot 2H_2O$ in ultrapure water. Sorption was the most effective in a strongly acidic environment (pH = 3), where maximal adsorption of MoO_4^{2-} and minimal adsorbent dissolution (Fe concentrations–0.34 mg/L, Al concentrations—0.012 mg/L). The adsorption capacity for molybdenum in optimal conditions (pH = 3 and 25 °C) is 17.9 mg per gram of adsorbent. In the absence of molybdenum the point zero charge of the obtained composite lies around the pH = 4, so below that value its surface is charged positively. In the presence of Mo the point zero charge is no longer observed and magnetite-zeolite surface is negatively charged into whole studied pH range (2–10). That facts indicated a chemical adsorption of MoO_4^{2-}, which causes the shift of pH_{pzc} to a lower value. Negatively charged Mo species ($HMoO_4^-$ and MoO_4^{2}) are pulled toward composite surface via electrostatic forces and bind chemically to the Fe(III) present on magnetite surface. Such a connection results in a negative charge of the composite, even below pH 4, and lowering of the zeta potential. In pH 4–6.5 where the composite surface is negatively charged, the attractive specific sorption forces still overcome the repulsive electrostatic forces, which indicate that molybdenum adsorption onto the magnetite-zeolite surface is a specific chemisorption process. The maximum adsorption capacity increases from 13.6 mg/g at 4 °C to 20.2 mg/g at 40 °C, which indicates the endothermic adsorption. Kinetic data fits well to pseudo second order equation. Experimental results indicate that Mo adsorption is better fitted by the Langmuir isotherm, which indicates chemical bonding between the adsorbent and adsorbate, assuming the formation of a Mo monolayer onto the magnetite-supported zeolite surface. In conclusion, the molybdenum is adsorbed onto the magnetite-supported zeolite composite via formation of an inner-sphere $FeOMoO_2(OH) \cdot 2H_2O$ complex. Such a sorption mechanism was verified for different iron oxides, like goethite and feriihydrite, by Goldberg et al. [124] and Gustafsson [117]. In their later research Verbinnen et al. [130] had tested magnetite supported zeolite sorption abilities in real industrial wastewater coming from the wet treatment of flue gases from a rotary kiln for industrial waste combustion. The samples were obtained from Indaver, a waste treatment company in Belgium. However, though cations were removed by coagulation and flocculation, and mercury by precipitation with trimercaptotriazine (TMT) by Indaver, the effluent contained oxyanion-forming elements (like Mo and Sb). Anions like chloride and sulphate are hardly affected by the precipitation/flocculation treatment. Researchers increased pH from 1 to optimal sorption value—3.5. The initial concentration of Mo oxyanions was 872 µg/L and after treatment it decreased to 7 µg/L while the adsorbent concentration was 20 g/L. The adsorption order on zeolite-supported magnetite is Mo(VI) > Sb(V) > Se(VI) in both synthetic and real systems.

Molybdenum sorption onto metal oxide-based adsorbents has not been a subject of research of many scientists. Adsorption data gained by literature survey is gathered in Table 6. Adsorption of molybdates is favored in acidic pH in room temperature.

Table 6. Sorption properties of metal oxide-based adsorbents for molybdenum oxyanions removal.

Adsorbent	Surface Area (m²/g)	Mo Concentration (mg/L)	Adsorption Capacity (mg/g)	Temperature (°C)	Contact Time (h)	pH	Ref.
Fe₃O₄ embedded hydrolyzed triazine polymer	-	2.5	0.213	25	2.5	2.5	[120]
zeolite-supported-Fe₃O₄	74.5	1	17.92	25	24	3	[119]
Goethite	-	1	1.76	25	24	3	[119]
	43.96	0–32	25.9	room	17	4.0 ± 0.1	[33,123]
Hematite	-	1	1.43	25	24	3	[119]

7. Conclusions

In spite of the fact that metal and metalloid oxyanion pollution is a real phenomenon and their elimination is the subject of increased interest of scientific communities, due to the lack of data in published papers the results are incomparable to each other. Authors are unwilling to share exact results of their work, which cause replication of the same studies over and over again by different research teams. In addition most of the research is carried out only on model solutions, so proposed methods might not be applicable as such for the removal of oxyanions from industrial wastewaters, because they are more complex systems containing other oxyanions and compounds, that can compete and interact with each other. Some research in the field of competitive adsorption was held, but the exact mechanisms of oxyanion behavior in solution are still unknown. Thus single sorbate and single mineral adsorption studies in the laboratory may not be directly applicable [131].

The presented literature review unambiguously indicates the complexity of the process of removing metal oxyanions from aqueous systems via adsorption. In contrast to metal cations, which have been extensively studied and described in scientific papers, oxyanions studies are not entirely clear. Analysis of the M^{n+} ion removal process indicates the optimal conditions for its realization, which depend on many factors such as the type of cation, the type of sorption material and its physicochemical parameters, as well as the parameters of the adsorption process. The key element seems to be the pH of the adsorption environment—for metal cations, in most cases analyzed, the optimal pH at which the highest removal efficiency is observed is pH = 5–6. This is related to the Pourbaix diagrams, indicating pH, beyond which precipitation of appropriate forms of metal hydroxides will occur. In this aspect, one should also remember the influence of pH on the nature of the sorbent functional groups, which can be protonated (at low pH values) or deprotonated (at higher pH values). This is important when defining the adsorption mechanism, which in the case of M^{n+} is based on electrostatic interactions—attraction when the charge of functional groups is different from the M^{n+} charge or repulsion when these charges are the same. In this respect, the oxyanion adsorption is definitely more complex. The probable mechanism of oxyanions adsorption on the surface of (hydr)oxides has been presented in Figure 9.

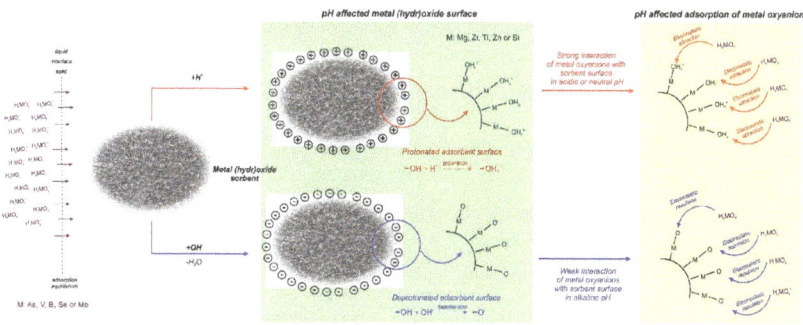

Figure 9. Mechanism of metal oxyanions adsorption onto metal (hydr)oxide-based sorbents.

Moreover, a significant majority of papers pertain to arsenic removal, which is visible on the statistic presented in Figure 10. In comparison with arsenite and arsenate, research related to other oxyanions is infinitesimal and so is its significance in scientific work.

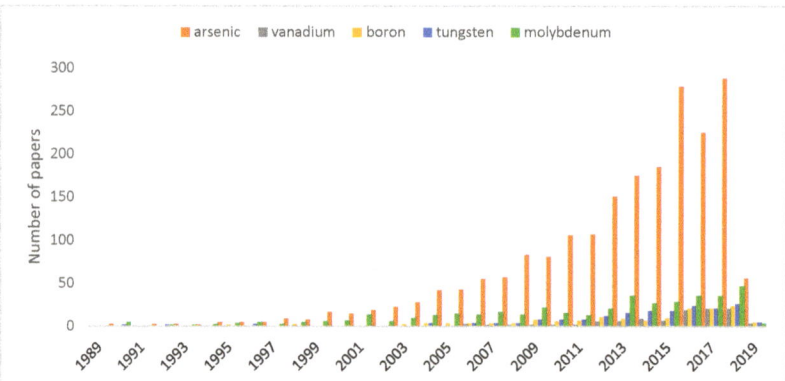

Figure 10. Bar chart of the number of articles per year about oxyanions adsorption for the 1989–28 January 2019. The statistical data was obtained by searching "adsorption metal oxide arsenite and arsenate/vanadate/borate/tungstate/molybdate" in the Scopus data base as title and keywords.

The environment pH influences as well metal chemistry in solution (occurrence of various oxyforms) as the metal (hydr)oxide surface's protonation/deprotonation [68]. Thus in most cases it is the determining factor of adsorption effectiveness, but unfortunately for research results it is neglected.

The spectrum of speciation forms of oxyanions of the relevant metal is very wide, which translates into their random behavior in aqueous solutions, especially those with varying pH. Unfortunately, the aspect of the effect of pH on the efficiency of oxyanion-binding by various sorbents, e.g., those presented in the literature review, is treated fairly generally, and in some cases even overlooked. This hinders the interpretation of the dependencies obtained and introduces confusion regarding the comparison of the behavior of various sorption materials towards the removal of these types of impurities from aqueous solutions. The presented comparisons unequivocally do not indicate optimal conditions for the removal of a particular oxyanion group using the available sorption materials. It seems that the mechanism of their binding should be at least similar for different sorbents, and as a result it is completely random. Analyzing at least the effect of the pH discussed earlier, one would expect such an environment pH, which would enable a strong attraction between the positively charged surface of the sorbent and the negative form of most metal oxyanions. On the other hand, paying attention to the type of sorption material, it would be important to use one which after synthesis or preparation would exhibit a significant positive charge that could ensure strong interaction with oxyanions. These issues should not be problematic when conducting research on model solutions, which are different to real wastewaters as their composition is diverse (high concentrations of components and their diversity) and can affect the selective adsorption of individual components. All oxyanions presented in this review, excluding borates, are preferably adsorbed in acidic media, due to the positive electrical charge present on adsorbents' surfaces. Electrostatic attraction is one that the most important mechanism of oxyanions bonding.

The following issues seem to be worth analyzing here:

- The role of sorption material, and in fact its physicochemical parameters designed at the synthesis stage—the presence of functional groups exhibiting a positive charge, facilitating the binding of negatively charged oxyanions;
- The influence of the pH of the adsorption environment on the nature of functional groups of the sorption material and the form of oxyanions in aqueous solutions, so important analyzing their potential interactions;

- Selectivity tests of sorption materials towards various metal oxyanions, which in the scientific papers published so far are effectively omitted, inversely as in case of sorption of metal cations,
- The effect of the presence of other components of wastewater on the sorption efficiency of a particular oxyanion group.

In these aspects, the presented scientific papers leave considerable dissatisfaction, and on the other hand, they open a wide range of activities to optimize the process of removal of these particularly "uncomfortable" inorganic impurities present in water systems. This all is more justified when analyzing the number of works published on this subject.

Funding: This work was supported by the National Science Centre Poland under research project no. 2018/29/B/ST8/01122.

Conflicts of Interest: The authors declare no conflict of interest.

References

1. Adegoke, H.I.; Adekola, F.A.; Fatoki, O.S.; Ximba, B.J. Sorptive interaction of oxyanions with iron oxides: A review. *Pol. J. Environ. Stud.* **2013**, *22*, 7–24.
2. Cornelis, G.; Johnson, C.A.; Van Gerven, T.; Vandecasteele, C. Leaching mechanisms of oxyanionic metalloid and metal species in alkaline solid wastes: A review. *Appl. Geochem.* **2008**, *23*, 955–976. [CrossRef]
3. Kailasam, V.; Rosenberg, E. Oxyanion removal and recovery using silica polyamine composites. *Hydrometallurgy* **2012**, *129–130*, 97–104. [CrossRef]
4. Verbinnen, B.; Block, C.; Van Caneghem, J.; Vandecasteele, C. Recycling of spent adsorbents for oxyanions and heavy metal ions in the production of ceramics. *Waste Manag.* **2015**, *45*, 407–411. [CrossRef]
5. Hajji, S.; Montes-Hernandez, G.; Sarret, G.; Tordo, A.; Morin, G.; Ona-nguema, G.; Bureau, S.; Turkia, T.; Mzoughi, N. Arsenite and chromate sequestration onto ferrihydrite, siderite and goethite nanostructured minerals: Isotherms from flow-through reactor experiments and XAS measurements. *J. Hazard. Mater.* **2019**, *362*, 358–367. [CrossRef]
6. Wang, Y.; Ding, S.; Shi, L.; Gong, M.; Xu, S.; Zhang, C. Simultaneous measurements of cations and anions using diffusive gradients in thin films with a ZrO-Chelex mixed binding layer. *Anal. Chim. Acta* **2017**, *972*, 1–11. [CrossRef] [PubMed]
7. Mon, M.; Bruno, R.; Ferrando-Soria, J.; Armentano, D.; Pardo, E. Metal-organic framework technologies for water remediation: Towards a sustainable ecosystem. *J. Mater. Chem. A* **2018**, *6*, 4912–4947. [CrossRef]
8. Baouab, M.H.V.; Gauthier, R.; Bernard, C. Sorption of chromium oxy-anions onto cationized lignocellulosic material. *J. Appl. Polym. Sci.* **2003**, *87*, 1660–1665.
9. Hristovski, K.D.; Markovski, J. Science of the total environment engineering metal (hydr)oxide sorbents for removal of arsenate and similar weak-acid oxyanion contaminants: A critical review with emphasis on factors governing sorption processes. *Sci. Total Environ.* **2017**, *598*, 258–271. [CrossRef]
10. Ungureanu, G.; Filote, C.; Santos, S.C.R.; Boaventura, R.A.R.; Volf, I.; Botelho, C.M.S. Antimony oxyanions uptake by green marine macroalgae. *J. Environ. Chem. Eng.* **2016**, *4*, 3441–3450. [CrossRef]
11. Chung, J.; Ahn, C.H.; Chen, Z.; Rittmann, B.E. Bio-reduction of arsenate using a hydrogen-based membrane biofilm reactor. *Chemosphere* **2006**, *65*, 24–34. [CrossRef]
12. Gupta, V.K.; Fakhri, A.; Kumar, A.; Agarwal, S.; Naji, M. Optimization by response surface methodology for vanadium(V) removal from aqueous solutions using PdO-MWCNTs nanocomposites. *J. Mol. Liq.* **2017**, *234*, 117–123. [CrossRef]
13. Kołodyńska, D.; Budnyak, T.M.; Hubicki, Z.; Tertykh, V.A. Sol–gel derived organic—Inorganic hybrid ceramic materials for heavy metal removal. In *Sol-Gel Based Nanoceramic Materials: Preparation, Properties and Applications*; Mishra, A.K., Ed.; Springer International Publishing: Berlin, Germany, 2017; pp. 253–274.
14. Atia, A.A. Adsorption of chromate and molybdate by cetylpyridinium bentonite. *Appl. Clay Sci.* **2008**, *41*, 73–84. [CrossRef]
15. Deng, B.; Caviness, M.; Gu, Z. Arsenic removal by activated carbon-based materials. In *Advances in Arsenic Research*; O'Day, P.A., Vlassopoulos, D., Meng, X., Benning, L.G., Eds.; American Chemical Society: Washington, DC, USA, 2005; pp. 284–293.

16. Santonastaso, G.F.; Erto, A.; Bortone, I.; Chianese, S.; Di Nardo, A.; Musmarra, D. Experimental and simulation study of the restoration of a thallium (I)-contaminated aquifer by Permeable Adsorptive Barriers (PABs). *Sci. Total Environ.* **2018**, *630*, 62–71. [CrossRef]
17. Melendres, C.A.; Hahn, F.; Bowmaker, G.A. Oxyanion adsorption and competition at a gold electrode. *Electrochim. Acta* **2000**, *46*, 9–13. [CrossRef]
18. Cumberland, S.L.; Strouse, G.F. Analysis of the nature of oxyanion adsorption on gold nanomaterial surfaces. *Langmuir* **2002**, *18*, 269–276. [CrossRef]
19. Stevenson, K.J.; Gao, X.; Hatchett, D.W.; White, H.S. Voltammetric measurement of anion adsorption on Ag(111). *J. Electroanal. Chem.* **1998**, *447*, 43–51. [CrossRef]
20. Zhu, H.; Jia, Y.; Wu, X.; Wang, H. Removal of arsenic from water by supported nano zero-valent iron on activated carbon. *J. Hazard. Mater.* **2009**, *172*, 1591–1596. [CrossRef]
21. Figueiredo, H.; Quintelas, C. Tailored zeolites for the removal of metal oxyanions: Overcoming intrinsic limitations of zeolites. *J. Hazard. Mater.* **2014**, *274*, 287–299. [CrossRef]
22. Bissen, M.; Frimmel, F.H. Arsenic—A review. Part II: Oxidation of arsenic and its removal in water treatment. *Acta Hydrochim. Hydrobiol.* **2003**, *31*, 97–107. [CrossRef]
23. Baccile, N.; Falco, C.; Titirici, M.-M. Characterization of biomass and its derived char using 13 C-solid state nuclear magnetic resonance. *Green Chem.* **2014**, *16*, 4839–4869. [CrossRef]
24. Chao, H.P.; Lee, C.K.; Juang, L.C.; Han, Y.L. Sorption of organic compounds, oxyanions, and heavy metal ions on surfactant modified titanate nanotubes. *Ind. Eng. Chem. Res.* **2013**, *52*, 9843–9850. [CrossRef]
25. Hemming, G.; Reeder, R.J.; Hart, S.R. Growth-step-selective incorporation of boron on the calcite surface. *Geochim. Cosmochim. Acta* **1998**, *62*, 2915–2922. [CrossRef]
26. Hiemstra, T.; Van Riemsdijk, W.H. Fluoride adsorption on goethite in relation to different types of surface sites. *J. Colloid Interface Sci.* **2000**, *225*, 94–104. [CrossRef]
27. Iwai, T.; Hashimoto, Y. Adsorption of tungstate (WO_4) on birnessite, ferrihydrite, gibbsite, goethite and montmorillonite as affected by pH and competitive phosphate (PO_4) and molybdate (MoO_4) oxyanions. *Appl. Clay Sci.* **2017**, *143*, 372–377. [CrossRef]
28. Manning, B.A.; Fendorf, S.E.; Goldberg, S. Surface structures and stability of arsenic(III) on goethite: Spectroscopic evidence for inner-sphere complexes. *Environ. Sci. Technol.* **1998**, *32*, 2383–2388. [CrossRef]
29. Hiemstra, T.; Van Riemsdijk, W.H. Surface structural adsorption modeling of competitive binding of oxanions by metal (hydr)oxides. *J. Colloid Interface Sci.* **1999**, *210*, 182–193. [CrossRef]
30. Matis, K.A.; Zouboulis, A.I.; Zamboulis, D.; Valtadoru, A.V. Sorption of As(V) by goethite particles and study of their flocculation. *Water Air Soil Pollut.* **1999**, *111*, 297–316. [CrossRef]
31. Li, Z.; Bowman, R.S. Retention of inorganic oxyanions by organo-kaolinite. *Water Res.* **2001**, *35*, 3771–3776. [CrossRef]
32. Sherlala, A.I.A.; Raman, A.A.A.; Bello, M.M. Synthesis and characterization of magnetic graphene oxide for arsenic removal from aqueous solution. *Environ. Technol.* **2018**, 1–9. [CrossRef]
33. Xu, N.; Christodoulatos, C.; Braida, W. Modeling the competitive effect of phosphate, sulfate, silicate, and tungstate anions on the adsorption of molybdate onto goethite. *Chemosphere* **2006**, *64*, 1325–1333. [CrossRef]
34. Laudadio, E.D.; Bennett, J.W.; Green, C.M.; Mason, S.E.; Hamers, R.J. Impact of phosphate adsorption on complex cobalt oxide nanoparticle dispersibility in aqueous media. *Environ. Sci. Technol.* **2018**, *52*, 10186–10195. [CrossRef]
35. Davis, S.A.; Misra, M. Transport model for the adsorption of oxyanions of selenium(IV) and arsenic(V) from water onto lanthanum-and aluminum-based oxides. *J. Colloid Interface Sci.* **1997**, *188*, 340–350. [CrossRef]
36. Ravenscroft, P.; Brammer, H.; Richards, K. *Arsenic Pollution*; Environmental Chemistry: West Sussex, UK, 2009.
37. Smedley, P.L.; Kinniburgh, D.G. A review of the source, behaviour and distribution of arsenic in natural waters. *Appl. Geochem.* **2002**, *17*, 517–568. [CrossRef]
38. Peng, F.F.; Di, P. Removal of arsenic from aqueous solution by adsorbing colloid flotation. *Ind. Eng. Chem. Res.* **1994**, *33*, 922–928. [CrossRef]
39. Van Halem, D.; Bakker, S.A.; Amy, G.L.; Van Dijk, J.C. Arsenic in drinking water: A worldwide water quality concern for water supply companies. *Drink. Water Eng. Sci.* **2009**, *2*, 29–34. [CrossRef]
40. World Health Organization. Arsenic. 2019. Available online: https://www.who.int/news-room/fact-sheets/detail/arsenic (accessed on 27 January 2018).

41. Su, C.; Puls, R.W. Arsenate and arsenite removal by zerovalent iron: Effects of phosphate, silicate, carbonate, borate, sulfate, chromate, molybdate, and nitrate, relative to chloride. *Environ. Sci. Technol.* **2001**, *35*, 4562–4568. [CrossRef]
42. Pena, M.E.; Korfiatis, G.P.; Patel, M.; Lippincott, L.; Meng, X. Adsorption of As(V) and As(III) by nanocrystalline titanium dioxide. *Water Res.* **2005**, *39*, 2327–2337. [CrossRef]
43. Meng, X.; Bang, S.; Korfiatis, G.P. Effects of silicate, sulfate, and carbonate on arsenic removal by ferric chloride. *Water Res.* **2000**, *34*, 1255–1261. [CrossRef]
44. Dambies, L. Existing and prospective sorption technologies for the removal of arsenic in water. *Sep. Sci. Technol.* **2005**, *39*, 603–627. [CrossRef]
45. Jeong, Y.; Fan, M.; Singh, S.; Chuang, C.L.; Saha, B.; van Leeuwen, J. Evaluation of iron oxide and aluminum oxide as potential arsenic(V) adsorbents. *Chem. Eng. Process. Process Intensif.* **2007**, *46*, 1030–1039. [CrossRef]
46. Lin, T.F.; Wu, J.K. Adsorption of arsenite and arsenate within activated alumina grains: Equilibrium and kinetics. *Water Res.* **2001**, *35*, 2049–2057. [CrossRef]
47. Manna, B.R.; Dey, S.; Debnath, S.; Ghosh, U.C. Removal of arsenic from groundwater using crystalline hydrous ferric oxide (CHFO). *Water Qual. Res. J. Can.* **2003**, *38*, 193–210. [CrossRef]
48. Mohan, D.; Pittman, C.U. Arsenic removal from water/wastewater using adsorbents—A critical review. *J. Hazard. Mater.* **2007**, *142*, 1–53. [CrossRef]
49. Ren, Z.; Zhang, G.; Paul Chen, J. Adsorptive removal of arsenic from water by an iron-zirconium binary oxide adsorbent. *J. Colloid Interface Sci.* **2011**, *358*, 230–237. [CrossRef]
50. Oscarson, D.W.; Huang, P.M.; Defosse, C.; Herbillon, A. Oxidative power of Mn(IV) and Fe(III) oxides with respect to As(III) in terrestrial and aquatic environments. *Nature* **1981**, *291*, 50–51. [CrossRef]
51. Su, C.; Puls, R.W. Arsenate and arsenite sorption on magnetite: Relations to groundwater arsenic treatment using zerovalent iron and natural attenuation. *Water Air Soil Pollut.* **2008**, *193*, 65–78. [CrossRef]
52. Bissen, M.; Vieillard-Baron, M.M.; Schindelin, A.J.; Frimmel, F.H. TiO_2-catalyzed photooxidation of arsenite to arsenate in aqueous samples. *Chemosphere* **2001**, *44*, 751–757. [CrossRef]
53. Zhang, F.S.; Itoh, H. Photocatalytic oxidation and removal of arsenite from water using slag-iron oxide-TiO_2 adsorbent. *Chemosphere* **2006**, *65*, 125–131. [CrossRef]
54. Gupta, K.; Ghosh, U.C. Arsenic removal using hydrous nanostructure iron(III)-titanium(IV) binary mixed oxide from aqueous solution. *J. Hazard. Mater.* **2009**, *161*, 884–892. [CrossRef]
55. Gładysz-Płaska, A.; Skwarek, E.; Budnyak, T.M.; Kołodyńska, D. Metal ions removal using nano oxide Pyrolox™ material. *Nanoscale Res. Lett.* **2017**, *12*, 1–9. [CrossRef]
56. Oscarson, D.W.; Huang, P.M.; Hammer, U.T.; Liaw, W.K. Oxidation and sorption of arsenite by manganese dioxide as influenced by surface coatings of iron and aluminum oxides and calcium carbonate. *Water Air Soil Pollut.* **1982**, *20*, 233–244. [CrossRef]
57. Lei, M.; Qin, P.; Peng, L.; Ren, Y.; Sato, T.; Zeng, Q.; Yang, Z.; Chai, L. Using Fe-Mn binary oxide three-dimensional nanostructure to remove arsenic from aqueous systems. *Water Sci. Technol. Water Supply* **2016**, *16*, 516–524. [CrossRef]
58. Gupta, K.; Biswas, K.; Ghosh, U.C. Nanostructure iron(III)-zirconium(IV) binary mixed oxide: Synthesis, characterization, and physicochemical aspects of arsenic(III) sorption from the aqueous solution. *Ind. Eng. Chem. Res.* **2008**, *47*, 9903–9912. [CrossRef]
59. Gupta, K.; Basu, T.; Ghosh, U.C. Sorption characteristics of arsenic(V) for removal from water using agglomerated nanostructure iron(III)-zirconium(IV) bimetal mixed oxide. *J. Chem. Eng. Data* **2009**, *54*, 2222–2228. [CrossRef]
60. Erdoğan, H.; Yalçinkaya, Ö.; Türker, A.R. Determination of inorganic arsenic species by hydride generation atomic absorption spectrometry in water samples after preconcentration/separation on nano ZrO_2/B_2O_3 by solid phase extraction. *Desalination* **2011**, *280*, 391–396. [CrossRef]
61. Oram, B. The pH of Water. Water Research Center Website. 2019. Available online: https://www.water-research.net/index.php/ph (accessed on 27 January 2019).
62. Kwon, O.H.; Kim, J.O.; Cho, D.W.; Kumar, R.; Baek, S.H.; Kurade, M.B.; Jeon, B.H. Adsorption of As(III), As(V) and Cu(II) on zirconium oxide immobilized alginate beads in aqueous phase. *Chemosphere* **2016**, *160*, 126–133. [CrossRef]
63. Youngran, J.; Maohong, F.; Van Leeuwen, J.; Belczyk, J.F. Effect of competing solutes on arsenic(V) adsorption using iron and aluminum oxides. *J. Environ. Sci.* **2007**, *19*, 910–919. [CrossRef]

64. Mojiri, A.; Hui, W.; Arshad, A.K.; Ruslan, A.; Ridzuan, M.; Hamid, N.H.A.; Farraji, H.; Gholami, A.; Vakili, A.H. Vanadium(V) removal from aqueous solutions using a new composite adsorbent (BAZLSC): Optimization by response surface methodology. *Adv. Environ. Res.* **2017**, *6*, 173–187.
65. Wright, M.T.; Stollenwerk, K.G.; Belitz, K. Assessing the solubility controls on vanadium in groundwater, northeastern San Joaquin Valley, CA. *Appl. Geochem.* **2014**, *48*, 41–52. [CrossRef]
66. Sharififard, H.; Soleimani, M.; Zokaee Ashtiani, F. Application of nanoscale iron oxide-hydroxide-impregnated activated carbon (Fe-AC) as an adsorbent for vanadium recovery from aqueous solutions. *Desalin. Water Treat.* **2016**, *57*, 15714–15723. [CrossRef]
67. Omidinasab, M.; Rahbar, N.; Ahmadi, M.; Kakavandi, B.; Ghanbari, F.; Kyzas, G.Z.; Martinez, S.S.; Jaafarzadeh, N. Removal of vanadium and palladium ions by adsorption onto magnetic chitosan nanoparticles. *Environ. Sci. Pollut. Res.* **2018**, *25*, 34262–34276. [CrossRef]
68. Naeem, A.; Westerhoff, P.; Mustafa, S. Vanadium removal by metal (hydr)oxide adsorbents. *Water Res.* **2007**, *41*, 1596–1602. [CrossRef]
69. Larsson, M.A.; Hadialhejazi, G.; Petter, J. Vanadium sorption by mineral soils: Development of a predictive model. *Chemosphere* **2017**, *168*, 925–932. [CrossRef]
70. Tracey, A.S.; Galeffi, B.; Mahjour, S. Vanadium (V) oxyanions. The dependence of vanadate alkyl ester formation on the p K a of the parent alcohols. *Can. J. Chem.* **1988**, *66*, 2294–2298. [CrossRef]
71. Vega, E.D.; Pedregosa, J.C.; Narda, G.E.; Morando, P.J. Removal of oxovanadium(IV) from aqueous solutions by using commercial crystalline calcium hydroxyapatite. *Water Res.* **2003**, *37*, 1776–1782. [CrossRef]
72. Leiviskä, T.; Khalid, M.K.; Sarpola, A.; Tanskanen, J. Removal of vanadium from industrial wastewater using iron sorbents in batch and continuous flow pilot systems. *J. Environ. Manag.* **2017**, *190*, 231–242. [CrossRef]
73. Ghazvini, M.P.T.; Ghorbanzadeh, S.G. Bioresource technology effect of salinity on vanadate biosorption by Halomonas sp. GT-83: Preliminary investigation on biosorption by micro-PIXE technique. *Bioresour. Technol.* **2009**, *100*, 2361–2368. [CrossRef]
74. Huang, J.; Huang, F.; Evans, L.; Glasauer, S. Vanadium: Global (bio) geochemistry. *Chem. Geol.* **2015**, *417*, 68–89. [CrossRef]
75. Baes, C.F.; Mesmer, R.S. *The Hydrolysis of Cations*; John Wiley & Sons: New York, NY, USA, 1976.
76. Kunz, R.G.; Giannelli, J.F.; Stensel, H.D. Vanadium removal from industrial wastewaters. *J. Water Pollut. Control Fed.* **2016**, *48*, 762–770.
77. Salvestrini, S. Analysis of the Langmuir rate equation in its differential and integrated form for adsorption processes and a comparison with the pseudo first and pseudo second order models. *React. Kinet. Mech. Catal.* **2018**, *123*, 455–472. [CrossRef]
78. Su, T.; Guan, X.; Gu, G.; Wang, J. Adsorption characteristics of As(V), Se(IV), and V(V) onto activated alumina: Effects of pH, surface loading, and ionic strength. *J. Colloid Interface Sci.* **2008**, *326*, 347–353. [CrossRef]
79. Golob, J.; Kosta, L.; Modic, R. Über die adsorptive trennung des vanadiums von arsen, phosphor und fluor mit aktiven aluminiumoxyd. *Vestn. Slov. Kem. Društva* **1971**, *18*, 21–25.
80. Agnoli, S.; Castellarin-cudia, C.; Sambi, M.; Surnev, S.; Ramsey, M.G.; Granozzi, G.; Netzer, F.P. Vanadium on TiO_2 (110): Adsorption site and sub-surface migration. *Surf. Sci.* **2003**, *546*, 117–126. [CrossRef]
81. Kantcheva, M.M.; Hadjiivanov, K.I. Adsorption of vanadium-oxo species on pure and peroxide-treated TiO_2 (anatase). *J. Chem. Soc. Chem. Commun.* **1991**, *02*, 1057–1058. [CrossRef]
82. Davydov, S.Y.; Pavlyk, A.V. Adsorption of vanadium on rutile: A change in the electron work function. *Tech. Phys. Lett.* **2003**, *29*, 500–501. [CrossRef]
83. Zhang, S.; Liu, S.; Hu, W.; Zhu, X.; Qu, R.; Wu, W.; Zheng, C.; Gao, X. New insight into alkali resistance and low temperature activation on vanadia-titania catalysts for selective catalytic reduction of NO. *Appl. Surf. Sci.* **2018**, *466*, 99–109. [CrossRef]
84. Erto, A.; Chianese, S.; Lancia, A.; Musmarra, D. On the mechanism of benzene and toluene adsorption in single-compound and binary systems: Energetic interactions and competitive effects. *Desalin. Water Treat.* **2017**, *86*, 259–265. [CrossRef]
85. Li, X.; Deng, G.; Zhang, Y.; Wang, J. Rapid removal of copper ions from aqueous media by hollow polymer nanoparticles. *Colloids Surf. A* **2019**, *568*, 345–355. [CrossRef]
86. Kołodyńska, D.; Bąk, J.; Kozioł, M.; Pylychuk, L.V. Investigations of heavy metal ion sorption using nanocomposites of iron-modified biochar. *Nanoscale Res. Lett.* **2017**, *12*, 1–13. [CrossRef]

87. Jansson-Charrier, M.; Guibal, E.; Roussy, J.; Delanghe, B.; Le Cloirec, P. Vanadium(IV) sorption by chitosan: Kinetics and equilibrium. *Water Res.* **1996**, *30*, 465–475. [CrossRef]
88. Guzma, J.; Saucedo, I.; Navarro, R.; Revilla, J.; Guibal, E. Vanadium interactions with chitosan: Influence of polymer protonation and metal speciation. *Langmuir* **2018**, *18*, 1567–1573. [CrossRef]
89. Navarro, R.; Guzmán, J.; Saucedo, I.; Revilla, J.; Guibal, E. Recovery of metal ions by chitosan: Sorption mechanisms and influence of metal speciation. *Macromol. Biosci.* **2003**, *3*, 552–561. [CrossRef]
90. Padilla-Rodríguez, A.; Hernández-Viezcas, J.A.; Peralta-Videa, J.R.; Gardea-Torresdey, J.L.; Perales-Pérez, O.; Román-Velázquez, F.R. Synthesis of protonated chitosan flakes for the removal of vanadium(III, IV and V) oxyanions from aqueous solutions. *Microchem. J.* **2015**, *118*, 1–11. [CrossRef]
91. Guibal, E.; Saucedo, I.; Jansson-Charrier, M.; Delanghe, B.; Le Cloirec, P. Uranium and vanadium sorption by chitosan derivatives. *Water Sci. Technol.* **1994**, *30*, 183–190. [CrossRef]
92. Zdarta, J.; Antecka, K.; Jędrzak, A.; Synoradzki, K.; Łuczak, M.; Jesionowski, T. Biopolymers conjugated with magnetite as support materials for trypsin immobilization and protein digestion. *Colloids Surf. B Biointerfaces* **2018**, *169*, 118–125. [CrossRef]
93. Jędrzak, A.; Grześkowiak, B.F.; Coy, E.; Wojnarowicz, J.; Szutkowski, K.; Jurga, S.; Jesionowski, T.; Mrówczyński, R. Dendrimer based theranostic nanostructures for combined chemo- and photothermal therapy of liver cancer cells in vitro. *Colloids Surf. B Biointerfaces* **2019**, *173*, 698–708. [CrossRef]
94. Talebzadeh, F.; Zandipak, R.; Sobhanardakani, S. CeO_2 nanoparticles supported on $CuFe_2O_4$ nanofibers as novel adsorbent for removal of Pb(II), Ni(II), and V(V) ions from petrochemical wastewater. *Desalin. Water Treat.* **2016**, *57*, 28363–28377. [CrossRef]
95. Demetriou, A.; Pashalidis, I. Adsorption of boron on iron-oxide in aqueous solutions. *Desalin. Water Treat.* **2012**, *37*, 37–41. [CrossRef]
96. Yamahira, M.; Kikawada, Y.; Oi, T. Boron isotope fractionation accompanying formation of potassium, sodium and lithium borates from boron-bearing solutions. *Geochem. J.* **2007**, *41*, 149–163. [CrossRef]
97. De la Fuente García-Soto, M.M.; Muñoz Camacho, E. Boron removal by means of adsorption processes with magnesium oxide—Modelization and mechanism. *Desalination* **2009**, *249*, 626–634. [CrossRef]
98. Liu, H.; Qing, B.; Ye, X.; Li, Q. Boron adsorption by composite magnetic particles. *Chem. Eng. J.* **2009**, *151*, 235–240. [CrossRef]
99. World Health Organization. Boron in drinking-water: Background document for development of WHO guidelines for drinking-water quality. *Guidel. Drink. Water Qual.* **1998**, *2*, 1–12.
100. World Health Organization. *Boron in Drinking-Water*; World Health Organization: Geneva, Switzerland, 2009.
101. Hinz, K.; Altmaier, M.; Gaona, X.; Rabung, T.; Schild, D.; Richmann, M.; Reed, D.T.; Alekseev, E.V.; Geckeis, H. Interaction of Nd(III) and Cm(III) with borate in dilute to concentrated alkaline NaCl, $MgCl_2$ and $CaCl_2$ solutions: Solubility and TRLFS studies. *New J. Chem.* **2015**, *39*, 849–859. [CrossRef]
102. Kameda, T.; Oba, J.; Yoshioka, T. Use of Mg-Al oxide for boron removal from an aqueous solution in rotation: Kinetics and equilibrium studies. *J. Environ. Manag.* **2016**, *165*, 280–285. [CrossRef]
103. Wang, H.; Yang, Y.; Guo, L. Renewable-biomolecule-based electrochemical energy-storage materials. *Adv. Energy Mater.* **2017**, *7*, 1700663. [CrossRef]
104. Peak, D.E.P.; Luther, G.E.W.L.; Sparks, D.O. ATR-FTIR spectroscopic studies of boric acid adsorption on hydrous ferric oxide. *Geochim. Cosmochim. Acta* **2003**, *67*, 2551–2560. [CrossRef]
105. Pylypchuk, I.V.; Kołodyńska, D.; Gorbyk, P.P. Gd(III) adsorption on the DTPA-functionalized chitosan/magnetite nanocomposites. *Sep. Sci. Technol.* **2018**, *53*, 1006–1016. [CrossRef]
106. Xu, H.; Yuan, H.; Yu, J.; Lin, S. Study on the competitive adsorption and correlational mechanism for heavy metal ions using the carboxylated magnetic iron oxide nanoparticles (MNPs-COOH) as efficient adsorbents. *Appl. Surf. Sci.* **2019**, *473*, 960–966. [CrossRef]
107. Öztürk, N.; Kavak, D. Boron removal from aqueous solutions by batch adsorption onto cerium oxide using full factorial design. *Desalination* **2008**, *223*, 106–112. [CrossRef]
108. Irawan, C.; Liu, J.C.; Wu, C. Removal of boron using aluminum-based water treatment residuals (Al-WTRs). *Desalination* **2011**, *276*, 322–327. [CrossRef]
109. Afkhami, A.; Aghajani, S.; Mohseni, M.; Madrakian, T. Effectiveness of $Ni_{0.5}Zn_{0.5}Fe_2O_4$ for the removal and preconcentration of Cr(VI), Mo(VI), V(V) and W(VI) oxyanions from water and wastewater samples. *J. Iran. Chem. Soc.* **2015**, *12*, 1–7. [CrossRef]

110. Rakshit, S.; Sallman, B.; Davantes, A.; Lefevre, G. Tungstate(VI) sorption on hematite: An in situ ATR-FTIR probe on the mechanism. *Chemosphere* **2017**, *168*, 685–691. [CrossRef]
111. Ding, S.; Xu, D.; Wang, Y.; Wang, Y.; Li, Y.; Gong, M.; Zhang, C. Simultaneous measurements of eight oxyanions using high-capacity diffusive gradients in thin films (Zr-Oxide DGT) with a high-efficiency elution procedure. *Environ. Sci. Technol.* **2016**, *50*, 7572–7580. [CrossRef]
112. Strigul, N.; Koutsospyros, A.; Arienti, P.; Christodoulatos, C.; Dermatas, D.; Braida, W. Effects of tungsten on environmental systems. *Chemosphere* **2005**, *61*, 248–258. [CrossRef]
113. Kashiwabara, T.; Kubo, S.; Tanaka, M.; Senda, R.; Iizuka, T.; Tanimizu, M.; Takahashi, Y. Stable isotope fractionation of tungsten during adsorption on Fe and Mn (oxyhydr)oxides. *Geochim. Cosmochim. Acta* **2017**, *204*, 52–67. [CrossRef]
114. Hur, H.; Reeder, R.J. Tungstate sorption mechanisms on boehmite: Systematic uptake studies and X-ray absorption spectroscopy analysis. *J. Colloid Interface Sci.* **2016**, *461*, 249–260. [CrossRef]
115. Plattes, M.; Bertrand, A.; Schmitt, B.; Sinner, J.; Verstraeten, F.; Welfring, J. Removal of tungsten oxyanions from industrial wastewater by precipitation, coagulation and flocculation processes. *J. Hazard. Mater.* **2007**, *148*, 613–615. [CrossRef]
116. Mikko, D.U.S. Military "green bullet". *Assoc. Firearm Tool Mark Exam. J.* **1999**, *31*, 1–4.
117. Gustafsson, J.P. Modelling molybdate and tungstate adsorption to ferrihydrite. *Chem. Geol.* **2003**, *200*, 105–115. [CrossRef]
118. Halmi, M.I.E.; Ahmad, S.A. Chemistry, biochemistry, toxicity and pollution of molybdenum: A mini review. *J. Biochem. Microbiol. Biotechnol.* **2014**, *2*, 1–6.
119. Verbinnen, B.; Block, C.; Hannes, D.; Lievens, P.; Vaclavikova, M.; Stefusova, K.; Gallios, G.; Vandecasteele, C. Removal of molybdate anions from water by adsorption on zeolite-supported magnetite. *Water Environ. Res.* **2012**, *84*, 753–760. [CrossRef]
120. Keskİn, C.S. Arsenic and molybdenum ions removal by Fe_3O_4 embedded acrylamide-yirazine hybrid-polymer. *Rev. Roum. Chim.* **2017**, *62*, 139–148.
121. Torres, J.; Gonzatto, L.; Peinado, G.; Kremer, C.; Kremer, E. Interaction of molybdenum(VI) oxyanions with +2 metal cations. *J. Solut. Chem.* **2014**, *43*, 1687–1700. [CrossRef]
122. Oyerinde, O.F.; Weeks, C.L.; Anbar, A.D.; Spiro, T.G. Solution structure of molybdic acid from Raman spectroscopy and DFT analysis. *Inorg. Chim. Acta* **2008**, *361*, 1000–1007. [CrossRef]
123. Xu, N.; Christodoulatos, C.; Braida, W. Adsorption of molybdate and tetrathiomolybdate onto pyrite and goethite: Effect of pH and competitive anions. *Chemosphere* **2006**, *62*, 1726–1735. [CrossRef]
124. Goldberg, S.; Forster, H.S.; Godfrey, C.L. Molybdenum adsorption on oxides, clay minerals, and soils. *Soil Sci. Soc. Am. J.* **1996**, *60*, 425–432. [CrossRef]
125. Aulmann, M.A.; Siri, G.J.; Blanco, M.N.; Caceres, C.V.; Thomas, H.J. Molybdenum adsorption isotherms on γ-alumina. *Appl. Catal.* **1983**, *7*, 139–149. [CrossRef]
126. Luthraand, N.P.; Cheng, W.C. Molybdenum-95 NMR study of the adsorption molybdates on alumina. *J. Catal.* **1987**, *107*, 154–160. [CrossRef]
127. Spanos, N.; Vordonis, L.; Kordulis, C.; Koutsoukos, P.G.; Lycourghiotis, A. Molybdenum-oxo species deposited on alumina by adsorption: II. regulation of the surface Mo(VI) concentration by control of the protonated surface hydroxyls. *J. Catal.* **1990**, *124*, 315–323. [CrossRef]
128. Wu, C.; Lo, S.; Lin, C. Competitive adsorption of molybdate, chromate, sulfate, selenate, and selenite on γ-Al_2O_3. *Colloids Surf. A Physicochem. Eng. Asp.* **2000**, *166*, 251–259. [CrossRef]
129. Sbai, S.; Elyahyaoui, A.; Sbai, Y.; Bentayeb, F.; Bricha, M.R. Study of adsorption of molybdate ion by alumina. *J. Environ. Res. Develop.* **2017**, *11*, 452–460.
130. Verbinnen, B.; Block, C.; Lievens, P.; Van Brecht, A.; Vandecasteele, C. Simultaneous removal of molybdenum, antimony and selenium oxyanions from wastewater by adsorption on supported magnetite. *Waste Biomass Valor* **2013**, *4*, 635–645. [CrossRef]
131. Swedlund, P.J.; Webster, J.G. Adsorption and polymerisation of silicic acid on ferrihydrite, and its effect on arsenic adsorption. *Water Res.* **1999**, *33*, 3413–3422. [CrossRef]

© 2019 by the authors. Licensee MDPI, Basel, Switzerland. This article is an open access article distributed under the terms and conditions of the Creative Commons Attribution (CC BY) license (http://creativecommons.org/licenses/by/4.0/).

Article

Laccase Immobilized onto Zirconia–Silica Hybrid Doped with Cu^{2+} as an Effective Biocatalytic System for Decolorization of Dyes

Katarzyna Jankowska, Filip Ciesielczyk, Karolina Bachosz, Jakub Zdarta, Ewa Kaczorek and Teofil Jesionowski *

Institute of Chemical Technology and Engineering, Faculty of Chemical Technology, Poznan University of Technology, Berdychowo 4, PL-60965 Poznan, Poland; katarzyna.a.antecka@doctorate.put.poznan.pl (K.J.); filip.ciesielczyk@put.poznan.pl (F.C.); karolina.h.bachosz@doctorate.put.poznan.pl (K.B.); jakub.zdarta@put.poznan.pl (J.Z.); ewa.kaczorek@put.poznan.pl (E.K.)
* Correspondence: Teofil.Jesionowski@put.poznan.pl; Tel.: +48-616-65-3720

Received: 21 March 2019; Accepted: 15 April 2019; Published: 16 April 2019

Abstract: Nowadays, novel and advanced methods are being sought to efficiently remove dyes from wastewaters. These compounds, which mainly originate from the textile industry, may adversely affect the aquatic environment as well as living organisms. Thus, in presented study, the synthesized ZrO_2–SiO_2 and Cu^{2+}-doped ZrO_2–SiO_2 oxide materials were used for the first time as supports for laccase immobilization, which was carried out for 1 h, at pH 5 and 25 °C. The materials were thoroughly characterized before and after laccase immobilization with respect to electrokinetic stability, parameters of the porous structure, morphology and type of surface functional groups. Additionally, the immobilization yields were defined, which reached 86% and 94% for ZrO_2–SiO_2–laccase and ZrO_2–SiO_2/Cu^{2+}–laccase, respectively. Furthermore, the obtained biocatalytic systems were used for enzymatic decolorization of the Remazol Brilliant Blue R (RBBR) dye from model aqueous solutions, under various reaction conditions (time, temperature, pH). The best conditions of the decolorization process (24 h, 30 °C and pH = 4) allowed to achieve the highest decolorization efficiencies of 98% and 90% for ZrO_2–SiO_2–laccase and ZrO_2–SiO_2/Cu^{2+}–laccase, respectively. Finally, it was established that the mortality of *Artemia salina* in solutions after enzymatic decolorization was lower by approx. 20% and 30% for ZrO_2–SiO_2–laccase and ZrO_2–SiO_2/Cu^{2+}–laccase, respectively, as compared to the solution before enzymatic treatment, which indicated lower toxicity of the solution. Thus, it should be clearly stated that doping of the oxide support with copper ions positively affects enzyme stability, activity and, in consequence, the removal efficiency of the RBBR dye.

Keywords: inorganic oxide materials; surface functionalization; enzyme immobilization; laccase; dyes decolorization

1. Introduction

Laccases are oxidoreductases which catalyse the oxidation of a wide variety of organic compounds, including mono-, di- and polyphenols as well as aliphatic and aromatic amines [1]. These enzymes are widespread in nature and most commonly extracted from white or red rot fungi, such as *Trametes versicolor* or *Trametes vilosa* [2]. Four adjacent copper atoms are located at the active sites of laccase which correspond to the blue color to the enzyme molecule, hence the protein is often called "blue oxidase". The mechanism of reactions catalysed by laccases involves the oxidation of the substrate molecule to radicals and simultaneous reduction of the oxygen molecule into two water molecules [3]. Due to wide substrate specificity, laccases are used in many industrial processes, such as wood pulp delignification, purification of contaminated water and wastewater treatment as well as in bioremediation and decolorization of textile dye effluents [4].

The high amount of pollutants results in increasing interest in searching for new methods of their removal. Methods such as adsorption, sedimentation, coagulation, membrane techniques or photocatalysis are used for this purpose. However, these techniques are characterized by insufficient efficiency [5,6]. Therefore, the microbiological and enzymatic degradation of harmful compounds is increasingly employed. Unfortunately, due to the properties of native enzymes, such as low stability, narrow pH and temperature range of high catalytic activity, there is a need to use methods for their improvement [7]. It should be emphasized that various strategies are used to improve the thermal and chemical stability of the biocatalysts as well as to prolong their activity and facilitate their reusability, such as protein engineering, chemical modification or enzyme immobilization (which is used most frequently) [8,9]. Immobilization enhances the feasibility of the process, its economy as well as purity of the products [10,11]. Moreover, it is possible to carry out one-step immobilization and simultaneous purification of enzymes [12]. Immobilized metal ion affinity chromatography (IMAC) is also an interesting method of enzyme purification, which is based on covalent binding of the chelating compounds onto chromatographic support. The entrapment of metal ions and their high affinity to enzyme facilitate rapid peptide purification and production of heterogeneous biocatalysts. Due to its simplicity and relatively low costs, this method is used with increasing frequency in various branches of the industry [13,14]. Nevertheless, it should be clearly stated that only a proper realization of immobilization process increases enzyme rigidity or generates a stabilizing and protective microenvironment for the biomolecules. In this regard, multipoint enzyme immobilization which improves the biocatalytic properties of biomolecules by providing stable enzyme–matrix interactions and reducing subunit dissociation is crucial, particularly for multimeric enzymes such as laccases and most of the dehydrogenases [15]. Additionally, other enzyme features such as selectivity, specificity, purity, resistance to inhibitors and stability at harsh environmental conditions may also be improved [16–18]. Furthermore, the immobilization of biocatalysts allows to practically use the enzymes for degradation of pollutants, which has recently gained great scientific interest [19]. For instance, Koloti et al. [20] immobilized laccase onto hyperbranched polyethyleneimine/polyethersulfone (HPEI/PES) electrospun nanofibrous membrane and used this system for removal of bisphenol A. An approach for the removal of the same compound, which included immobilization of laccase onto *Hippospongia communis* spongin scaffolds, was proposed by Zdarta et al. [21]. Nevertheless, it should be noted that not only phenol and its derivatives may be degraded using immobilized laccases. Interesting examples have been presented by Bayramoglu et al. and Gioia et al., which were focused on the immobilization of laccase onto poly(hydroxyethyl methacrylate-co-vinylene carbonate) and thiolsulfinate-agarose, respectively. The obtained systems were used for degradation of Cibacron Blue 3GA, Acid Red 88 and Acid Black 172 [22,23]. It should also be underlined that proper selection of a suitable support has a significant impact on the effectiveness of the immobilization as well as on the final properties of the produced biocatalytic system. Moreover, the industrial processes carried out at various conditions, e.g., at high temperature and pressure or in the presence of salts and organic solvents, require stabilization of multimeric enzymes [24,25]. Therefore, novel, multifunctional supports for enzyme immobilization, which are characterized by numerous different functional groups on their surfaces, are highly required. The presence of different moieties improves enzyme stability and reusability by formation of covalent bonds between the biomolecules and the support. Particularly, hydroxyl, epoxy, carbonyl and divinylsulfone groups are of high importance [26]. However, it should be mentioned that heterofunctional supports also possess disadvantages, as a specific pH value (approx. 10) is required in case of carbonyl groups to form covalent bonds [16]. Thus, hybrid oxide systems should be distinguished among the support materials due to the presence of numerous –OH groups onto their surface. Research carried out by Pezella et al., which included the use of perlite (mixture of SiO_2, Al_2O_3, Na_2O, K_2O, Fe_2O_3, MgO and CaO) for immobilization of laccase and verification of this biocatalytic system in the process of decolorization of Remazol Brilliant Blue

R dye, confirm the previously mentioned statement [27]. Nevertheless, only a few literature reports concerning the use of the aforementioned materials for immobilization of laccase and removal of dyes can be found to date. Hybrid oxide materials, such as SiO_2–Fe_3O_4 [28,29], TiO_2–ZrO_2–SiO_2 [30] or graphene oxide enriched by inorganic additives [31,32], were used for the removal of dyes from wastewater. It should also be mentioned that nanoparticles of ZnO/MnO_2 chelated with Cu^{2+} were used as a support for laccase, and allowed to achieve more than 85% degradation of the Alizarin Red S dye [33]. Despite numerous studies regarding the immobilization of laccase on various supports, new support materials are still being sought for improvement of the effectiveness of biocatalytic processes. This fact is caused by the necessity to deeply understand the kinetics of pollutants degradation, in case of which the substrates consist of various phenolic compounds at different concentrations and various origins [34].

Taking the aforementioned information into account, the decolorization efficiency of Remazol Brilliant Blue R by laccase immobilized onto ZrO_2–SiO_2 and ZrO_2–SiO_2/Cu^{2+} materials was determined in the presented study. The use of zirconia–silica hybrid supports is determined by the need to search for more effective support materials and methods of enzyme immobilization, and hence, more efficient biocatalytic systems. Moreover, laccase immobilization onto the obtained porous oxide materials by adsorption did not alter the three-dimensional structure of the enzyme and may provide a stabilizing microenvironment for the biomolecules, and thus increase their catalytic activity and stability. Nevertheless, it should be mentioned that enzyme stabilization due to adsorption is challenging. This is associated with the blocking of pores of the support as well as the formation of physical interactions which lead to the partial inactivation of the enzyme, as suggested by dos Santos et al. [35]. Furthermore, the presence of –OH moieties and hydrophilic character of the produced supports may additionally positively affect enzyme–support interactions. In addition, the use of a laccase inducer, such as copper ions [36], allowed to compare the efficiency of the processes catalysed by this oxidoreductase with and without the presence of the metal ions. During the investigation, the solution of RBBR dye at a concentration of 50 mg/L was used in order to reflect the actual concentration of this dye in wastewater [37]. The effect of process duration, temperature and pH was investigated in order to determine the most suitable decolorization conditions. The obtained materials were extensively characterized using scanning electron microscopy, Fourier transform infrared spectroscopy, low-temperature N_2 sorption and electrophoretic mobility measurements. Kinetic parameters, such as Michaelis–Menten constant (K_m) and the maximum reaction rate (V_{max}), were calculated. Furthermore, the amount of immobilized enzyme as well as reusability and storage stability of the novel biocatalytic systems were determined. The median lethal concentration (LC_{50}) of *Artemia salina* microorganisms and their mortality were evaluated to compare the toxicity of dye solution before and after the decolorization processes.

2. Materials and Methods

2.1. Chemicals and Materials

Zirconium(IV) isopropoxide, tetraetoxysilane, copper(II) sulfate pentahydrate, laccase from *Trametes versicolor* (EC 1.10.3.2.), Remazol Brilliant Blue R (RBBR), sodium acetate, ammonium and phosphate buffer solutions, potassium chloride, sodium chloride, Coomassie Brilliant Blue G-250 (CBB G-250), and 2,2-azinobis-3-ethylbenzothiazoline-6-sulfonate (ABTS) were obtained from Sigma-Aldrich (St. Louis, MO, USA). Isopropyl and ethyl alcohol, ammonia solution, hydrochloric acid (35–38%) and glacial acetic acid were purchased from the Chempur Company (Piekary Śląskie Poland). The chemical structure of the Remazol Brilliant Blue R dye is presented in Figure 1.

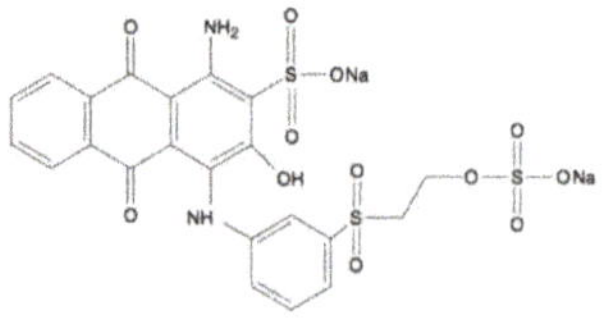

Figure 1. Chemical structure of the Remazol Brilliant Blue R dye (RBBR).

2.2. Synthesis of ZrO_2–SiO_2 and ZrO_2–SiO_2/Cu^{2+} Oxide Systems

The ZrO_2–SiO_2 hybrid, with a molar ratio of precursors equal to 1.5:1, was synthesized via a modified sol–gel route, according to the previously published research [38]. The reactor (1 L) was filled with 500 mL of isopropyl alcohol. Simultaneous dosing of 120 mL of zirconium(IV) isopropoxide and 90 mL of tetraetoxysilane using ISM833A peristaltic pumps (ISMATEC, Wertheim, Germany) was the crucial step of the synthesis. After 1 h of stirring (Eurostar Digital stirrer, Ika Werke GmbH, Staufen im Breisgau, Germany), 60 mL of 25% solution of ammonia (promoter of hydrolysis), was added dropwise. The mixture was additionally stirred within 1 h. After the specified reaction time, the reactor content was placed in a fume hood until a gel was obtained (24 h). Then, the obtained material was slowly dried at 105 °C within 24 h. In order to obtain ZrO_2–SiO_2/Cu^{2+}, the classified material was modified with copper (II) ions. Briefly, 2.5 g of ZrO_2–SiO_2 hybrid was placed into a round bottom flask together with 10 mL of 10% copper (II) sulfate solution. After 1 h of stirring, the mixture was placed in a vacuum evaporator in order to remove the solvent. The obtained materials were comprehensively analyzed and used as supports in the immobilization process of laccase.

2.3. Immobilization of Laccase

In order to immobilize laccase onto the obtained materials, 100 mg of the ZrO_2–SiO_2 or ZrO_2–SiO_2/Cu^{2+} hybrids were placed into vials (20 mL). In the next step, 10 mL of laccase solution at pH 5 (sodium acetate buffer) and concentration of 1 mg/mL was added. The immobilization was carried out using an IKA KS 4000i control incubator (Ika Werke GmbH, Staufen im Breisgau, Germany) at 25 °C for 1 h. After incubation, the obtained biocatalytic systems were centrifuged using a LLG uniCFUGE 5 (LLG Labware, Dublin, Ireland) at 4000 rpm and washed several times with sodium acetate buffer (pH 5).

2.4. Storage Stability and Kinetic Measurements of Free and Immobilized Laccase

In order to define the storage stability of free and immobilized enzyme and to calculate the Michaelis–Menten constant (K_m) and maximum rate of reaction (V_{max}), investigations were conducted according to the previously published study [30]. The measurements were carried out using ABTS (maximum absorbance at $\lambda = 420$ nm) as the model substrate. Briefly, free and immobilized biocatalysts were stored in sodium acetate buffer solution (pH 5) at 4 °C within 20 days in order to examine their storage stability. The V-750 spectrophotometer (Jasco, Tokio, Japan) was used to investigate changes in substrate concentration after the catalytic reaction. Based on the results, storage stability was estimated and relative activity was calculated using the following Equation (1):

$$Relative\ activity\ (\%) = \frac{A_I}{A_0} \cdot 100\% \qquad (1)$$

where A_0 denotes the initial activity of the laccase, and A_I denotes the activity of the immobilized enzyme.

The oxidation reaction of ABTS was used to calculate the Michaelis–Menten constant (K_m) and the maximum rate of reaction (V_{max}), based on Hanes–Wolf plot. Kinetic parameters were evaluated under optimal reaction conditions using substrate solutions at various concentrations.

2.5. Decolorization of Remazol Brilliant Blue R Dye

In order to determine the sorption properties of oxide materials prior to dye decolorization, the experiments were carried out using both biocatalytic systems with a thermally inactivated enzyme. The obtained heterogeneous biocatalysts were placed in an IKA KS 4000i control incubator (Ika Werke GmbH, Staufen im Breisgau, Germany) at 80 °C for 2 h, in order to deactivate the enzyme. Furthermore, 100 mg of each of the materials with the inactivated enzyme were placed in the vials together with 10 mL of Remazol Brilliant Blue R dye at a concentration of 50 mg/L (pH 4, 30 °C, 24 h). After a specified period of time, the absorbance of the resulting solution was measured.

To establish the effect of process duration on decolorization efficiency, the experiments were performed at 30 °C and pH 4 using dye solution at the concentration of 50 mg/L. The samples were collected after 0.5, 1, 3, 5, 8, 12 and 24 h of the decolorization process. The influence of temperature, ranging from 10 to 70 °C, on decolorization efficiency was examined using 10 mL of dye solution at the concentration of 50 mg/L (pH 4, 24 h). The effect of pH on the dye removal was also examined using 10 mL of dye solution at the concentration of 50 mg/L, at 30 °C and at wide pH range from 2 to 10.

Due to the fact that immobilized enzymes should be characterized by good reusability, the biocatalytic systems produced were tested over seven consecutive catalytic cycles. Briefly, after 24 h of decolorization reaction, the obtained heterogeneous biocatalysts were centrifuged using a 5810 R centrifuge (Eppendorf, Hamburg, Germany) at 4000 rpm and washed with sodium acetate buffer at pH 4 to remove dye molecules. The materials prepared in this way were used once again. The reusability study was performed under optimal process conditions which allowed for the highest removal efficiency of the dye (pH 4, 30 °C and 24 h). Each experiment was carried out in triplicate and the results are presented as an average value.

2.6. Analytical Techniques

In order to describe the morphology of the ZrO_2–SiO_2 and ZrO_2–SiO_2/Cu^{2+} oxide materials before and after immobilization of laccase, SEM images were obtained using EVO40 apparatus (Zeiss, Berlin, Germany). FTIR spectra were obtained using a Vertex 70 spectrometer (Bruker, Billerica, MA, USA). Samples were prepared in the form of pellets by mixing 1.5 mg of the analyzed material with 200 mg of anhydrous KBr. ASAP 2020 physisorption analyzer (Micromeritics Instrument Co., Norcross, GA, USA) was used in order to determine the parameters of the porous structure of synthesized materials, such as Brunauer–Emmett–Teller (BET) surface area, meanwhile mean pore size and total pore volume were calculated based on the Barrett–Joyner–Halenda (BJH) method. Before measurement, ZrO_2–SiO_2 and ZrO_2–SiO_2/Cu^{2+} hybrids and biocatalytic systems were degassed at 120 °C within 4 h. Furthermore, they were subjected to analysis using low-temperature (−196 °C) sorption of N_2. The electrokinetic stability of the materials was investigated using an Zetasizer Nano ZS instrument equipped with an MPT-2 autotitrator (Malvern Instruments Ltd., Malvern, United Kingdom). The samples were prepared by dispersing 10 mg of the oxide material in 25 mL of a 0.001 M NaCl solution. The amount of laccase immobilized onto ZrO_2–SiO_2 and ZrO_2–SiO_2/Cu^{2+} oxide materials was calculated based on the Bradford method [39]. The UV-Vis spectroscopy was used to measure the changes in the absorbance of the RBBR during the decolorization process and to calculate the degradation efficiency. The measurements were carried out at wavelength equal to 592 nm (λ_{max} of RBBR dye) using a V-750 spectrophotometer (Jasco, Tokio, Japan). The decolorization efficiencies were calculated using Equation (2):

$$DDE\ (\%) = \frac{C_B - C_A}{C_B} \cdot 100\% \tag{2}$$

where DDE (%) denotes RBBR dye decolorization efficiency, C_B and C_A denote RBBR dye concentration before and after decolorization process, respectively.

The idea of the investigations is presented in Figure 2.

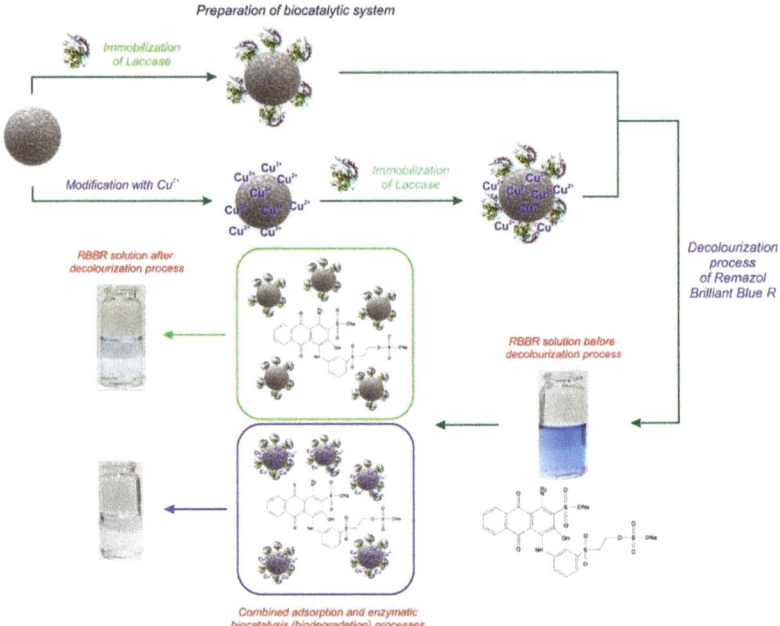

Figure 2. Schematic diagram of preparation of biocatalytic systems and decolorization process of Remazol Brilliant Blue R (RBBR) dye.

Toxicity study of the RBBR dye solution before and after decolorization was carried out using the *Artemia salina* test microorganism, according to the previously published study [40]. Briefly, 0.5 g of *Artemia salina* eggs were incubated in 500 mL of NaCl solution at a concentration of 25 g/L (25 °C with exposure to permanent lighting for 24 h). After that time, 10 larvae were placed into a specified sample and left for 24 h at 25 °C. The percentage of mortality of *Artemia salina* was calculated for the reaction media before and after the enzymatic treatment of dyes. The calculation of number of dead larvae was conducted and median lethal concentration (LC_{50}) was defined. The tests were carried out in triplicate.

3. Results and Discussion

3.1. Characterization of the Oxide Materials before and after Immobilization of Laccase

SEM and FTIR spectral analyses were performed in order to confirm both the effective synthesis of ZrO_2–SiO_2 and ZrO_2–SiO_2/Cu^{2+} oxide systems and immobilization of laccase onto the obtained materials (Figure 3). The morphology of materials was evaluated based on SEM images, presented in Figure 3a,b, respectively. The obtained materials are characterized by irregular particles, 5 μm in diameter, which tend to aggregate. It is worth noticing that there were no significant changes in the structure of oxide material before and after doping with Cu^{2+}.

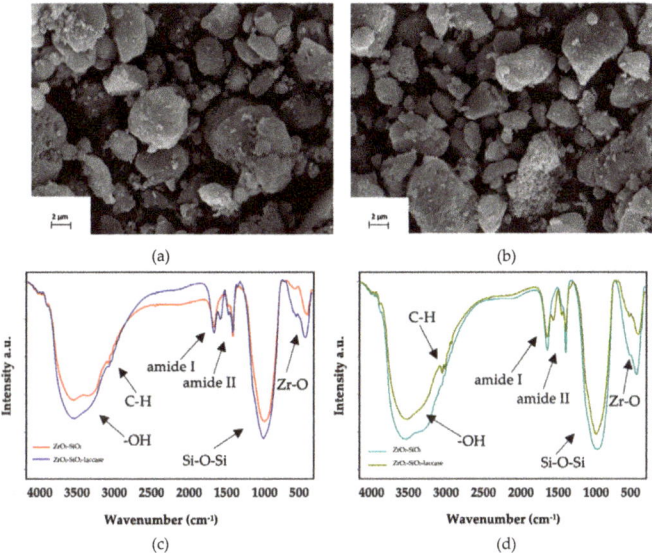

Figure 3. Scanning electron microscope (SEM) images of: (**a**) ZrO_2–SiO_2 and (**b**) ZrO_2–SiO_2/Cu^{2+} and FTIR spectra of: (**c**) ZrO_2–SiO_2, ZrO_2–SiO_2–laccase and (**d**) ZrO_2–SiO_2/Cu^{2+} and ZrO_2–SiO_2/Cu^{2+}–laccase systems.

The FTIR spectra of ZrO_2–SiO_2 material and ZrO_2–SiO_2/Cu^{2+} before and after immobilization of laccase are presented in Figure 3c,d. The spectrum of ZrO_2–SiO_2 possessed a wide band assigned to stretching vibrations of –OH groups between 3650 and 3350 cm^{-1} and a signal at 1630 cm^{-1} characteristic for bending vibrations of physically adsorbed water. The signal at 1480 cm^{-1}, which is characteristic for deformational vibrations of –NH group, presumably resulted from the use of ammonia during synthesis of the oxide material. The most characteristic groups in the structure of the obtained material are represented by the peaks with maxima at 1330 cm^{-1} (Zr-OH bond), 1200–950 cm^{-1} (Si-O-Si and Zr-O-Zr bonds), 675 cm^{-1} (Si-O bond) and 600 cm^{-1} (Zr-O bond) [38,41]. The FTIR spectrum of the ZrO_2–SiO_2–laccase system included signals, which can be seen at 1625, 1555 and 1250 cm^{-1}, assigned to the hybrid oxide system as well as laccase. These signals were attributed to the stretching vibrations of amide I, II and III bonds, respectively, and their presence indicates an effective deposition of enzyme molecules onto oxide material surface [21]. The FTIR spectra of ZrO_2–SiO_2/Cu^{2+} and ZrO_2–SiO_2/Cu^{2+}–laccase materials (Figure 3d), included the same signals as those presented in the spectra of ZrO_2–SiO_2 and ZrO_2–SiO_2–laccase systems. However, particular attention should be paid to the signal at wavenumber of 601 cm^{-1}, which corresponds to vibrations of Cu-O bonds and confirms efficient doping of oxide material with Cu^{2+} ions [42]. Thus, based on the FTIR results, the effective synthesis and functionalization of the oxide materials as well as successful enzyme immobilization have been confirmed.

The results of low-temperature N_2 adsorption/desorption for materials before and after immobilization process are presented in Figure 4. Each of the illustrated isotherms (for ZrO_2–SiO_2, ZrO_2–SiO_2–laccase, ZrO_2–SiO_2/Cu^{2+} and ZrO_2–SiO_2/Cu^{2+}–laccase systems) was classified as type IV with type H4 hysteresis loops. The calculated values of BET surface area, the total pore volume and the mean pore diameter are presented in Table 1. The parameters of the porous structure of the materials before laccase immobilization showed that the analyzed samples possessed higher surface area and total pore volume compared to the ZrO_2–SiO_2–laccase and ZrO_2–SiO_2/Cu^{2+}–laccase systems. These results confirm the effective enzyme immobilization onto the surface and inside the pores of the synthesized oxide materials.

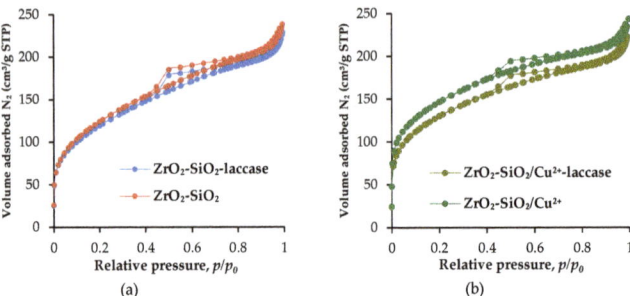

Figure 4. Low temperature N_2 adsorption/desorption isotherms of: (**a**) ZrO_2–SiO_2, ZrO_2–SiO_2–laccase and (**b**) ZrO_2–SiO_2/Cu^{2+} and ZrO_2–SiO_2/Cu^{2+}–laccase systems.

Table 1. Parameters of the porous structure of the obtained materials before and after immobilization of laccase.

Examined System	Parameter		
	A_{BET} (m^2/g)	V_p (cm^3/g)	S_p (nm)
ZrO_2–SiO_2	440.2	0.369	3.4
ZrO_2–SiO_2–laccase	419.9	0.354	3.4
ZrO_2–SiO_2/Cu^{2+}	498.3	0.376	3.0
ZrO_2–SiO_2/Cu^{2+}–laccase	445.7	0.344	3.0

The zeta potential is an important feature for the evaluation of the surface properties of the oxide materials [43]. The obtained results (zeta potential vs. pH) for both pure and Cu^{2+}-doped materials before and after the immobilization process are presented in Figure 5. It can be observed that the zeta potential values of all samples are negative in the analyzed pH range. In addition, it can be concluded that the zeta potential strongly depends on the pH value. Beyond the isoelectric point (IEP), which is almost the same for each of the samples and equal to 3, the zeta potential values become positive. Moreover, it should be underlined that the zeta potential after functionalization and enzyme immobilization is almost unaltered, due to the relatively low amount of Cu^{2+} ions or laccase which were deposited onto the surface of oxide system. Nevertheless, values of IEP close to 3 were mainly caused by the presence of silica in the oxide material, the IEP of which is located between 2 and 3 [44,45]. However, the slight shift of isoelectric point values towards higher pH for samples after laccase immobilization indicated that the attached enzyme also affected the electrokinetic properties of the analyzed samples as laccase possesses two IEP, the first one at approx. pH 3 and the second in a pH range of 4.6–6.8, as previously reported by Jolivalt et al. [46]. Moreover, each of tested systems possessed higher stability in neutral and alkaline environment (pH above 6).

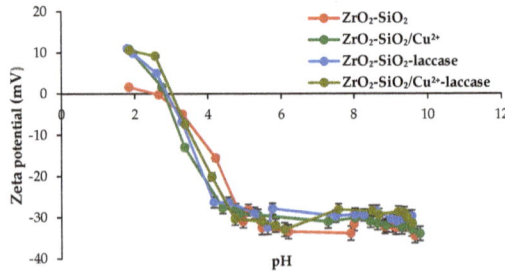

Figure 5. Electrokinetic curves of the synthesized materials and biocatalytic systems.

The kinetic parameters of free and immobilized enzymes as well as efficiency of laccase immobilization onto ZrO_2–SiO_2 and ZrO_2–SiO_2/Cu^{2+} were also calculated. The immobilization yield obtained for hybrid and Cu^{2+}-doped oxide systems reached 86% and 94%, respectively, which corresponds to effective enzyme immobilization onto both supports (Table 2). The high immobilization yields were mainly associated with the high porosity of the supports and the presence of numerous hydroxyl groups on their surface, which facilitate the formation of hydrogen bonds and electrostatic interactions between the biocatalyst and surface of the oxide systems [47].

Kinetic parameters, such as the Michaelis–Menten constant (K_m) and the maximum reaction rate (V_{max}), were also calculated and presented in Table 2. The higher values of K_m and lower values of V_{max} noticed for immobilized laccase (in comparison to free laccase) indicated its lower substrate affinity and lower reaction rate, respectively. Lower substrate affinity of the immobilized enzymes might be related to the conformational changes in the enzyme structure upon immobilization, as lower maximum reaction rate could be explained by the formation of substrate diffusional limitations after enzyme binding. However, blocking of the enzymes active sites by the substrate or products of the reaction cannot be excluded [48]. It is worth noticing that the oxide matrix doped with copper ions showed a slight decrease of K_m value and a slight increase of V_{max} value, in comparison to the system without Cu^{2+}, which implies the positive effect of these ions on the kinetic parameters of immobilized laccase. The obtained data are in agreement with the results of catalytic activity retention, as laccase immobilized onto ZrO_2–SiO_2 and ZrO_2–SiO_2/Cu^{2+} retained 65 and 76% of free enzyme activity, respectively.

Table 2. Kinetic parameters of free and immobilized enzyme, amount of immobilized enzyme and immobilization yield.

Kinetic Parameters and Immobilization Data	Free Laccase	ZrO_2–SiO_2–Laccase	ZrO_2–SiO_2/Cu^{2+}–Laccase
K_m (mM)	0.049 ± 0.002	0.132 ± 0.009	0.098 ± 0.008
V_{max} (U/mg)	0.041 ± 0.006	0.029 ± 0.007	0.037 ± 0.009
Amount of enzyme (mg/g)	-	86 ± 3.8	94 ± 3.2
Immobilization yield (%)	-	86 ± 3.9	94 ± 3.1

Storage stability is one of the crucial parameters which indicate the possibility to use the immobilized enzymes at an industrial scale. The investigation of the storage stability of the free and immobilized laccases was carried out over 20 days (Figure 6). After 20 days of storage, laccase deposited onto ZrO_2–SiO_2 and ZrO_2–SiO_2/Cu^{2+} exhibited definitely higher catalytic activity (over 80%), as compared to the native enzyme (43%). However, a slight difference between the relative activity of laccase immobilized onto initial and Cu^{2+}-doped oxide system can be observed. The enzyme deposited onto material doped with copper ions retained approx. 90% activity, as laccase immobilized onto ZrO_2–SiO_2 possessed only 82% of its initial activity after 20 days of storage. The higher activity of the enzyme immobilized onto both oxide systems might be explained by the stabilization of the biocatalyst structure upon immobilization and protective effect of the support material, which was also shown in previously published studies [30]. Nevertheless, the higher relative activity of the laccase immobilized onto Cu^{2+}-doped oxide material is directly associated with the additional reinforcement of the enzyme structure by the metal ions [48]. The presented results correspond well with the data published by Fu et al., regarding the degradation of bisphenol A using laccase–copper phosphate hybrid nanoflowers. After 60 days of storage, the immobilized biocatalysts retained 90% of its initial activity [49]. In another study, presented by Batule et al., the laccase nanoflowers synthesized with copper phosphate significantly enhanced enzyme stability, which exceeded 95% after 1 month of storage. Moreover, after that time free laccase demonstrated only approx. 20% of its initial activity [50]. These results confirm the stabilizing effect of copper ions towards laccase structure and its activity.

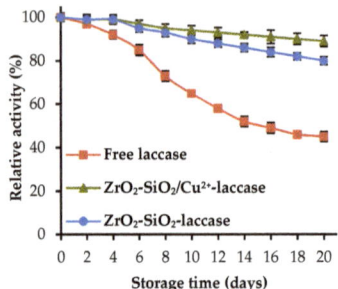

Figure 6. Storage stability of the free and immobilized enzyme over 20 days of storage.

3.2. Decolorization of Dye

3.2.1. Decolorization of Dye Using Oxide Materials with Inactivated Enzyme

The first step of investigations concerning the dye decolorization process involved testing of the sorption capacities of ZrO_2–SiO_2 and ZrO_2–SiO_2/Cu^{2+} oxide systems toward RBBR. For this purpose, supports with thermally inactivated enzyme were placed into solution of the dye at concentration of 50 mg/L. The results showed that sorption efficiencies were equal to 6.38% and 0.52% for the ZrO_2–SiO_2–laccase and ZrO_2–SiO_2/Cu^{2+}–laccase, respectively. Since the error values of the measurements were estimated at 3%, it can be concluded that sorption of the dye onto the obtained materials was negligible. These results can be compared with our previous study [30], in which the sorption efficiencies of the RBBR dye from solution at concentration of 10 mg/L for oxide materials such as TiO_2–ZrO_2 and TiO_2–ZrO_2–SiO_2 were equal to 46% and 28%, respectively.

3.2.2. Effect of Process Duration on Decolorization Efficiency

After determination of the sorption capacities of oxide materials with inactivated enzymes, the effect of the process duration on decolorization efficiency of RBBR dye by the immobilized enzyme was investigated. The obtained results are presented in Figure 7.

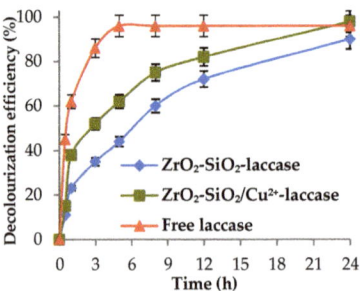

Figure 7. Effect of process duration on decolorization efficiency of the RBBR dye using free laccase and synthesized biocatalytic systems.

It can be observed that application of the biocatalytic system doped with copper ions resulted in higher decolorization efficiency of the RBBR dye over the analyzed process duration, as compared with biocatalytic systems without metal ions. Clearly, after 1 h of process, the decolorization efficiency reached 39% and 22% using the ZrO_2–SiO_2–laccase system with and without Cu^{2+}, respectively. After 24 h, the efficiency of the decolorization process of the RBBR dye by biocatalytic system with copper ions was equal to 98%, as the removal rate of the dye by the ZrO_2–SiO_2–laccase system reached 90%. The solutions of the RBBR dye before and after enzymatic decolorization processes were presented in Figure 8. Based on the results, it was confirmed that 24 h is the most suitable time, because its prolongation does not increase the biodegradation efficiencies, irrespectively of the biocatalytic system used. It should be noted that the decolorization efficiency reached 96% using a native enzyme after 24 h of the process at pH 5 and temperature of 30 °C. The difference between decolorization efficiencies using ZrO_2–SiO_2 and ZrO_2–SiO_2/Cu^{2+} systems resulted from the presence of Cu^{2+} ions, which positively affected the activity and stability of the laccase [51,52]. In another publication, Bayramoglu et al. showed that the presence of transition metal ions, such as Cu^{2+}, on the surface of the support obtained from poly(4-vinylpyridine) molecules on the magnetic beads is responsible for strong binding between pyridine ring/Cu^{2+} and laccase [53]. This may explain the higher efficiency of decolorization by laccase immobilized onto ZrO_2–SiO_2/Cu^{2+} support. Similar observations were made by Olajuyigbe et al. which indicated that laccase immobilized onto Cu–alginate beads shows higher relative activity, as compared to native enzyme and calcium–alginate support [54,55].

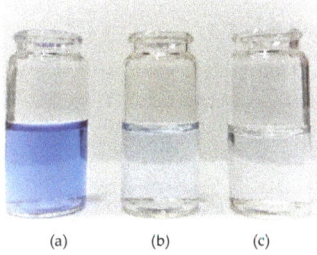

Figure 8. The solutions of the RBBR dye: (**a**) at initial concentration of 50 mg/L, (**b**) after decolorization using ZrO_2–SiO_2–laccase and (**c**) after decolorization using ZrO_2–SiO_2/Cu^{2+}–laccase. Each of experiment was conducted at 30 °C and pH 4 within 24 h.

3.2.3. Effect of pH and Temperature on Decolorization Efficiency

The next step of investigations was focused on the evaluation of the influence of the pH and temperature of dye solution on the decolorization efficiency. These parameters play a crucial role due to the relatively low stability of enzymes under harsh reaction conditions [34]. The effect of temperature was examined in the range from 10 to 70 °C at pH 4, whereas the effect of pH was evaluated in the range from 2 to 10 at 30 °C using the RBBR dye solution at the concentration of 50 mg/L (Figure 9).

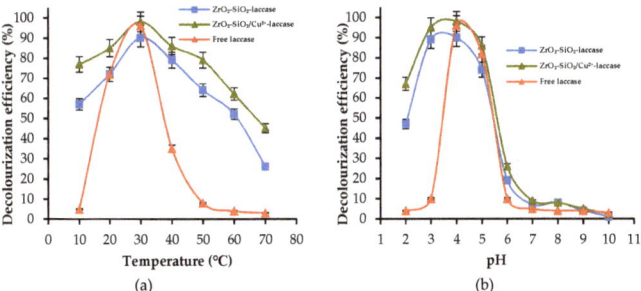

Figure 9. Effect of: (**a**) temperature and (**b**) pH of dye solution on decolorization efficiencies.

In the analyzed temperature range, the decolorization process occurred with higher efficiencies when the dye was degraded in a system containing copper ions. It is worth mentioning that the most significant difference between the removal rates of the dye obtained using both types of materials was equal to approx. 25% at 10 °C. Nevertheless, the highest values of dye removal were obtained at 30 °C, irrespectively of the biocatalytic system used. It should also be added that even at 60 °C both systems after immobilization allowed for the removal of more than 50% of the dye. However, increase of the temperature to 70 °C caused a decrease of the decolorization efficiency, which reached 26% for ZrO_2–SiO_2–laccase and 45% for ZrO_2–SiO_2/Cu^{2+}–laccase, respectively. This fact may be related to the thermal inactivation of the enzyme. The obtained results were compared with the data presented by Tapia-Orozco et al., which were focused on laccase immobilized onto TiO_2 particles. The obtained biocatalytic system was tested at 80 °C and showed a lack of enzymatic activity [56]. Conclusion can be drawn based on the analysis of the results which show effect of pH on the decolorization efficiency (Figure 9b). Over the tested pH range, the removal rates were higher when the ZrO_2–SiO_2/Cu^{2+}–laccase system was used. Moreover, the highest decolorization efficiency was achieved at pH 4 for both systems. Significant decrease of the decolorization efficiency was noticed when the process was realized at the pH range of 6–10. This may result from the reduction of catalytic activity of laccase in the near-neutral and basic environment due to conformational rearrangements in the structure of biomolecule, caused by changes in the nature of aminoacid groups [57].

3.2.4. Reusability of the Biocatalytic Systems

Reusability as well as storage stability are important parameters which need to be considered when immobilized enzymes are used at an industrial scale. The possibility to use the biocatalytic systems multiple times possesses many advantages, the most important among which is the reduction of process costs [58]. In the presented study, the decolorization efficiency of the RBBR dye by ZrO_2–SiO_2–laccase and ZrO_2–SiO_2/Cu^{2+}–laccase systems was tested over seven successive decolorization cycles (Figure 10).

The decolorization efficiency constantly decreased over the catalytic cycles. However, the more prominent decrease of the removal rates can be observed from the 4th catalytic cycle, particularly in case of ZrO_2–SiO_2–laccase system (decrease from 77% in the 3rd cycle to 50% in the 4th cycle). It may be explained by the elution of laccase from the support, which has been also shown in our previous study [30]. After seven consecutive catalytic cycles, the results showed that approx. 40% and 10% of the RBBR dye was removed by the ZrO_2–SiO_2/Cu^{2+}–laccase system and system without copper ions, respectively. The decrease in enzyme activity might be related to the inhibition of the enzyme by the reaction products and partial deactivation of the biocatalyst. It is worth noticing that significant differences between the degradation efficiency of the RBBR dye after treatment with ZrO_2–SiO_2–laccase and ZrO_2–SiO_2/Cu^{2+}–laccase over seven catalytic cycles might be related to the stabilizing effect of copper ions on laccase activity as well as the fact that Cu^{2+} ions act as an inducer for the transfer of electrons which helps overcome enzyme inactivation [36]. In another study, Chen et al. immobilized laccase onto magnetic graphene oxide and used this system for degradation of dyes such as Crystal Violet, Malachite Green and Brilliant Green. After 10 catalytic cycles, the relative activity of the obtained biocatalytic system reached approx. 60% [59]. On the other hand, Le et al. showed that laccase encapsulated in core–shell magnetic copper alginate beads removed only approx. 20% of the RBBR dye after three consecutive decolorization cycles [60]. These results clearly show that type of the support material strongly affects both laccase activity and decolorization efficiency.

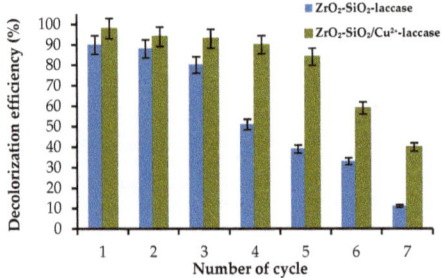

Figure 10. Reusability of the obtained biocatalytic systems.

3.3. Toxicity Study

In order to determine the lethal concentration of the RBBR dye, solutions of the dye at different concentrations, ranging from 0.5 to 50 mg/L, were tested. It was evaluated that the LC_{50} of the RBBR dye before enzymatic treatment towards *Artemia salina* larvae as a model microorganism is equal to 16 mg/L. Moreover, the mortality of the tested microorganisms was lower after the decolorization process (pH 4, 30 °C and 24 h) by 20 and 30% for ZrO_2–SiO_2–laccase and ZrO_2–SiO_2/Cu^{2+}–laccase, respectively, as compared to dye solution before enzymatic treatment (Figure 11).

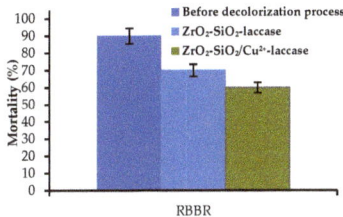

Figure 11. Mortality of *Artemia salina* in the solution before and after enzymatic treatment.

In another study by da Silva et al., horseradish peroxidase was used for removal of RBBR from an aqueous solution. Authors noticed that enzymatic treatment caused a reduction of mortality of *Artemia salina* up to 45% as compared to the solution before the decolorization [61]. The higher mortality data presented in this work might indicate the formation of RBBR degradation products which exhibit higher toxicity as compared to the RBBR degradation products generated using laccase [62]. That fact is closely related to the different mechanism of catalytic behaviour of laccase and horseradish peroxidase.

4. Conclusions

In this study the hybrid ZrO_2–SiO_2 and ZrO_2–SiO_2/Cu^{2+} oxide systems were used as supports for laccase immobilization and further as an effective biocatalytic system for the decolorization processes of the Remazol Brilliant Blue R dye. Results of scanning electron microscopy and Fourier transform infrared spectroscopy confirmed the effective synthesis of the oxide systems and efficient laccase immobilization. Analysis of the effect of various decolorization parameters on its efficiency showed that the highest removal rates of RBBR, equal 98% and 90% for ZrO_2–SiO_2–laccase and ZrO_2–SiO_2/Cu^{2+}–laccase, respectively, were observed after 24 h of the process at 30 °C and pH 4. Thus, the obtained results show that addition of copper ions into the support material has a positive impact on laccase stability and activity. Additionally, the toxicity of the dye solution after enzymatic treatment using ZrO_2–SiO_2–laccase and ZrO_2–SiO_2/Cu^{2+}–laccase was reduced by approx. 20% and 30%, respectively, as compared to initial RBBR dye solution. However, due to slight decrease of toxicity of the solution after laccase treatment, the use of another enzyme, such as horseradish peroxidase, could be considered. Nevertheless, analysis of the results allows to conclude that effective enzymatic systems were obtained, which may potentially be applied for the removal of dyes. Moreover, it should be noted that the obtained oxide materials might be used as supports for the immobilization of a wide range of catalytic proteins. These biocatalytic systems may be also used for the degradation processes of other contaminants, e.g., bisphenols or pharmaceuticals.

Author Contributions: K.J. planned the studies, carried out the experiments, evaluated the enzyme immobilization yield and wrote the manuscript. F.C. obtained support materials, carried out the experiments and participated in discussion. K.B. carried out the immobilization experiments and participated in discussion. J.Z. and E.K. planned the studies and participated in discussion and interpretation of the results. T.J. edited the manuscript, coordinated project tasks and carried out the discussion.

Funding: This work was funded by research grant funds from the National Science Centre Poland (DEC-2018/29/N/ST8/01026).

Conflicts of Interest: The authors declare no conflict of interest.

References

1. Fatarella, E.; Spinelli, D.; Ruzzante, M.; Pogni, R. Nylon 6 film and nanofiber carriers: Preparation and laccase immobilization performance. *J. Mol. Catal. B–Enzym.* **2014**, *102*, 41–47. [CrossRef]
2. Geng, A.; Wu, J.; Xie, R.; Li, X.; Chang, F.; Sun, J. Characterization of a laccase from a wood-feeding termite, *Coptotermes formosanus*. *Insect Sci.* **2018**, *25*, 251–258. [CrossRef]
3. Mayer, A.; Staples, R. Laccase: New functions for an old enzyme. *Phytochemistry* **2002**, *60*, 551–565. [CrossRef]
4. Cristovao, R.O.; Silverio, S.C.; Tavares, A.P.M.; Brigida, A.I.S.; Loureiro, J.M.; Boaventura, R.A.R.; Macedo, E.A.; Coelho, M.A.Z. Green coconut fiber: A novel carrier for immobilization of commercial laccase by covalent attachment for textile dyes decolorization. *World J. Microbiol. Biotechnol.* **2012**, *28*, 2827–2838. [CrossRef]
5. Robinson, T.; McMullan, G.; Marchant, R.; Nigam, P. Remediation of dyes in textile effluent: A critical review on current treatment technologies with a proposed alternative. *Bioresour. Technol.* **2001**, *77*, 247–255. [CrossRef]
6. Goscianska, J.; Ciesielczyk, F. Lanthanum enriched aminosilane-grafted mesoporous carbon material for efficient adsorption of tartrazine azo dye. *Microporous Mesoporous Mater.* **2019**, *280*, 7–19. [CrossRef]
7. Jesionowski, T.; Zdarta, J.; Krajewska, B. Enzyme immobilization by adsorption: A review. *Adsorption* **2014**, *20*, 801–821. [CrossRef]
8. Bolivar, J.M.; Rocha-Martin, J.; Mateo, C.; Guisan, J.M. Stabilization of a highly active but unstable alcohol dehydrogenase from yeast using immobilization and post-immobilization techniques. *Process Biochem.* **2012**, *47*, 679–686. [CrossRef]
9. Bilal, M.; Cui, J.; Iqbal, H.M.N. Tailoring enzyme microenvironment: State-of-art strategy to fulfil the quest for efficient bio-catalysis. *Int. J. Biol. Macromol.* **2019**, *130*, 186–196. [CrossRef] [PubMed]
10. Datta, S.; Christena, L.R.; Rajaram, Y.R.S. Enzyme immobilization: An overview on techniques and support materials. *3 Biotech* **2013**, *3*, 1–9. [CrossRef] [PubMed]
11. D'Souza, S.F. Immobilized enzymes in bioprocess. *Curr. Sci.* **1999**, *77*, 69–79.
12. Barbosa, O.; Ortiz, C.; Berenguer-Murcia, A.; Torres, R.; Rodrigues, R.C.; Fernandez-Lafuente, R. Strategies for the one-step immobilization–purification of enzymes as industrial biocatalysts. *Biotechnol. Adv.* **2015**, *33*, 435–456. [CrossRef] [PubMed]
13. Gaberc-Porekar, V.; Menart, V. Perspectives of immobilized-metal affinity chromatography. *J. Biochem. Biophys. Methods* **2001**, *49*, 335–360. [CrossRef]
14. Porath, J. Immobilized metal ion affinity chromatography. *Protein Expr. Purif.* **1992**, *3*, 263–281. [CrossRef]
15. Bilal, M.; Asgher, M.; Cheng, H.; Yan, Y.; Iqbal, H.M.N. Multi-point enzyme immobilization, surface chemistry, and novel platforms: A paradigm shift in biocatalyst design. *Crit. Rev. Biotechnol.* **2019**, *39*, 202–219. [CrossRef] [PubMed]
16. Bilal, M.; Iqbal, H.M.N. Chemical, physical and biological coordination: An interplay between materials and enzymes as potential platforms for immobilization. *Coord. Chem. Rev.* **2019**, *388*, 1–23. [CrossRef]
17. Bilal, M.; Asgher, M.; Iqbal, H.M.; Hu, H.; Zhang, X. Gelatin-immobilized manganese peroxidase with novel catalytic characteristics and its industrial exploitation for fruit juice clarification purposes. *Catal. Lett.* **2016**, *146*, 2221–2228. [CrossRef]
18. Sheldon, R.A.; Pereira, P.C. Biocatalysis engineering: The big picture. *Chem. Soc. Rev.* **2017**, *46*, 2678–2691. [CrossRef] [PubMed]
19. Bilal, M.; Iqbal, H.M.N. Naturally-derived biopolymer: Potential platforms for enzyme immobilization. *Int. J. Biol. Macromol.* **2019**, *130*, 462–482. [CrossRef]
20. Koloti, L.E.; Gule, N.P.; Arotiba, O.A.; Malinga, S.P. Laccase-immobilized dendritic nanofibrous membranes as a novel approach towards the removal of bisphenol A. *Environ. Technol.* **2018**, *39*, 392–404. [CrossRef] [PubMed]
21. Zdarta, J.; Antecka, K.; Frankowski, R.; Zgoła-Grześkowiak, A.; Ehrlich, H.; Jesionowski, T. The effect of operational parameters on the biodegradation of bisphenols by *Trametes versicolor* laccase immobilized on *Hippospongia communis* spongin scaffolds. *Sci. Total Environ.* **2018**, *615*, 784–795. [CrossRef] [PubMed]
22. Bayramoglu, G.; Salih, B.; Akbulut, A.; Arica, M.Y. Biodegradation of Cibacron Blue 3GA by insolubilized laccase and identification of enzymatic byproducts using MALDI-ToF-MS: Toxity assessment studies by *Daphnia magna* and *Chlorella vulgaris*. *Ecotoxicol. Environ. Saf.* **2019**, *170*, 453–460. [CrossRef]

23. Gioia, L.; Ovsejevi, K.; Manta, C.; Miguez, D.; Menendez, P. Biodegradation of acid dyes by an immobilized laccase: An ecotoxicological approach. *Environ. Sci. Water Res. Technol.* **2018**, *4*, 2125–2135. [CrossRef]
24. Iyer, P.V.; Ananthanarayan, L. Enzyme stability and stabilization–Aqueous and non-aqueous environment. *Process Biochem.* **2008**, *43*, 1019–1032. [CrossRef]
25. Fernandez-Lafuente, R. Stabilization of multimeric enzymes: Strategies to prevent subunit dissociation. *Enzym. Microb. Technol.* **2009**, *45*, 405–418. [CrossRef]
26. Barbosa, O.; Torres, R.; Ortiz, C.; Berenguer-Murcia, A.; Rodrigues, R.C.; Fernadez-Lafuente, R. Heterofunctional supports in enzyme immobilization: From traditional immobilization protocols to opportunities in tuning enzyme properties. *Biomacromolecules* **2013**, *14*, 2433–2462. [CrossRef]
27. Pezzella, C.; Russo, M.E.; Marzocchella, A.; Salatino, P.; Sannia, G. Immobilization of a *Pleurotus ostreatus* laccase mixture on perlite and its application to dye decolorisation. *BioMed Res. Int.* **2014**, *2014*, 308613. [CrossRef]
28. Amin, R.; Khorshidi, A.; Shojaei, A.F.; Rezaai, S.; Faramarzi, M.A. Immobilization of laccase on modified Fe_3O_4@SiO_2@Kit-6 magnetite nanoparticles for enhanced delignification of olive pomace bio-waste. *Int. J. Biol. Macromol.* **2018**, *114*, 106–113. [CrossRef]
29. Li, Q.Y.; Wang, P.Y.; Zhou, Y.L.; Nie, Z.R.; Wei, Q. A magnetic mesoporous SiO_2/Fe_3O_4 hollow microsphere with a novel network-like composite shell: Synthesis and application on laccase immobilization. *J. Sol-Gel Sci. Technol.* **2016**, *78*, 523–530. [CrossRef]
30. Antecka, K.; Zdarta, J.; Siwińska-Stefańska, K.; Sztuk, G.; Jankowska, E.; Oleskowicz-Popiel, P.; Jesionowski, T. Synergistic degradation of dye wastewaters using binary or ternary oxide systems with immobilized laccase. *Catalysts* **2018**, *8*, 402. [CrossRef]
31. Kashefi, S.; Borghei, S.M.; Mahmoodi, N.M. Covalently immobilized laccase onto graphene oxide nanosheets: Preparation, characterization, and biodegradation of azo dyes in colored wastewater. *J. Mol. Liq.* **2019**, *276*, 153–162. [CrossRef]
32. Kashefi, S.; Borghei, S.M.; Mahmoodi, N.M. Superparamagnetic enzyme-graphene oxide magnetic nanocomposite as an environmentally friendly biocatalyst: Synthesis and biodegradation of dye using response surface methodology. *Microchem. J.* **2019**, *145*, 547–558. [CrossRef]
33. Rani, M.; Shanker, U.; Chaurasia, A.K. Catalytic potential of laccase immobilized on transition metal oxides nanomaterials: Degradation of Alizarin Red S dye. *J. Environ. Chem. Eng.* **2017**, *5*, 2730–2739. [CrossRef]
34. Zdarta, J.; Meyer, A.S.; Jesionowski, T.; Pinelo, M. Developments in support materials for immobilization of oxidoreductases: A comprehensive review. *Adv. Colloids Interf.* **2018**, *258*, 1–20. [CrossRef] [PubMed]
35. Dos Santos, J.C.; Barbosa, O.B.; Ortiz, C.; Berenguer-Murcia, A.; Rodrigues, R.C.; Fernadez-Lafuente, R. Importance of the suport properties for immobilization or purification of enzymes. *ChemCatChem* **2015**, *7*, 2413–2432. [CrossRef]
36. Gomma, O.M.; Momtaz, O.A. Copper induction and differential expression of laccase in *Aspergillus flavus*. *Braz. J. Microbiol.* **2015**, *46*, 285–292. [CrossRef] [PubMed]
37. Yaseen, D.A.; Scholz, M. Treatment of synthetic textile wastewater containing dye mixtures with microcosms. *Environ. Sci. Pollut. Res.* **2018**, *25*, 1980–1997. [CrossRef]
38. Ciesielczyk, F.; Goscianska, J.; Zdarta, J.; Jesionowski, T. The development of zirconia/silica hybrids for the adsorption and controlled release of active pharmaceutical ingredients. *Colloids Surf. A* **2018**, *545*, 39–50. [CrossRef]
39. Bradford, M.M. A rapid and sensitive method for the quantitation of microgram quantities of protein utilizing the principle of protein-dye binding. *Anal. Biochem.* **1976**, *72*, 248–254. [CrossRef]
40. Lu, J.; Zhu, X.; Tian, S.; Lv, X.; Chen, Z.; Jiang, Y.; Liao, X.; Cai, Z.; Chen, B. Graphene oxide in the marine environment: Toxicity to *Artemia salina* with and without the presence of Phe and Cd^{2+}. *Chemosphere* **2018**, *211*, 390–396. [CrossRef]
41. Sidhu, G.K.; Kaushik, A.K.; Rana, S.; Bhansali, S.; Kumar, R. Photoluminescence quenching of zirconia nanoparticle by surface modification. *Appl. Surf. Sci.* **2015**, *334*, 216–221. [CrossRef]
42. Radhakrishnan, A.; Beena, B. Structural and optical absorption analysis of CuO nanoparticles. *Indian J. Adv. Chem. Sci.* **2014**, *2*, 158–161.
43. Kadhom, M.; Deng, B. Thin film nanocomposite membranes filled with bentonite nanoparticles for brackish water desalination: A novel water uptake concept. *Microporous Mesoporous Mater.* **2019**, *279*, 82–91. [CrossRef]

44. Alves, J.A., Jr.; Baldo, J.B. The behaviour of zeta potential of silica suspensions. *New J. Glass Ceram.* **2014**, *4*, 29–37. [CrossRef]
45. Bousse, L.; Mostarshed, S.; van der Shoot, B.; de Rooij, N.F.; Gimmel, P.; Gopel, W. Zeta potential measurements of Ta_2O_5 and SiO_2 thin films. *J. Colloid Interf. Sci.* **1991**, *147*, 22–32. [CrossRef]
46. Jolivalt, C.; Brenon, S.; Caminade, E.; Mougin, C.; Pontie, M. Immobilization of laccase from *Trametes versicolor* on a modified PVDF microfiltration membrane: Characterization of the grafted support and application in removing a phenylurea pesticide in wastewater. *J. Membr. Sci.* **2000**, *180*, 103–113. [CrossRef]
47. Arica, M.Y.; Senel, S.; Alaeddinoglu, N.G.; Patir, S.; Denizli, A. Invertase immobilized on spacer-arm attached poly(hydroxyethyl metacrylate) membrane: Preparation and properties. *J. Appl. Polym. Sci.* **2000**, *75*, 1685–1692. [CrossRef]
48. Xu, R.; Cui, J.; Tang, R.; Li, F.; Zhang, B. Removal of 2,4,6-trichlorophenol by laccase immobilized on nano-copper incorporated electrospun fibrous membrane-high efficiency, stability and reusability. *Chem. Eng. J.* **2017**, *326*, 647–655. [CrossRef]
49. Fu, M.; Xing, J.; Ge, Z. Preparation of laccase-loaded magnetic nanoflowers and their recycling for efficient degradation of bisphenol A. *Sci. Total Environ.* **2019**, *651*, 2857–2865. [CrossRef]
50. Batule, B.S.; Park, K.S.; Kim, M.I.; Park, H.G. Ultrafast sonochemical synthesis of protein-inorganic nanoflowers. *Int. J. Nanomed.* **2015**, *10*, 137–142. [CrossRef]
51. Stajic, M.; Persky, L.; Hadar, Y.; Friesem, D.; Duletic-Lausevic, S.; Wasser, S.P.; Nevo, E. Effect of copper and manganese ions on activities of laccase and peroxidases in three *Pleurotus* species grown on agricultural wastes. *Appl. Biochem. biotechnol.* **2006**, *128*, 87–96. [CrossRef]
52. Baldrian, P.; Gabriel, J. Copper and cadmium increase laccase activity in *Pleurotus ostreatus*. *FEMS Microbiol. Lett.* **2002**, *206*, 69–74. [CrossRef]
53. Bayramoglu, G.; Yilmaz, M.; Arica, M.Y. Reversible immobilization of laccase to poly(4-vinylpyridine) grafted and Cu(II) chelated magnetic beads: Biodegradation of reactive dyes. *Bioresour. Technol.* **2010**, *101*, 6615–6621. [CrossRef]
54. Wang, X.; Hu, J.; Liang, Y.; Zhan, H. Effects of metal ions on laccase activity. *Asian J. Chem.* **2011**, *23*, 5422–5424.
55. Olajuyigbe, F.M.; Adetuyi, O.Y.; Fatokun, C.O. Characterization of free and immobilized laccase from *Cyberlindnera fabianii* and application in degradation of bisphenol A. *Int. J. Biol. Macromol.* **2019**, *125*, 856–864. [CrossRef]
56. Tapia-Orozco, N.; Melendez-Saavedra, F.; Figueroa, M.; Gimeno, M.; Garcia-Arrazola, R. Removal of bisphenol A in canned liquid food by enzyme-based nanocomposites. *Appl. Nanosci.* **2018**, *8*, 427–434. [CrossRef]
57. Zhang, X.; Wang, M.; Lin, L.; Xiao, G.; Tang, Z.; Zhu, X. Synthesis of novel laccase-biotitania biocatalysts for malachite green decolorization. *J. Biosci. Bioeng.* **2018**, *126*, 69–77. [CrossRef]
58. Patel, S.K.S.; Kalia, V.C.; Choi, J.H.; Haw, J.R.; Kim, I.W.; Lee, J.K. Immobilization of laccase on SiO_2 nanocarriers improves its stability and reusability. *J. Microб. Biotechnol.* **2014**, *24*, 639–647. [CrossRef]
59. Chen, J.; Leng, J.; Yang, X.; Liao, L.; Liu, L.; Xiao, A. Enhanced performance of magnetic graphene oxide-immobilized laccase and its application for the decolorization of dyes. *Molecules* **2017**, *22*, 221. [CrossRef]
60. Le, T.T.; Murugesan, K.; Lee, C.-S.; Vu, C.H.; Chang, Y.-S.; Jeon, J.-R. Degradation of synthetic pollutants in real wastewaters using laccase encapsulated in core-shell magnetic copper alginate beads. *Bioresour. Technol.* **2016**, *216*, 203–210. [CrossRef]
61. Da Silva, M.R.; da Sa, L.R.V.; Russo, C.; Scio, E.; Ferreira-Leitao, V.S. The use of HRP in decolorization of reactive dyes and toxicological evaluation of their products. *Enzym. Res.* **2010**, *2010*, 703824. [CrossRef] [PubMed]
62. Osma, J.F.; Toca-Herra, J.L.; Rodriguez-Couto, S. Transformation pathway of Remazol Brilliant Blue R by immobilised laccase. *Bioresour. Technol.* **2010**, *101*, 8509–8514. [CrossRef] [PubMed]

© 2019 by the authors. Licensee MDPI, Basel, Switzerland. This article is an open access article distributed under the terms and conditions of the Creative Commons Attribution (CC BY) license (http://creativecommons.org/licenses/by/4.0/).

Article

Modification of Ti6Al4V Titanium Alloy Surface Layer in the Ozone Atmosphere

Mariusz Kłonica * and Józef Kuczmaszewski

Department of Production Engineering, Faculty of Mechanical Engineering, Lublin University of Technology, 20-618 Lublin, Poland
* Correspondence: m.klonica@pollub.pl; Tel.: +48-815-384-231

Received: 7 May 2019; Accepted: 26 June 2019; Published: 30 June 2019

Abstract: The paper reports the results of a study on the Ti6Al4V titanium alloy involving the XPS (X-ray photoelectron spectroscopy) photoelectron spectroscopy method. The position of bands in the viewing spectrum serves as a basis for the qualitative identification of atoms forming the surface layer, while their intensity is used to calculate the aggregate concentration of these atoms in the analyzed layer. High-resolution spectra are used to determine the type of chemical bonds based on characteristic numerical values of the chemical shift. The paper also presents the 3D results of surface roughness measurements obtained from optical profiling, as well as the results of energy state measurements of the Ti6Al4V titanium alloy surface layer after ozone treatment. It was shown that the ozone treatment of the Ti6Al4V titanium alloy removes carbon and increases concentrations of Ti and V ions at higher oxidation states at the expense of metal atoms and lower valence ions. The modification of the surface layer in ozone atmosphere caused a 30% increase in the Ti element concentration in the surface layer compared to the samples prior to ozone treatment. The carbon removal rate from the Ti6Al4V titanium alloy samples amounted to 35%, and a 13% increase was noted in oxides. The tests proved that the value of the surface free energy of the Ti6Al4V titanium alloy increased as a result of ozone treatment. The highest increase in the surface free energy was observed for Variant 4 samples, and amounted to 17% compared to the untreated samples, while the lowest increase was equal to 14%. For the analyzed data, the maximum value of standard deviation was 0.99 [mJ/m^2].

Keywords: Ti6Al4V titanium alloy; ozone treatment; surface layer; surface free energy; adhesive joint

---

1. Introduction

Titanium alloys are difficult to use in adhesive joining or other technologies where adhesive phenomena play a key role. This is due to a very wide variation in the properties of titanium oxides, depending on the oxidation conditions [1]. In industrial applications, particularly in the aerospace industry, the well-established electrochemical treatment methods for titanium alloys have been employed for years. However, these methods are cost-ineffective, also considering bath deposition. Chemical or electrochemical treatment technologies develop a properly constituted oxide layer. Although the obtained oxides typically exhibit high energy characteristics, they tend to be loosely attached to the substrate surface, which proves problematic considering adhesion processes. The growing interest in ozone treatment as an alternative method for imparting favorable energy conditions to the surface layer is an effect of high reactivity of ozone. The promising potential of this method may eventually lead to abandoning the well-established surface treatments, which, however, bear heavy environmental cost.

Searching for effective alternative methods for modifying adhesive properties [2] of titanium alloys is crucial from both the scientific and the practical perspective.

The energy state [3–6] of structural materials, especially titanium alloys in industrial applications, is particularly important in technologies where adhesion is crucial to process efficiency. Such technologies include adhesive joints, construction encapsulation, coating, printing, and sintering [1,7]. Proper bonding of structural materials requires not only suitable preparation of the substrate surface, but also developing adequate adhesion properties of adhesives or sealants [8–10].

Studies indicate that ozone treatment exhibits a promising potential [1,9] due to the fact that it increases the surface free energy. It is difficult to account for the phenomena that cause these changes as they take place at the nanoscale. Such changes may be observed at the macroscale only to a small extent by means of surface microscopy [11].

Chemical methods [12,13] and electrochemical pretreatment of the surface layer of structural materials are effective when removal of strongly bound material is required. They generate a suitable surface texture and provide high physicochemical activity of the surface with respect to the adhesive used.

The authors of the paper [14] used ozone to obtain titanium oxide. The results showed that this method produces a surface layer of the required properties.

This study set out to provide a more thorough chemical and topographic analysis of the titanium alloy surface following ozone treatment. The results offer an interesting insight into the changes occurring in the fusion layer and partly also in the surface layer of the cohesion zone of the Ti6Al4V alloy following the surface treatment in question.

2. Materials and Methods

The study was performed on the Ti6Al4V (Grade 5) titanium alloy according to the standards (AMS4911, ASTM B265). This alloy is widely used in the aviation, space and marine industries. The 25 mm × 100 mm × 1.6 mm Ti6Al4V titanium alloy samples were adhesively joined with the two-component epoxy adhesive, Loctite Hysol 9466. The adhesive was prepared by means of a static flow stirrer and applied to the adherend surfaces according to the manufacturer's recommendations. Uniform thickness of the adhesive was ensured by the application of appropriate pressure during cure.

Figure 1 shows a schematic diagram of a single-lap adhesive joint.

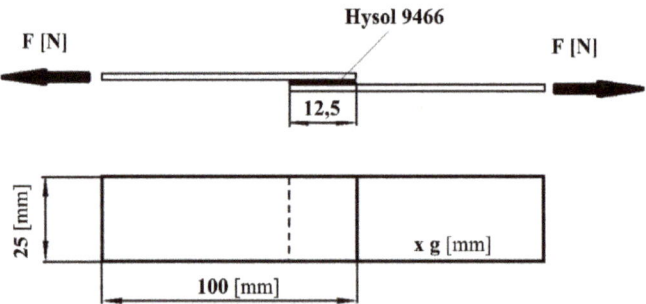

Figure 1. Single-lap adhesive joint: g-adherend thickness, F-force of the joint.

The dimensional accuracy of the adherends was 0.1 mm, while that the length of the adhesive joint, 0.5 mm, with adhesive thickness gk = 0.1 ± 0.01 mm.

The adhesive was cured at an ambient temperature of 19 °C–22 °C, and relative humidity of 38–45%. The cure time was set to 120 h, and the unit pressure applied to the surface of the samples during bonding was equal to 0.2 MPa.

The layout diagram illustrating the test stand for modification of the surface layer of construction materials in ozone atmosphere is shown in Figure 2.

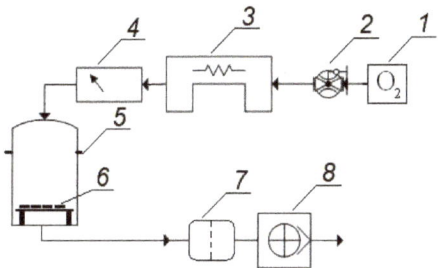

Figure 2. Layout diagram of test stand for ozone treatment of construction material samples: 1—oxygen concentrator, 2—adjustable flow meter, 3—ozone generator, 4—ozone concentration meter, 5—reaction chamber, 6—samples subjected to surface layer modification, 7—ozone destroyer, 8—suction pump.

The ozone flow in the ozone treatment process of samples was 0.9 dm³/min. The Ozone ANALYZER BMT 964 (BMT MESSTECHNIK GMBH, Stahnsdorf, Germany) was used for measuring ozone concentration.

The specimens were subjected to one of the four variants of ozone treatment, i.e.,

No. 1—samples after machining prior to ozone treatment,
No. 2—samples after machining and ozone treatment: 50g O_3/m^3 for 10 min,
No. 3—samples after machining and ozone treatment: 50g O_3/m^3 for 30 min,
No. 4—samples after machining and ozone treatment: 50g O_3/m^3 for 45 min.

The surface free energy (SFE) of the specimens was determined by means of the indirect method, based on the contact angle measurement. One of the most commonly used methods for determining the SFE of structural materials [3–6] is the Owens–Wendt method. This method assumes that the value of the SFE is the sum of its two constituents: The dispersive γ_s^d and the polar γ_s^p components of the SFE, and, in addition, that there is an additive relationship between these Equation (1):

$$\gamma_s = \gamma_s^d + \gamma_s^p, \tag{1}$$

where, γ_s—the SFE of solids,
γ_s^d—the dispersive component of the SFE of tested substrates,
γ_s^p—the polar component of the SFE of tested substrates.

To calculate the dispersive and polar components of the SFE, of the tested substrate materials it is necessary to know the value of their contact angles. This is determined using the measuring liquids whose surface free energy, as well as polar and dispersive components, are known. In the presented tests, the measuring liquids were distilled water, as a polar liquid, and diiodomethane, as an apolar liquid. The dispersion component of the surface free energy is obtained from the given Equation (2):

$$\left(\gamma_s^d\right)^{0.5} = \frac{\gamma_d(\cos\Theta_d + 1) - \sqrt{\frac{\gamma_d^p}{\gamma_w^p}}\gamma_w(\cos\Theta_w + 1)}{2\left(\sqrt{\gamma_d^d} - \sqrt{\gamma_d^p \frac{\gamma_w^d}{\gamma_w^p}}\right)} \tag{2}$$

and the polar component of the surface free energy according is given by Equation (3):

$$\left(\gamma_s^p\right)^{0.5} = \frac{\gamma_w(\cos\Theta_w + 1) - 2\sqrt{\gamma_s^d \gamma_w^d}}{2\sqrt{\gamma_w^p}}, \tag{3}$$

where, γ_d—the SFE of diiodomethane,
  $\gamma_d{}^d$—the dispersive component of the SFE of diiodomethane,
  $\gamma_d{}^p$—the polar component of the SFE of diiodomethane,
  γ_w—the SFE of water,
  $\gamma_w{}^d$—the dispersive component of the SFE of water,
  $\gamma_w{}^p$—the polar component of the SFE of water,
  θ_w—the contact angle of water.

3. Results

Photographs of the Ti6Al4V titanium alloy surface are shown in Figure 3a,b.

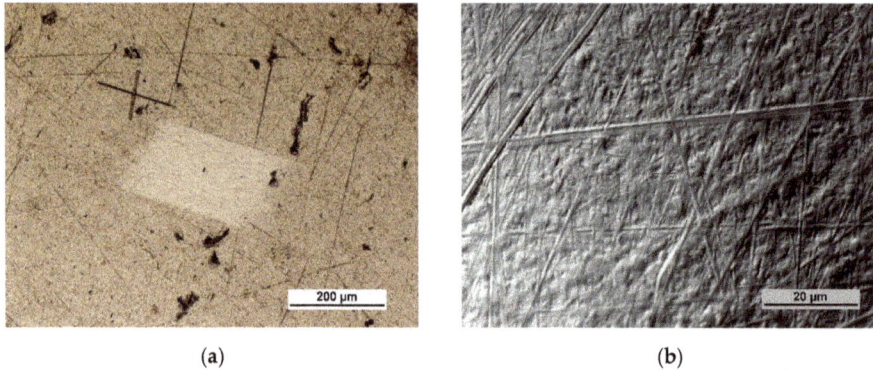

Figure 3. Ti6Al4V titanium alloy surface: (**a**) In polarized light at 50× magnification; (**b**) DIC image at 1000× magnification.

Figure 3a shows a fragment of the titanium alloy surface with a visible metallographic section produced by the ion abrasion technique obtained by means of a Quanta 3D FEG microscope (FEI, Hillsboro, OR, USA). A cross-shaped marker is also visible. The marker was used to identify the place of measurements before and after ozone treatment. Precise measurements enabled recording the changes occurring at the same place of the surface layer following ozone treatment.

3.1. Surface Roughness

The surface roughness was measured by means of optical profilometry using a Contour GT-K1 optical profilometer (Veeco, Bruker, the Netherlands). The purpose of the study was to determine the effect of surface layer modification in the ozone atmosphere on the surface layer of structural materials. The following parameters were measured in the experiments: Sa—arithmetic mean of the 3D profile ordinates, Sq—root mean square of 3D profile ordinates, Sp—value of the highest elevation of 3D profile, Sv—value of the lowest 3D profile depth and Sz—maximum height of the 3D profile.

The paper [10] pointed to the significance of surface topography on the strength of lap adhesive joints and investigated the surface roughness parameters that affect the strength of adhesive joints.

Figure 4 shows the topography of Ti6Al4V titanium alloy samples before and after ozone treatment. The size of the scanned surface is 156 μm × 117 μm.

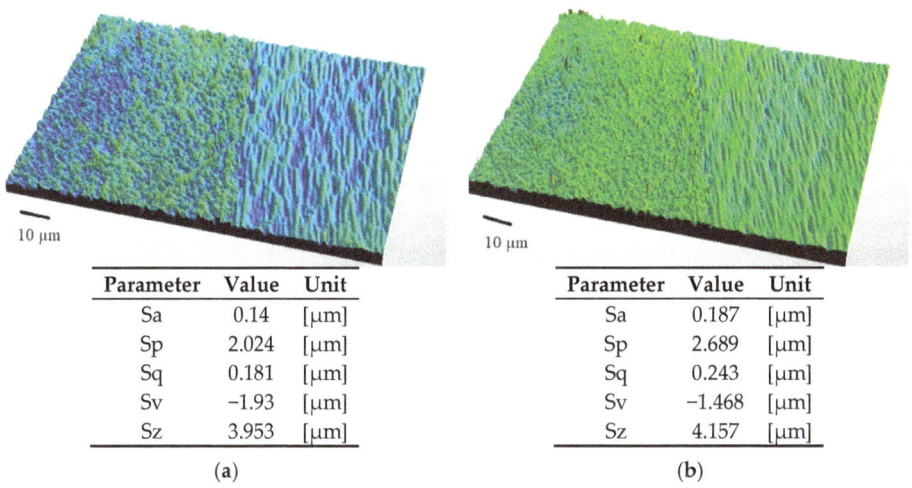

Figure 4. Images of Ti–6Al–4V titanium alloy samples surface with clear limit after ionic polishing: (**a**) Before ozone treatment; (**b**) after ozone treatment (samples after ozone treatment: 50 g O_3/m^3 for 45 min).

The results indicate an over 25% increase in the surface roughness of the entire joint area after ozone treatment compared to the untreated surface. The value of the roughness parameter Sq before pretreatment, 0.181 µm, increases to 0.243 µm after the ozone treatment. Similar changes are observed in other parameters after treatment.

The experimental results demonstrate that ozone "interacts" with the smooth surface of this alloy.

3.2. Examination of the Surface Layer with Electron Scanning Microscopy

Figure 5 shows photographs of untreated Ti6Al4V titanium alloy surface and after modification of the surface layer in ozone atmosphere. The figures show examples of areas after ionic polishing, prior to and following ozone treatment.

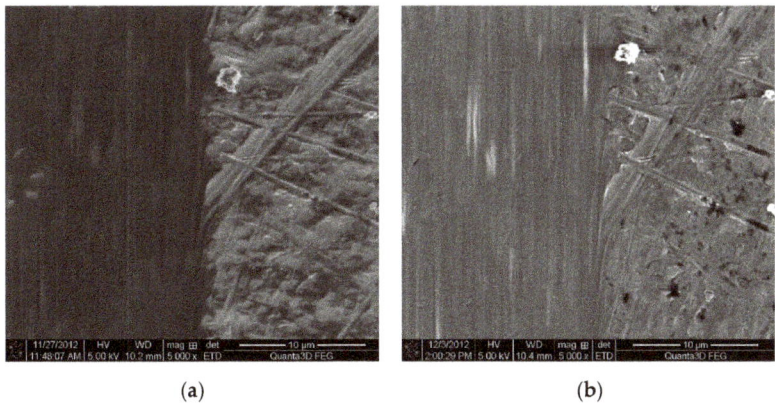

Figure 5. Topography of Ti6Al4V titanium alloy surface after ionic polishing with FIB (focus ion beam) technique: (**a**) Before ozone treatment; (**b**) after ozone treatment (samples after ozone treatment: 50 g O_3/m^3 for 45 min).

The surface layer of the samples was examined using a high-contrast scanning electron microscope (Quanta 3D FEG; FEI, Hillsboro, OR, USA), which enables registration of secondary electrons (SE). The maximum magnification was 5000×.

The obtained increase in the values of surface roughness parameters following ozone treatment is notable on nano-scale. The effect is complex, certain micro-irregularities of the surface may become leveled out, whereas new micro-cavities may be formed on the treated surface, which effects in the rise in surface roughness parameters.

The results of the surface examination are shown in Table 1. Table 1 shows the results for Ti6Al4V titanium alloy before and after ozone treatment. The scanning area was set to 1 mm^2.

Table 1. Results of EDX (energy dispersive X-ray spectroscopy) analysis of Ti6Al4V titanium alloy sample before and after ozone treatment.

Element	Sample before Ozone Treatment		Sample after Ozone Treatment: 50 g O_3/m^3 for 45 min	
	wt % (weight)	at % (atom)	wt % (weight)	at % (atom)
Al	6.13	10.20	6.08	10.13
Ti	89.14	83.53	89.11	83.65
V	3.61	3.18	3.73	3.29
Fe	0.22	0.17	0.25	0.21
Other	0.9	2.92	0.84	2.72

The ozone treatment of Ti6Al4V titanium alloy effectively removes carbon from the surface of contaminated samples. The removal of atmospheric carbon leads to a percentage increase of other Ti6Al4V titanium alloy components.

3.3. Examination of the Surface Layer with XPS Photoelectron Spectroscopy

The results of XPS and FTIR (Fourier Transform Infrared Spectroscopy) analyses [15] reveal the existence of an oxidized polymer layer in ozone-treated samples and the effect of ozone on polar bonds such as C-O or C=O.

The XPS results reported in the paper [16] show irreversible changes in the surface layer of titanium alloy under the impact of ozone. The results analyzed in [16], however, concern the human body environment, which exhibits totally different conditions. The decrease in the surface carbon content is a result of the cleaning effect of ozone, while the increase in oxygen levels in the surface layer indicates the formation of oxides. Simultaneously, the analyses show that the content of pure titanium decreases.

Figures 6–12 show the results of Ti6Al4V titanium alloy surface composition examination. The surface layer of this alloy is highly sensitive to contamination. Despite washing and degreasing before ozone treatment, the alloy was contaminated with a layer of surface carbon and lead. The concentration of titanium alloys constitutes approx. 31% of its nominal weight value, 85% of which is in the oxidized form.

A comparative kit of XPS wideband spectra of the Ti6Al4V titanium alloy sample is shown in Figure 6.

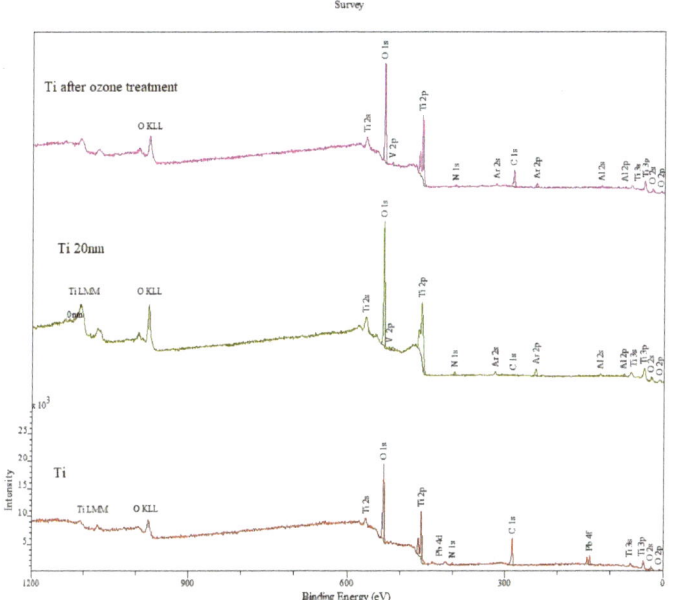

Figure 6. Comparative kit of XPS broadband spectra of Ti6Al4V titanium alloy samples in the following order: A spectrum of titanium alloy before ozone treatment (Ti), after plasma cleaning (Ti 20 nm) and after ozone treatment (Ti after ozone treatment).

What can be seen in the high-energy part of the XPS spectrum are wide KLL Auger bands of oxygen (EB ~980 eV) and LMM Ti (EB ~1120 eV). The titanium alloy surface is deficient in vanadium atoms, which remains even after ionic cleaning of the adherend. The cleaning of the adherend surface and removing 20 nm of the layer led to a considerable change in surface composition of the alloy. Lead was entirely removed from the surface layer of the specimen, and so was almost entire surface carbon. The removal of the carbon layer exposes the grains of aluminum and titanium. As a result of ozone treatment, C content on the surface of Ti6Al4V dropped by approx. 35%.

Table 2 shows the chemical composition of Ti6Al4V titanium alloy substrate prior to and following ozone treatment.

Table 2. The chemical composition of Ti6Al4V titanium alloy prior to and after ozone treatment.

Element	Ti6Al4V Titanium Alloy Prior to Ozone Treatment				Ti6Al4V Titanium Alloy after Ozone Treatment			
	%	at %	Fractions at %		%	at %	Fractions at %	
Ti	28.23	12.15	Ti (0) Ti (II) Ti (III) Ti (IV)	6.74 1.75 6.59 84.92	37.72	16.86	Ti (0) Ti (II) Ti (III) Ti (IV)	2.21 8.01 10.72 79.06
V	1.86	0.75	V (0) V (II) V (III) V (IV) V (V)	21.76 15.55 32.00 18.18 12.51	0.96	0.40	V (0) V (II) V (III) V (IV) V (V)	9.14 8.64 19.77 33.26 29.19
Al	4.89	3.73	Al (III)	100	4.44	3.52	Al (III)	100
C	23.78	40.78	-		15.56	27.72	-	
O	31.42	40.46	-		35.55	47.56	-	
Pb	8.98	0.89	-		-	-	-	

Pb content on the surface of specimens prior to treatment is an effect of ambient environment contamination. Ionic and ozone treatment remove PB entirely.

Figure 7 shows changes in the doublets at the valence band in the Ti2p spectrum following Ti6Al4V titanium alloy surface modification.

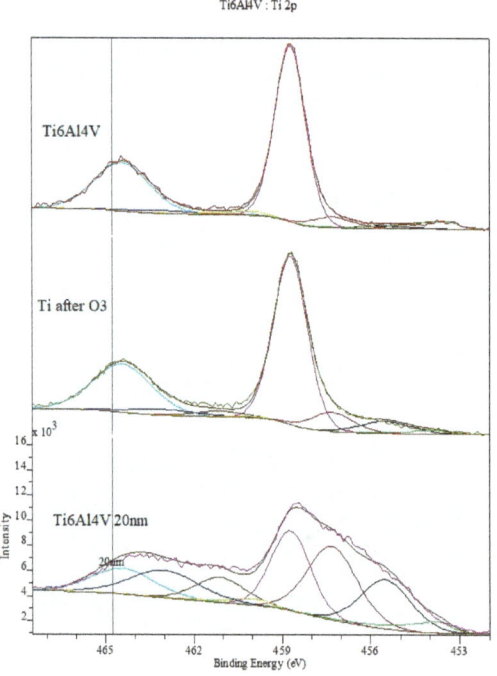

Figure 7. Changes in doublets at valence band in Ti2p spectrum.

The ozone treatment of the Ti6Al4V titanium alloy slightly changes the shape of the composite band and the intensity of its components. What can be noticed comparing the spectral band profiles of Ti2p alloy after ionic etching and ozone treatment is deep ozonation of the adherend surface layer. What is also notable is the drop in the intensity of metallic phase peaks in the spectrum as well as Ti (II) and Ti (III) ions in a low-energy XPS part of the spectrum. The peak located at EB = 459 eV bound to TiO2 titanium oxide at the highest oxidation state is the dominant band. Ti (IV) form amounts to 79% of total concentration after ozone treatment, which is more than a two-fold increase, compared to the 28% concentration after ionic cleaning.

Figure 8 shows the effect of ozone treatment on XPS spectrum based on V2p transitions.

The course and nature of the observed changes clearly indicate the oxidation process of the surface layer of the titanium alloy sample during ozone treatment. Unlike titanium ions, the ozone treatment process essentially changes the shape of the composite band and the distribution of matching components. The concentration of forms with a higher oxidation degree: V(IV) and V(V) significantly increases at the expense of metal and V(II) ions.

The deconvolution of the XPS spectrum for the Al2p band is shown in Figure 9.

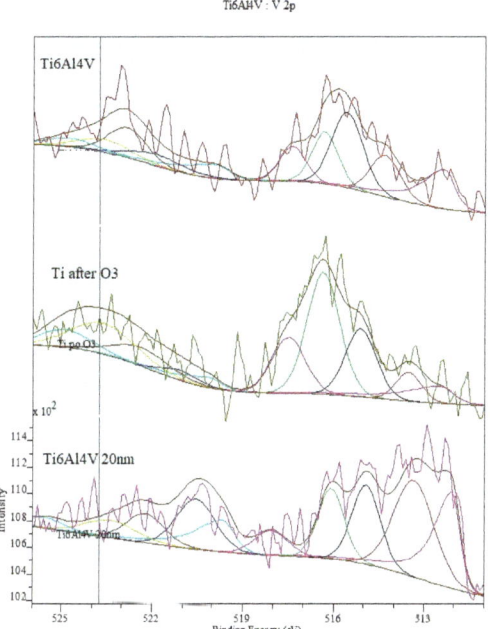

Figure 8. Effect of ozone treatment on XPS spectrum on the basis of V2p transitions.

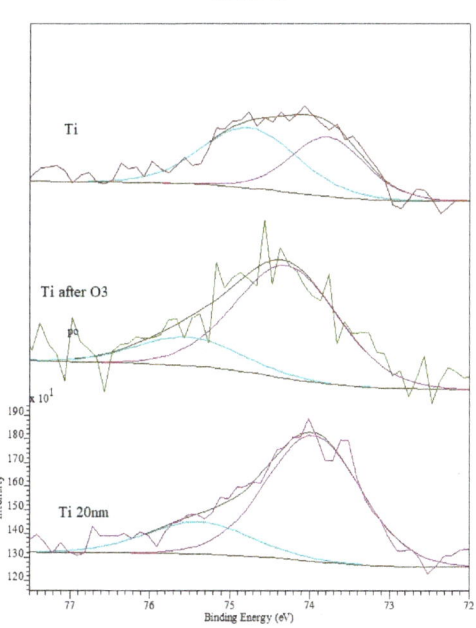

Figure 9. Deconvolution of XPS spectrum of Al2p band: Titanium alloy before ozone treatment (Ti), titanium alloy after ionic cleaning (Ti 20 nm) and titanium alloy after ozone treatment (Ti after ozone treatment).

The deconvolution procedure of the band reveals two phases of aluminum oxide on the alloy surface. After ozone treatment, aluminum oxide Al_2O_3 constitutes the main phase with the 80% content of Al.

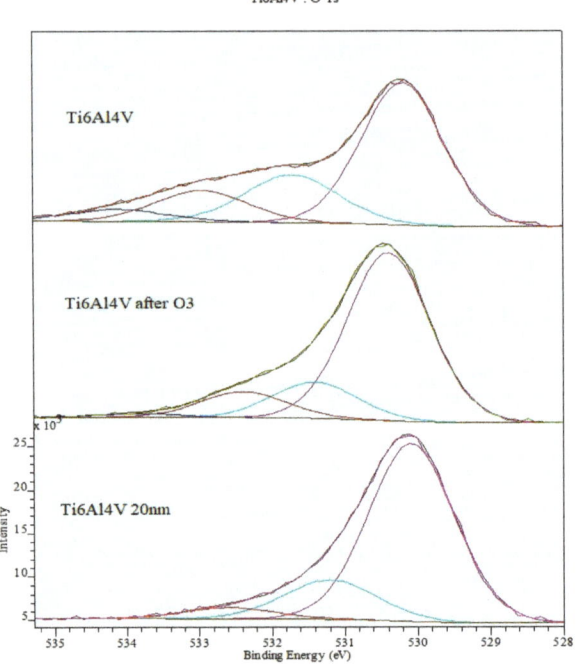

Figure 10. Deconvolution of XPS spectrum of O1s band.

The deconvolution procedure of the XPS spectrum for the O1s band (Figure 10) confirms the coexistence of various oxides and a high degree of oxidation of the surface layer of the alloy after ozone treatment. The intensity axis was normalized.

Figure 11 depicts the deconvolution of the XPS spectrum of the C1s band.

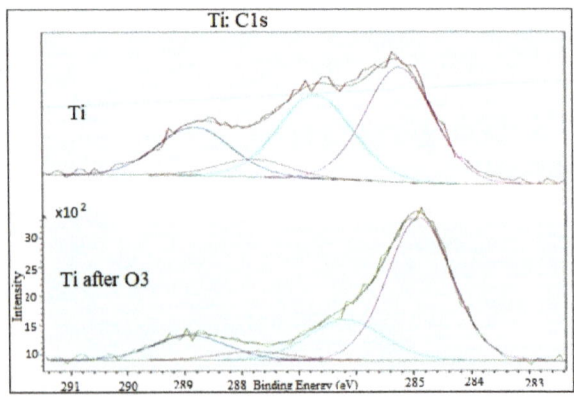

Figure 11. Matching XPS spectrum to C1s band.

The deconvolution procedure of the band confirms the coexistence of oxidized and non-oxidized forms of surface carbon.

Figure 12 shows the XPS spectrum of photoelectrons in the Ti6Al4V titanium alloy sample.

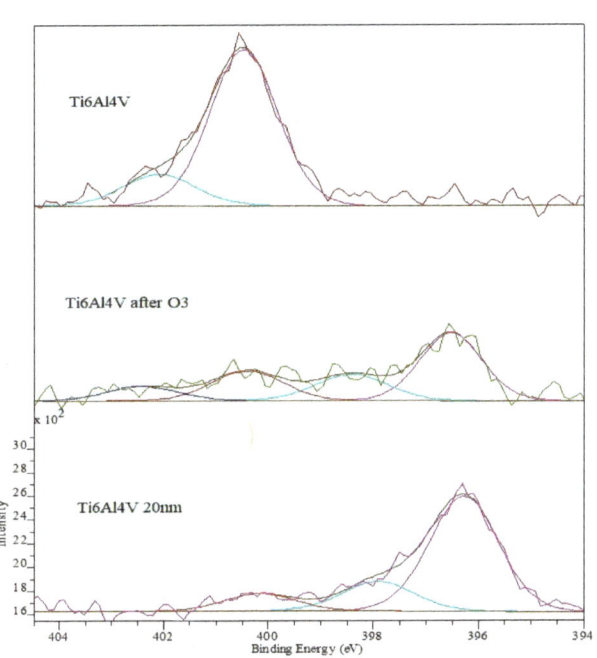

Figure 12. XPS spectrum of photoelectrons in titanium alloy sample.

After the deconvolution of the peaks on the spectrum, we may distinguish bands of N1s photoelectrons that come from three chemically different phases.

The results from the chemical analysis show that the surface layer of the alloy is enriched with aluminum atoms, as their concentration, over 6 wt %., exceeds the nominal value, and which occur in the entirely oxidized form as Al_2O_3 oxide, $Al(OH)_3$ hydroxides and AlO(OH) boehmite. The spectral analysis results, furthermore, reveal that argon permanently penetrates the crystal lattice and is not removed in high vacuum of the UHV system. The adherend also contains nitrogen which can form various chemical bonds.

The results clearly show that the surface layer of the titanium alloy before ozone treatment is practically entirely covered with a layer of metal oxides, showing different degrees of oxidation and contaminated with the carbon layer. The tests show that the ozone treatment of titanium alloy results in the removal of carbon and the increase in the concentration of Ti and V ions in an advanced stage of oxidation at the expense of metal atoms and ions with lower valence. The modification of the surface layer in the ozone atmosphere resulted in a 30% increase of Ti element concentration in the sample surface layer compared to the samples prior to the ozone treatment. The removal of carbon from Ti6Al4V titanium alloy samples during the ozone treatment process is estimated at 35%. A 13% increase in oxides after ozone treatment was also observed.

3.4. Surface Free Energy

Figure 13 shows the surface free energy of Ti6Al4V titanium alloy after machining using a density tool with a grit size of P320.

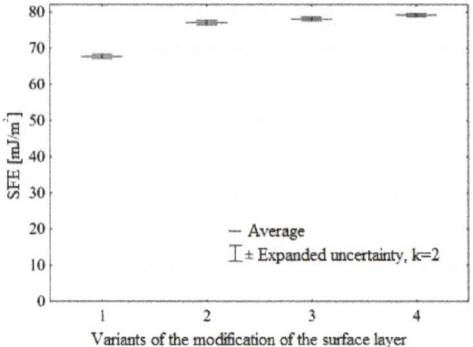

Figure 13. Surface free energy of Ti6Al4V titanium alloy for different ozone treatment conditions (modification variants of surface layer are described in "Materials and Methods").

The results demonstrate that the surface free energy of Ti6Al4V titanium alloy increased as a result of the ozone treatment. The highest increase in the surface free energy was observed for the samples prepared in accordance with the fourth variant of surface pretreatment, amounting to 17% compared to the samples before ozone treatment. The lowest observed increase was 14%. For the analyzed data, the maximum value of standard deviation was 0.99 [mJ/m^2].

Compared to the samples prior to ozone treatment, the highest increase in the polar component of the SFE of the Ti6Al4V titanium alloy after pretreatment (Figure 14) was 51%, while the lowest increase was by 44%.

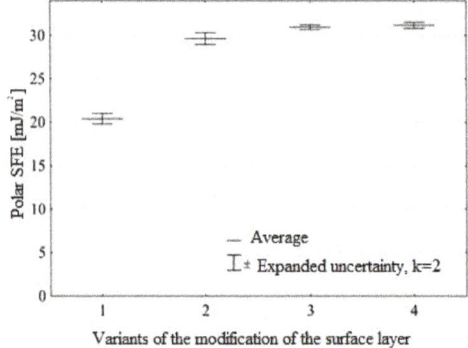

Figure 14. Polar component SFE of Ti6Al4V titanium alloy for different ozone treatment conditions (modification variants of surface layer are described in "Materials and Methods").

The results clearly demonstrate that modification of the surface layer of the tested construction materials in ozone atmosphere has a considerable effect on increasing the surface free energy and its polar component.

3.5. Strength of the Adhesive Joint

Figure 15 shows mean values of shear stresses in a single-lap adhesive joint obtained in experimental studies for the Ti6Al4V titanium alloy after machining and ozone treatment.

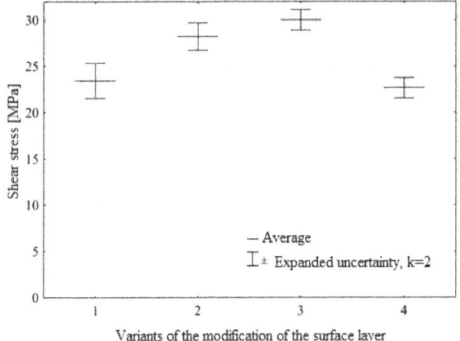

Figure 15. Shear stress at obtained in experimental studies for Ti6Al4V titanium alloy samples made with Hysol 9466 adhesive, for various ozone treatment conditions (modification variants of surface layer are described in "Materials and Methods").

The ozone treatment process of Ti6Al4V titanium alloy samples increases to some extent the strength of the single-lap adhesive joint made with the Hysol 9466 adhesive. The highest increase in strength of the adhesive joint, amounting to 28%, was observed for ozone samples with a concentration of 50 g O_3/m^3 in 30 minutes. In this case, the extreme has been reached. The decrease in strength observed for Variant 4 confirms the significant effect of the ozone treatment process on the final effects.

4. Conclusions

The findings of this study have led to the following conclusions:

(1) Ozone treatment of the titanium alloy results in the removal of carbon and an increase in the concentration of Ti and V ions at higher oxidation states at the expense of metal atoms and ions with lower valence.
(2) The modification of the surface layer in ozone atmosphere caused a 25% increase in the Ti element of the surface layer of the samples compared to the samples before ozone treatment.
(3) Removal of carbon from Ti6Al4V titanium alloy samples in the ozone treatment process amounted to 35%, and the content of oxides increased by 13%.
(4) The increase in the surface free energy was observed in the Ti6Al4V titanium alloy as a result of the ozone treatment, the highest increase in the surface free energy was observed for the samples prepared according to Variant 4 and increased by 17%, compared to the untreated samples, while the smallest increase was 14%.
(5) There is no correlation between the highest increase in the surface free energy and an increase in the strength of adhesive joints. The knowledge on the surface free energy value is, therefore, a necessary yet insufficient condition for designing predictive models of adhesive joints strength of metal substrates.
(6) The highest increase in the strength of adhesive joint was noted for the samples subjected to ozone treatment with a concentration of 50 g O_3/m^3 over 30 minutes. The untreated samples exhibited an increase of 28%.

The study confirms that ozone treatment, if performed under appropriate conditions, can be used in bonding technologies to shape surface micro-topography and free energy, thus offering an alternative option to electrochemical methods.

Author Contributions: For research articles with several authors, a short paragraph specifying their individual contributions must be provided. The following statements should be used "conceptualization, M.K. and J.K.; methodology, M.K., J.K.; software, M.K., J.K.; validation, M.K., J.K.; formal analysis, M.K., J.K.; investigation, M.K., J.K.; resources, M.K., J.K.; data curation, M.K., J.K.; writing—original draft preparation, M.K., J.K.; writing—review and editing, M.K., J.K.; visualization, M.K., J.K.; supervision, M.K., J.K.; project administration, M.K.; funding acquisition, M.K.

Funding: The project/research was financed in the framework of the project Lublin University of Technology - Regional Excellence Initiative, funded by the Polish Ministry of Science and Higher Education (contract no. 030/RID/2018/19).

Conflicts of Interest: The authors declare no conflict of interest.

References

1. Kwiatkowski, M.P.; Kłonica, M.; Kuczmaszewski, J.; Satoh, S. Comparative analysis of energetic properties of Ti6Al4V titanium and EN-AW-2017A(PA6) aluminum alloy surface layers for an adhesive bonding application. *Ozone: Sci. Eng.* **2013**, *35*, 220–228. [CrossRef]
2. Kłonica, M. *Studies on Energy State of Surface Layer in Selected Construction Materials after Ozonisation*; Lublin University of Technology: Lublin, Poland, 2014.
3. Kłonica, M.; Kuczmaszewski, J. Determining the value of surface free energy on the basis of the contact angle. *Adv. Sci. Technol. Res. J.* **2017**, *11*, 66–74. [CrossRef]
4. Packham, D.E. Surface energy, surface topography and adhesion. *Int. J. Adhes. Adhes.* **2003**, *23*, 437–448. [CrossRef]
5. Żenkiewicz, M. New method of analysis of the surface free energy of polymeric materials calculated with Owens-Wendt and Neumann methods. *Polimery* **2006**, *51*, 584–587. [CrossRef]
6. Żenkiewicz, M. Comparative study on the surface free energy of a solid calculated by different methods. *Polym. Test.* **2007**, *26*, 14–19. [CrossRef]
7. Dingemans, M.; Dewulf, J.; Van Hecke, W.; Van Langenhove, H. Determination of ozone solubility in polymeric materials. *Chem. Eng. J.* **2008**, *138*, 172–178. [CrossRef]
8. Lemu, H.G.; Trzepieciński, T.; Kubit, A.; Fejkiel, R. Friction modeling of Al-Mg alloy sheets based on multiple regression analysis and neural networks. *Adv. Sci. Technol. Res. J.* **2017**, *11*, 48–57. [CrossRef]
9. Kłonica, M.; Kuczmaszewski, J.; Kwiatkowski, M.; Ozonek, J. Polyamide 6 surface layer following ozone treatment. *Int. J. Adhes. Adhes.* **2016**, *64*, 179–187. [CrossRef]
10. Zielecki, W.; Pawlus, P.; Perłowski, R.; Dzierwa, A. Surface topography effect on strength of lap adhesive joints after mechanical pretreatment. *Arch. Civ. Mech. Eng.* **2013**, *13*, 175–185. [CrossRef]
11. Olewnik-Kruszkowska, E.; Nowaczyk, J.; Kadac, K. Effect of ozone exposure on thermal and structural properties of polylactide based composites. *Polym. Test.* **2016**, *56*, 299–307. [CrossRef]
12. Domingues, L.; Fernandes, J.C.S.; Da Cunha Belo, M.; Ferreira, M.G.S.; Guerra-Rosa, L. Anodising of Al 2024-T3 in a modified sulphuric acid/boric acid bath for aeronautical applications. *Corros. Sci.* **2003**, *45*, 149–160. [CrossRef]
13. Johnsen, B.B.; Lapique, F.; Bjørgum, A. The durability of bonded aluminium joints: A comparison of AC and DC anodizing pretreatments. *Int. J. Adhes. Adhes.* **2004**, *24*, 153–161. [CrossRef]
14. Fisher, P.; Maksimov, O.; Du, H.; Heydemann, V.D.; Skowronski, M.; Salvador, P.A. Growth, structure, and morphology of TiO_2 films deposited by molecular beam epitaxy in pure ozone ambients. *Microelectron. J.* **2006**, *37*, 1493–1497. [CrossRef]
15. Xu, J.; Yuan, Y.; Shan, B.; Shen, J.; Lin, S. Ozone-induced grafting phosphorylcholine polymer onto silicone film grafting 2-methacryloyloxyethyl phosphorylcholine onto silicone film to improve hemocompatibility. *Colloid. Surface B* **2003**, *30*, 215–223. [CrossRef]
16. Linderbäck, P.; Harmankaya, N.; Askendal, A.; Areva, S.; Lausmaa, J.; Tengvall, P. The effect of heat- or ultra violet ozone-treatment of titanium on complement deposition from human blood plasma. *Biomaterials* **2010**, *31*, 4795–4801. [CrossRef] [PubMed]

© 2019 by the authors. Licensee MDPI, Basel, Switzerland. This article is an open access article distributed under the terms and conditions of the Creative Commons Attribution (CC BY) license (http://creativecommons.org/licenses/by/4.0/).

MDPI
St. Alban-Anlage 66
4052 Basel
Switzerland
Tel. +41 61 683 77 34
Fax +41 61 302 89 18
www.mdpi.com

Materials Editorial Office
E-mail: materials@mdpi.com
www.mdpi.com/journal/materials

www.ingramcontent.com/pod-product-compliance
Lightning Source LLC
LaVergne TN
LVHW071947080526
838202LV00064B/6696